Similarity and Modeling in Science and Engineering

Josef Kuneš

Similarity and Modeling in Science and Engineering

CISP

Springer

Josef Kuneš
Plzeň
Czech Republic

ISBN 978-1-907343-77-3 e-ISBN 978-1-907343-78-0
DOI 10.1007/978-1-907343-78-0
Cambridge International Science Publishing Ltd

Library of Congress Control Number: 2012935674

Printed on acid-free paper

Springer is part of Springer Science+Business Media (www.springer.com)

I dedicate this book to the memory of my teachers Prof. Vladimír Marcelli and Prof. Josef Hošek. They took me and my workplace, half a century ago, on the path of modeling and experiment.

Josef Kuneš

If explaining science . . . ,
do not forget the practical
application of each theorem
lest your science should
become needless.

LEONARDO DA VINCI
(15.4.1452 Anchiano by Vinci – 2.5.1519 Cloux by Amboise)

A universal personality of the Renaissance who unified fine arts, science and engineering in himself. He was not only an outstanding painter, sculptor, architect, natural scientist, musician, and writer, but also the first engineer and designer from our contemporary point of view. He studied aerodynamics, physics, anatomy, meteorology, and astronomy. He leant on Pythagoras' mathematics, which became the basis for his research. Probably, he was the first who, at the end of his life, expressed two fundamental concepts of modeling: similarity and abstraction. He compared two different dynamic systems, the movement of a fish in water and a bird's flight in the air, based on a single description only, namely a mathematical model, or an equation of mathematical physics, as we would say today. Similarity and abstraction in the fine arts and modeling represent two fundamental tools in the continuous development of each of these fields. And it is just this high degree of abstraction in contemporary modeling, along with the application of practical experience, and intuition and even feeling, which changes the modeling to the unique form of the art of experiment.

Contents

Foreword

It is an unusual pleasure to present Professor Kuneš's book on Similarity and Modeling in Science and Engineering to a broader international community of interested readers. Professor Kuneš's leading place in the development of the area of modeling of various processes is well-known to those who have been following its growth.

The presentation of the enormous amount of material that has been accumulated in this field over the last few decades may follow two possible approaches: one can either concentrate on the experimentally observed phenomena and treat each of them as using a variety of theoretical methods, or one may put emphasis on the methodology and the technical aspects of the different theories that have been used to describe the relevant phenomena. Professor Kuneš has chosen the second approach – his book is in fact an attempt towards a unified view of models and modeling in various fields of science and engineering.

The power of modeling in giving us a better understanding of a variety of phenomena in nature is becoming increasingly evident. The book Similarity and Modeling in Science and Engineering deals with the general methods of modeling, the theory of similarity and the classification of models according to their similarity and their degree of abstraction. Nevertheless, distinctive properties of different models are pointed out as well.

The book can give readers an unconventional access to the methodology of modeling and model classification, and so to its applications in different scientific and engineering spheres.

Professor J. Vlček
Head of the Department of Physics
University of West Bohemia, Plzeň
Czech Republic

"Three paths lead to knowledge: the path of meditation – which is noblest, the path of imitation – which is lightest, and the path of experiment – which is driest."

Confucius *(28.9.551 BC–479 BC)*

Preface

The book *Similarity and Modeling in Science and Engineering* deals with the theory of similarity and modeling. The aim of the book is to interpret modeling in its breadth and diversity as far as possible. It also discusses the distinctive features of these diverse forms of modeling. The fundamental point of view which allows us to classify the various types of models relates to how much information is transferred from an object or process to its model. Thus, the essence of this classification of models is an information point of view.

This book concentrates on presenting a general view of various types of models, applied modeling methods and the main modeling tools. Among these tools is the theory of similarity, which accompanies all kinds of modeling. Dimensional analysis and methods of similarity analysis, following from analyses of physical phenomenological and asymptotic mathematical models, are also important. Following the continuity of development of various models leads to an understanding of two fundamental concepts which accompany their development. It is similarity and abstraction which enable a comprehensive view of the classification of models. Both these aspects are crucial for modeling, even though their role and importance has changed significantly during the development of modeling techniques. The emphasis has been shifting always toward more abstract models.

The utilisation of more abstract models can easily cause the loss of a model credibility. This is often due to a lack of feedback of more abstract model to the real object. This implies the importance of models with a low degree of abstraction. The natural and physical models are necessary because they are based on direct experiment with an object or its physical model. The accuracy of these models depends on that of measurement only. This is similar in the case of physical analogues, where the simplicity and clearness important to understanding the substance of physical problem is put in the foreground. The simplicity and clarity of the natural and physical models and physical analogues is significant in identification and education processes. The computer models, as the mathematical analogues, have a dominant position today, especially in solving complicated problems. They make use of the present state of software and computational methods and are distinguished by high accuracy of mathematical operations. However, the accuracy of the solution depends on the accuracy of, among other things, the physical parameters and conditions of unambiguity. The cybernetic functional models are unsubstitutable in solving the problems of control of complicated systems, their outer behavior and the intrinsic interaction among subsystems.

This book is an attempt to take an unified look at modeling which expresses the logical relationships among different sorts of models. As such, it has to draw attention to the importance of similarity and modeling theories in the contemporary period of computer modeling, characterized by, among other things, a great amount of information to be processed. In computer modeling, the use of dimensionless quantities enables one not only to decrease the number of variables significantly and the generalization of results, but also the scale factor assignment of the computer model to the examined object or process. Not least, the book deals with the application of dimensionless quantities in the formation of the phenomenological mathematical models where the asymptotic mathematical models cannot be used or are not known. The results of the theory and application of phenomenological mathematical models or, in other words, of experimental mathematical models, offer new possibilities for solving complicated processes in diverse systems.

This book highlights the present state and importance of modeling not only in conventional but also in recently developing fields.

The next book, *Dimensionless Physical Quantities in Science and Engineering* (Elsevier-Insights, 2012) continues presenting a comprehensive summary of dimensionless quantities in modeling. It incorporates examples from many of the fields in which the application of modeling in exceptionally important.

Acknowledgements

The author expresses his thanks to collaborators in the scientific team working at the University over many years. They helped him to develop the subject of modeling and to build a workplace concentrated especially on applied as well as theoretical research of thermomechanical processes. His Ph.D. students are part of this team. All of them assisted in developing this subject at the Department of Physics of the Faculty of Applied Sciences and contributed to achieving results as part of the Research Centre for New Technologies at the University.

Those collaborators must be introduced who contributed substantially at the very beginning to laying the foundations for this workplace, the results of which led to the preparation of this book. Asst. Prof. B. Ulrych, who assisted in numerical methods development, V. Franta, Ph.D in the development of stochastic methods and O.Vavroch, Ph.D. in the field of hybrid methods. Some results of present day collaborators are part of this book as well. This concerns especially work in the area of non-equilibrium processes (Asst. Prof, M. Honner) and help in preparing the 8th Chapter (J. Martan, Ph.D., M. Švantner, Ph.D. and V. Lang, Ph.D., O.E. Cibulka, Ph.D.) and the 9th Chapter (J. Šroub, Ph.D). Co-worker Mr. Z. Veselý, Ph.D. also participated significantly in computer-aided processing of the book into electronic form and in helping with the preparation and modification of the text and figures.

Thanks must also be expressed to the Management of the University of West Bohemia in Plzeň for financial and other support for publication of this book. The same must be expressed to the Department of Physics. Significant help in correction of the English text was provided by Joan DeForeest, M.A. and PhDr A.Telinová.

This work was supported by the Ministry of Education of the Czech Republic under Project No. MSM 4977751302 (Processes in discharge plasmas and new thin materials with unique properties) and by the University of West Bohemia.

"I have often been impressed by the scanty attention paid even by original workers in physics to the great principle of similitude. It happens not infrequently that results in the form of 'laws' are put forward as novelties on the basis of elaborate experiments, which might have been predicted a priori after few minutes of consideration."

Lord Rayleigh (John William Strutt) *(12.11.1842–30.6.1919)*

1

Introduction

Similarity and *modeling* are closely connected concepts which cannot be separated from the concept of *information.* The aim of each experiment, therefore of a modeling one, is to gain information about an examined object which can be the object itself or its model. In the case of modeling, another important concept appears, namely *abstraction,* which is one of the forms of human thinking. It came into being in the human world in order to understand complex phenomena and laws of nature during our historical development, specifically in the form of three types of abstract models – dimensional, physical and mathematical. These abstract models led to the simulation model and to our verification of reality on it. No abstract model exists independently of people and the development of their thinking. The success of physical research depends on the considered selection of everything that can be observed as important and on the combination of the results of this observation with abstraction.

Mathematization and *cybernetization* became a natural display of abstraction in the development of scientific fields. In modeling, mathematization delivers first an objective general view of an examined phenomena. From this, it leads to the formalization of the knowledge and to the elaboration of new theories, due to which it decisively influences the development of all fields. Cybernetization can be understood as the natural enlargement of the concept of mathematization to explain complex systems of both animate and inanimate nature. The characteristic feature of modeling is the integral comprehension of diverse concepts and fields from the point of view of cybernetic similarity which, in an investigated complex object, appears as similarity of its outer behavior. Cybernetics delivers not only deeper

J. Kuneš, *Similarity and Modeling in Science and Engineering*,
DOI: 10.1007/978-1-907343-78-0_1, © Cambridge International Science Publishing 2012

knowledge of the substance of nature into modeling but even new generalizing and controlling elements. Thus, modeling becomes a means affecting deeply the way of scientific thinking. The high level of abstraction, and the application of practical experience, intuition and feeling transform modeling into a unique form of the *art of experiment.*

Based on the above-mentioned view, models can be divided into physical, physical–mathematical, mathematical and cybernetic ones. The development of modeling is closely connected to the progress of civilization and the development of knowledge. It begins with primitive empirical modeling or, in other words, with the imitation of phenomena in nature, from the 2nd millennium B.C., and culminates in today's computer and cybernetic modeling of complex and global dynamic systems. However, all types of models are characterized by two fundamental properties, namely the *level of abstraction* from a real object or process and the *similarity* between an object and a model, which can be obvious or concealed.

Among the *physical models* working on the principle of physical similarity, even a *natural model* can be incorporated which is the object of examination itself. The *physical–mathematical models* (the *physical analogues)* are characterized by mathematical similarity in the analogy. The *mathematical (computer) models* make use of the mathematical similarity between an abstract mathematical model and a process or an object. In *cybernetic models,* the mathematical similarity disappears in the inner system structure and appears in its outer behavior (the *functional model).* When comparing the above-mentioned models, not only the changing character of similarity between a model and an object but even of model information can be seen. The unified logic of nature and of developing scientific knowledge is hidden behind the comparison of modeling methods.

With the use of model abstraction increasing, the problem of model credibility arises in relation to an examined object. When it comes to model credibility, a natural, physical or other less abstract type of model appears to be the best. It is true that the more abstract the model being used is, the greater is the danger of separation from reality and loss of model credibility. In order to achieve modeling credibility, one's physical feeling for the modeling process and understanding of how it transforms into a mathematical and simulation model are also important.

The *physical analogy* is remarkable too, not only from the standpoint of the historical development of pre-computer modeling, but first and foremost from the point of view of the on-going development of knowledge and abstraction of thinking. With this, the observation of a natural phenomenon, either with a microscope or a telescope or other measuring device, often leads to the creation of a similar abstract model in a quite different area. An example is the Copernican heliocentric model of the solar system and the model of the atom, arising after four centuries, which resembles a super-miniature solar system. As with the physical analogy in both macro- and microworlds, a process of recognition and the abstraction coming out of it, are under way and result in an abstract mathematical model. However, one cannot forget that the credibility of this model depends in particular on the knowledge of the physical principle of the phenomenon and the analysis of it.

The *theory of similarity* represents a fundamental modeling tool which expresses unambiguous assignment of a process or an object to its model. It is based on the existence of similarity criteria which are expressed as dimensionless quantities, replacing fully the dimensional ones in the examined phenomenon. From the mathematical point of view, the determination of dimensionless quantities in modeling, by means of dimensional analysis or analyses of a mathematical or physical model, can be considered as one of the methods for applying the theory of groups which is developing within the bounds of mathematical physics.

There is an opinion that the significance of similarity theory has decreased with the arrival of computer models and is restricted mostly to physical or scale models. But the opposite is true. With the amount of modeling information increasing, and demands on its processing and generalization growing, a need arises for data compacting, generalizing of results, and receiving generalized functional dependencies. Also, the creation of experimental mathematical models is often necessary where an exact mathematical model cannot be used or is not known. The credibility of these models is higher, despite the fact that they can be less precise than computer ones. This is due to the fact that they are based on real object conditions which cannot be fully incorporated by a computer model.

One of the main goals of this book is to highlight the broader significance and consequences of similarity theory and its unique possibilities. In two chapters, the principles of *dimensional analysis* and of *similarity analysis methods* are presented. The division into two independent chapters is due to the fact that there are different ways to derive dimensionless similarity criteria.

One of the chapters of the book is devoted to the classification of models and connected methodological questions of modeling. The classification of models is based on the principles of abstraction and similarity in modeling, which reflects their historical development. One cannot forget that despite a sudden shift of emphasis to computer modeling, some important properties of the preceding models remain. In the case of physical modeling, this especially relates to the non-substitutability of using *physical models* to verify the credibility of some computer models. First and foremost in *physical analogues*, there is a clarity in recognizing the mathematical similarity and analogy which can be used to simply express physical laws, e.g. in education. The new thin-layer technologies for making miniature electrical elements enables the utilization of single-purpose physical electrical analogues, e.g. in systems of process controlling.

The computer models, as mathematical analogues, represent the most important contemporary means of modeling. Essentially, they can be divided into *numerical, analogous, and hybrid models*, of which numerical or possibly hybrid models, incorporating both analogous and numerical parts, are most important. Each of these models can be *deterministic or stochastic*. They are distinguished not only by their principles, but also by specific properties which can be utilized in modeling, especially in solving complex tasks. Since there are many publications about deterministic and stochastic methods and means, only brief characteristics of methods are presented here from the standpoint of their use in modeling. The *cybernetic functional models* of the black-box type, which follow outer system behavior or outer interactions among subsystems, represent the culmination of the development of computer models and the most advanced level of abstraction.

Dimensionless physical quantities, their physical importance and application in classical and new branches and technologies are described in the book[1].

[1]Kuneš J.: *Dimensionless Physical Quantities in Science and Engineering* (Elsevier-Insights, 2012).

"The model approximates the complex prototype asymptotically. The trend of the approximation is to become identical with the prototype. In the limit, model and prototype coincide. For a typical example, the best model of a cat is another cat, or indeed, the original cat proper. In other words, if the model realizes the purpose of the prototype in all its aspects, then it is possible to explore the prototype in all detail, and no model is needed".

Norbert Wiener *(26.11.1894–18.3.1964)*

2

Methodology of Modeling

Modeling, as a specific form of experiment, enables one to recognize the laws of Nature. Its broad development, numerous forms and many application areas have resulted in the fact that modeling has gradually become a relatively disorganized field. It follows from this that modeling methodology is actually itself a problem.

Apart from other things, the task of methodology is to find a common point of view in sorting models, which enables their classification and explains their specific position in a broader system of experimental means. Clearly, this could be a point of view from which to look at information and similarity.

2.1. Identification, Modeling and Simulation

Modeling represents one of the most general ways of representing the outer world and studying the objective laws existing in it. It is an experimental information process where another physical or abstract object, called a *model,* is assigned unambiguously to the examined system (original, object or work) according to certain criteria. Dynamic system modeling, with direct or indirect feedback on the examined object, is called *simulation.*

With modeling, diverse forms of similarity and analogy can be connected. In this sense, *similarity* can be understood as unambiguous mutual assignment between different systems in terms of their structure, properties and behavior. *Physical similarity* refers to the similarity between systems and processes, having the same physical substances,

J. Kuneš, *Similarity and Modeling in Science and Engineering*,
DOI: 10.1007/978-1-907343-78-0_2, © Cambridge International Science Publishing 2012

and involves, besides geometric similarity, the similarity of parameters and values of the system state. *Mathematical similarity* expresses the similarity between systems and processes having identical mathematical descriptions. It is called an *analogy* when it is about different systems and processes. Finally, *cybernetic (functional) similarity* expresses the mathematical similarity in the external behavior of systems.

In accordance with the above-mentioned three kinds of similarity, modeling can be classified as *physical, mathematical and cybernetic.* Cybernetic modeling makes use of *black box models.* This black box concept, introduced by N. Wiener, is understood to be a system which does not deliver any information on its inner structure but on its external behavior only. Opposite to a black box is a white or, better, a transparent box, which represents the inner structure of a system and a process under way in it. Similarly, a grey box concept can be used, delivering partial information about the inner structure of the system. The *experiment* is the general superior concept to the modeling one.

Essentially, there are two ways to generate a mathematical model (Fig. 2.1). In direct identification, one proceeds from a summary of knowledge about the behavior of the examined object which has been obtained either as a result of the object identification or as a result of the development of knowledge in a corresponding field. Often, the necessary information is obtained indirectly as a result of identification

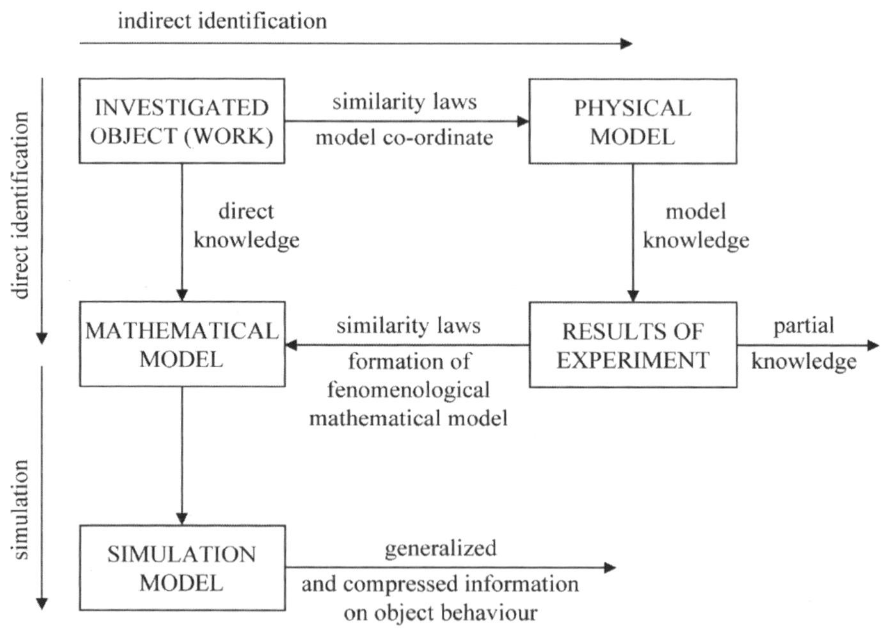

Fig. 2.1. Procedure diagram of direct and indirect identification of an object and generation of mathematical and simulation models.

of a physical model and this is called indirect identification. The *results of an experiment* are represented by a summary of knowledge expressed mostly in the form of a phenomenological description of a physical model or an examined object, because of assignment of the model to the examined object. To obtain a general mathematical model, a detailed analysis of the physical principles and of results of an experiment is necessary. This is the only way to create a simulation model and to get with it some credible information on system behavior.

2.1.1. Identification and Simulation

In modeling, reliable identification of the properties of the examined object is fundamentally significant. The credibility of a mathematical and simulation model and, therefore, even of modeling results, is closely connected with this. Therefore, the identification represents an important step for systems modeling and simulation on which the efficiency of the modeling depends to a large extent. *Identification* is an experimental way to determine important system and process characteristics enabling the construction of the mathematical model. The *system characteristics* include various physical properties, the structure and parameters of the system and the process. The *system diagnostic* is used to determine the state of the system and the process, especially its variations from a supposed state.

The procedure of identification of an examined dynamic system and consequential simulation process is in the block chart (Fig. 2.2).

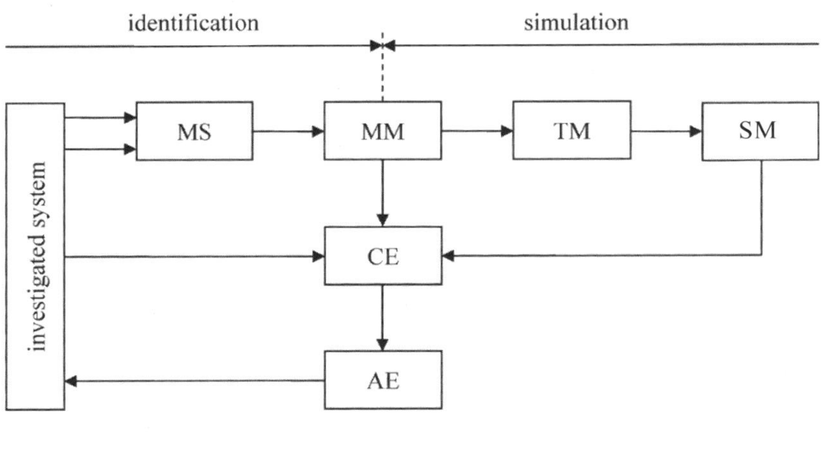

MS ... measurement system, MM ... mathematical model,
TM ... transformed mathematical model, SM ... simulation model,
CE ... comparative element, AE ... active element.

Fig. 2.2. Identification and simulation in the process of investigation of an examined system.

Substantial information on system behavior (about an examined object, original, prototype or work) are gained by direct measurement by means of a measuring system. The identification process results in the formulation of a mathematical model. The mathematical model is transformed into a shape which enables building a simulation model by selected technical means. Based on the comparison of the examined system with a simulation model, one can assume model credibility and utilize the obtained information to improve the original.

2.1.2. Cybernetics and Modeling

The contribution of cybernetics to the development of scientific knowledge is closely connected to improving traditional modeling methods and introducing computer aided modeling. This enrichment is based on changes in our way of thinking because *cybernetics is the theory of models in science without strong boundaries among physical, biological and economic processes* (*S. Bear*). It is based on the ability to solve quantitatively higher tasks of optimization and complex dynamic systems control by making use of the theory of cybernetic similarity and analogy. By removing the boundaries among disciplines, cybernetics enables the synthesis of knowledge from diverse fields, the mutual influence of fields and, as a result, even their accelerated development. System modeling and simulation play a crucial role in this process.

Since the beginning of the formulation of cybernetics as a discipline, it has been accompanied by the idea of constructing a computer working as a model analogous with the human brain. This idea steadily influences the development of cybernetic modeling, which represents a highly abstract modeling form. To a certain extent, the power of cybernetic modeling depends on utilizing functional analogies between various dynamic systems.

The *systems theory* is concerned with general questions of mathematical modeling, such as the analysis and synthesis of complicated technical, biological, economic, social, and other systems. Abstraction, modeling and analogy are fundamental tools in systems theory.

In the mathematical meaning of the word, *system* means a set of elements connected mutually and to their surroundings by means of interactions which create the properties enabling the fulfillment of their functions. The *structure* is the inner system arrangement expressed by mutual couplings and actions of components, the inner organization of the whole or the parts of which it is formed.

A formal description of a system can be in the form of $S(X, R)$, where

$$X = \{x_1, x_2, \ldots, x_n\} \tag{2.1}$$

is a set of elements and

$$R = \{r_1, r_2, \ldots, r_m\} \tag{2.2}$$

is a set of relations among elements where $n, m \in N$. In the simplest case, the relation can be binary $r_k(x_i, x_j)$. The simplest system consists of a pair of elements whereby the set R involves a single relation.

The concept of a system can be enlarged implicitly. In general, one can say that each system element is a system itself and is denoted as a *subsystem*. Analogously, by connecting systems, a *composed system* (*super-system*) is created. The system branching is called a *system hierarchy*. To model and simulate complex systems, the hierarchical approach has not only methodological but even practical significance.

In Fig. 2.3, the system consists of two subsystems, S_1 and S_2, where mutual couplings among the elements x_1 up to x_6 and couplings among the subsystems are displayed. This can concern, e.g. a bearing-base system. Individual elements of the subsystem S_1 represent parts of the bearing, whereas the elements of the subsystem S_2 are the parts of the base.

Dynamic system means a system $S(\tau)$ in which the state values occur in space and time τ. The limit state of a dynamic system is its initial or steady state. Dynamic systems can be divided into *deterministic* and *stochastic* ones, which is a point of view being applied significantly in

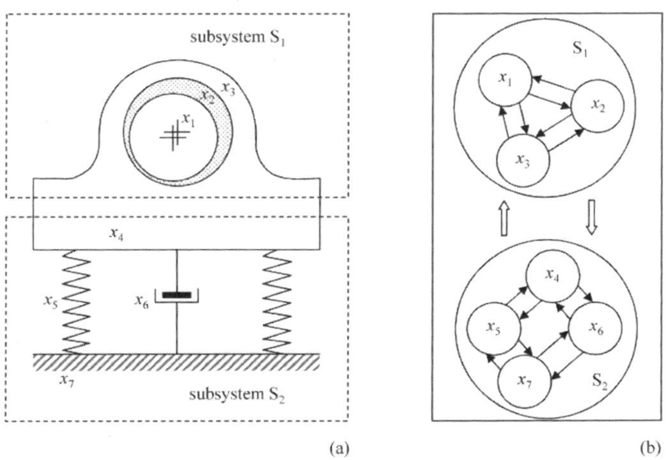

Fig. 2.3. Real system diagram (a) and its abstract model (b).

modeling as well. Further, a dynamic system can be classified as non-determined, adaptive, etc. Depending on the method of space and time expression, dynamic systems are divided into *continuous* or *discrete* ones in space and time. Dynamic systems can also be classified as *linear* land *non-linear, non-stationary* or *stationary*.

Composed dynamic system is a term used in studying the behavior of complicated sets (complexes). Every composed system can became an element (subsystem) of a more complex composed system. A *subsystem* forms an autonomic system part and its behavior need not correspond to the behavior of the system as a whole, but must be compatible with it by means of direct interactions or information flows. A *partial system* is a detached system part.

Adaptive dynamic systems are very important in simulation of controlled processes in situations where not all process parameters or properties are known at the beginning. The system parameters can be estimated only during the course of the process on the basis of information about previous behavior.

In simulating various controlled dynamic systems, the *state function* $F(U,\mathrm{Fo})$ of the following compound process, but not the function $U(M,\mathrm{Fo})$, appears often as a *controlled state quantity,* characterizing the fundamental process in place $M(x,y,z)$ and the time expressed by the Fourier number Fo. A *limiting condition* determines the extent of control of the boundary condition of the basic process. A simulator for starting up a turbo-set is an example, where the simulated basic thermal process is described by a mathematical model in the form of a partial 2nd-order differential equation; meanwhile, the following compound thermal stress process is described by a partial 4th-order differential equation. In this case, the temperature T corresponds to function U, the thermoelastic potential, and the thermal stress and deformation derived from it, corresponds to the function F.

2.1.3. Physical Similarity and Modeling

The physical similarity theory is based on similarity criteria. It is about dimensionless values replacing fully the dimensional ones in a studied phenomenon. In particular, the similarity criteria serve to state the modeling scales which state unambiguously the relations between a model and an object. The scales are determined from the equality of mutually corresponding similarity criteria for a model and an object. The determination of similarity criteria is the fundamental task in modeling, not only in simulation model creation but even for compressing and generalizing the modeling results and transferring them to other similar systems and processes. Therefore, they appear as *generalized variables.*

To determine the similarity criteria and, possibly, the dependence of criteria, similarity theory applies three methods of obtaining generalized variables:

- *dimensional analysis,*
- *analysis of a physical phenomenological model,*
- *analysis of a mathematical model.*

Sometimes, the last two methods are called the similarity analysis methods.

Because of the fact that similarity theory lowers the number of operations on a model and that of experiments and enables transferring the results to another similar problem, the efficiency of a model experiment is made higher. In Fig. 2.4, *four basic steps of modelling* are presented. By compacting the information and generalizing the variables in an experiment, the methods of similarity and modeling theory, the so-called *methods of generalized variables,* play the main role. The examined process is described by dimensional values x_1 to x_N, where N equals the total number of quantities in the process. Due to the conversion of dimensional values to dimensionless ones, by means of a Pi theorem and some of the methods of generalized variables, the total number of variables is reduced. The original process, described by dimensional values, can be expressed by a functional dependence of criteria. In the second phase, the process on a model is described analogically by the functional dependence of the dimensionless criteria of the model (M). By comparing the corresponding criteria for a model and an original, the modeling scales are determined and a simulation model (3rd step) is constructed. The fourth and last step represents the

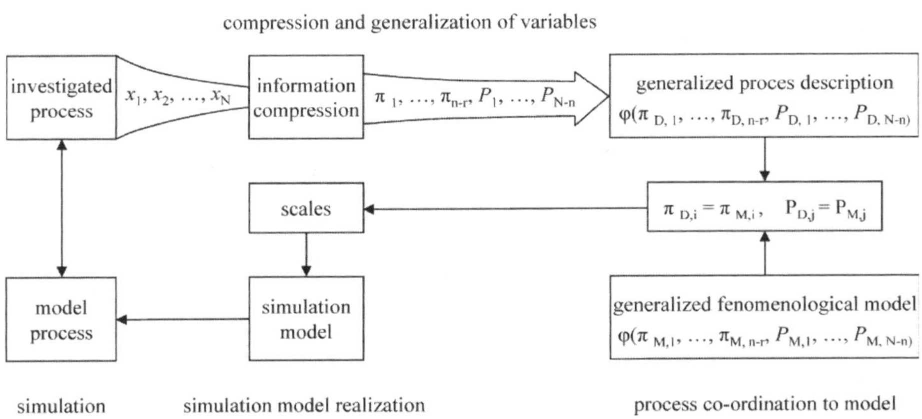

Fig. 2.4. Diagram of four fundamental modeling phases (π^i – composed criteria, P^i – simple criteria, N – total number of quantities, n – number of dimensionally different quantities, r – number of fundamental and complementary dimensions).

simulation process itself and comparison of results with the behavior of the object.

2.1.4 Overview of the Information Theory of Modeling

Essentially, modeling is an informational process in which information about the state and behavior of an examined object is obtained by means of a model. In modeling, the information increases and its *information entropy* is reduced at the same time due to the increasing knowledge of the object. The *extent of knowledge R* of an examined object can be expressed in the following form

$$R = 1 - \frac{H}{H_{max}}, \tag{2.3}$$

where H is the information entropy of the object and $H_{\max}$ is its maximum value where the amount of knowledge can become $R \in (0,1)$. The impossibility of reaching the boundary values $R = 0$ and $R = 1$ is among the basic modeling theorems. These boundaries express the ideal states. It follows from the above that there must be *a priori* and *a-posteriori information* about the object. The *amount of model information I* can be determined from the difference of *uncertainty* H_1 and *residual uncertainty* H_2

$$I = H_1 - H_2. \tag{2.4}$$

Information entropy is described by the relation derived by K.E. Shannon

$$H = -\sum_i p_i \log p_i, \tag{2.5}$$

where p_i is the probability that the studied system is in the i-state.

As opposed to some other disciplines, the model information is the main matter of interest of modeling theory. Generally, it has a quantitative as well as qualitative character. When examining it, one proceeds from three fundamental points of view – syntactic, semantic, and pragmatic ones. For modeling and system simulation, the use of all three informational standpoints is characteristic. The following is an example of the application of the above-mentioned standpoints in mathematical modeling:

A: Poisson stress equation *(syntactic access),*

B: torsion equation *(semantic access),*

C: torsion equation in a turbo-alternator shaft *(pragmatic access).*

When comparing these three cases, the quantitatively different information about the mathematical model of shaft torsion of a turbo-alternator is obvious. The syntactic standpoint (case A) is typical for classic information theory and does not take into account the content of a report that can be wrong or inexact, as it is in the cited example. Nevertheless, it must be technically and grammatically entirely correct. Case B is syntactically and semantically correct and defines a complete mathematical process model, but does not include the pragmatic part and does not answer the question about the purpose of the model. Only case C involves all three informational standpoints. In addition to the process and its mathematical model, the system in which the process is proceeding is also defined.

2.1.5 Historical Development of Modeling

Looking at the long history of its development, modeling can be divided into several shorter periods defined by substantial qualitative changes in the development of model abstraction. A more detailed explanation is presented in book [2.6]. At the beginning of its development, modeling involved the similarity the primitive man observed during his conscious activities and in contact with Nature, whereas now and in the foreseeable future, there is a complicated cybernetic model in which various hierarchically organized processors, with the structure approaching that of a brain, play the main role.

Most of the model development periods can be considered finished. However, this is not true for the *abstract model formulation period,* which culminates not only in technical fields, but also in economics, ecology, biology, sociology, and other disciplines. This period is the longest one, which began in history with the Greek astronomer and mathematician Ptolemaios, who built the first well-known mathematical model in the form of a set of cycles and hypercycles describing precisely the movement of planets he observed. The development continued in the Middle Ages, when Leonardo da Vinci assigned an abstract model to the movement of a fish in water and that of a bird in air. The development of *physical similarity and modeling theory* began towards the end of the 19th century and culminated in the 20th century. More recently, there has been a period of stagnation, due to the coming of computers and initial misunderstanding of the similarity theory in computer-aided modeling, especially in data compression, its treatment and the generalization of results. The development of modeling based on the *physical analogy* principle culminated in the 60's of the 20th century and was followed by a period of stagnation, due to the arrival of computers in modeling. Nowadays, physical analogy modeling can serve as the illustration way of task solutions.

Computer-aided modeling is undergoing intensive development which is not yet finished. The arrival of computers, the origin and development of cybernetics and the gradual mathematization and cybernetization of the sciences influence all significant fields of human activity. However, even today's computer-aided models will pass their dominant position in the system of modeling means to *cybernetic models.* The apex of development of computer-aided models can be presumed to occur in this century if one takes into account contemporary technical progress.

Technological modeling tasks still represent an important part of solving problems. But the time increment of the number of technological tasks is relatively smaller compared to the tasks in numerous other fields. Nevertheless, in technology, in addition to the nowadays tasks in classic fields such as machinery, power and electrical engineering, and technology, these tasks also appear in recently developing fields such as magneto-hydrodynamics, tribology, plasma technology, micro- and nanotechnology, geophysics, meteorology, non-equilibrium mechanics and many others.

2.2. Classification and Properties of Models

In compliance with the character of the model system, models can be divided into *abstract* (formal, ideal) and *physical* ones (objective, of material, of realization). The way to obtain the information is different in both models. The deductive process in an abstract model corresponds to experimental processing in a physical model. A *physical model* is built with natural or artificial materials. A physical model used for simulation of a dynamic system is called a *simulation model.* An *abstract model* consists of a non-material representation (by an idea, symbol or graphic expression, etc.) which describes an examined system. Abstract models are objective in terms of their content, which means that they express the laws of the real world and describe it. Nevertheless, their form is subjective because the same content can be expressed in many different forms. Among abstract models are *mathematical models,* especially, and models expressed, e.g., by programming languages, flow charts, and the like. In modeling, the use of natural language is inappropriate because it is polymorphic and uncertain.

From the model-to-object similarity point of view, models can be divided into physical models, physical analogues, mathematical analogues and cybernetic analogues. A *physical model* is based on physical similarity, a *physical analogue* is based on mathematical–physical similarity, a *mathematical analogue* is based on mathematical similarity, and a *cybernetic analogue* is based on mathematical functional similarity. In mathematical and cybernetic analogues, the term *analogue* is usually replaced with *model* as a superior term. A *natural model* is a limited case of a physical model, where the model is made identical with the object. *Physical analogues* are based on mathematical and physical similarity, where various phenomena are described by analogous equations; e.g., potential electric distribution and potential flow of a perfect liquid are described by a Laplace partial differential 2nd-order equation. *Mathematical models* are based on mathematical similarity and are usually computer-aided deterministic or stochastic ones. They have dominant positions in contemporary modeling. *Cybernetic models* are based on the principle of the functional input-to-output similarity of a system represented by a *black box*. Cybernetic models are contemporary ones, but first and foremost are a matter of the future, especially in solving global problems of mankind. In models based on mathematical similarity, the process of progressive transformation of mathematical models and the development of abstraction is remarkable.

There are other important terms connected to a model. The model-building process is called *composition* (*synthesis, construction*). In Fig. 2.5, the assignment of a model to an original object is suggested as an example. It is based on observing input X_i and output Y_i quantities. The functional relation $Y_i = F_M(X_i)$ is the model expression of quantity X_i behavior on an original. F_M denotes a *model operator. Verification* is the process of quantitative evaluation of a model's correctness. It is verified whether a model acceptably represents a system both from the structural and the behavioral point of view. The *credibility* (*validity*) expresses the quantitative degree of agreement of a model's behavior with that of the examined system. A model is credible if its behavior agrees with that of the original within acceptable limits under all the conditions considered in the investigation. This is not a precise check of hypotheses but a relatively subjective evaluation. *Validity* (*valence*) describes the extent of applications in which a model can be considered credible. The *variability* expresses the possibility to adapt a model according to changes of various factors, especially in connection with

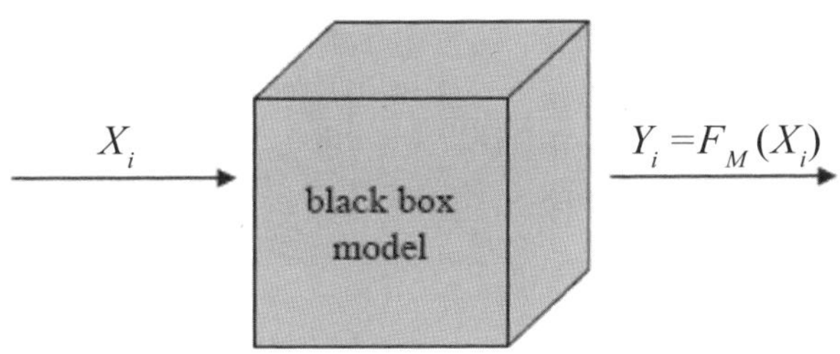

Fig. 2.5. Diagram of syntax and assignment of a functional model to the examined object.

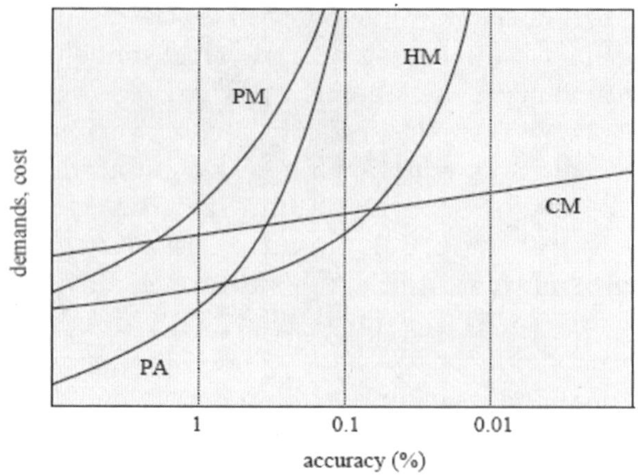

Fig. 2.6. The dependence of simulation demands on required solution precision for various simulation models (PM – physical model, PA – physical analogue, HM – hybrid model, CM – computer model).

the strategy of the simulation procedure. *Adaptability* is the capability of a model to collaborate as a subsystem with other model subsystems.

Time and *financial demands* are important characteristics of models. Usually, the costs are expressed as dependent on required *modeling precision* or on *task complexity.* This approximate dependence is expressed in Fig. 2.6, as a chart, but cannot be fully generalized. The task complexity concept involves such things as the dimensionality and shape of the solved region, the character of system parameters and boundary and initial conditions. The accuracy is a limited quantity, especially in physical models, physical analogues, analogue computer-aided models and hybrid computer-aided models, which is connected in particular with the limited accuracy of modeling elements. Sometimes, the term *credibility,* which is related to the real object, is more suitable to be used instead of the term accuracy of solution (e.g. the numerical one).

Finally, there is a *speculative model,* which is not really a model, in fact. It is called an insufficiently credible model, which is based on an excessive amount of speculations. It occurs in generating hypotheses or in case a system has not been identified. Speculative models often originate in examining complicated social, economic, ecological, and other systems. However, they also emerge if the model is irresponsibly formulated or realized.

2.2.1 Creation, Development and Transformation of a Model

Continuing from the definitions of basic modeling terms, an abstract model can be converted into a mathematical and simulation model, and various simulation models can be used.

In Fig. 2.7 the *development of an abstract model in generating a simulation model* is outlined. The first step is a thought model reflecting the real world object or system being examined.

Subsequent transformations of an abstract model are word models and then graphic or symbolic models, of which mathematical model is especially important. A simulation model originates with the solution's method and algorithm and further necessary transformations, depending on what means of simulation have been selected. The way back, indicated by the dotted line from a simulation model to an object in the drawing, only highlights the purposefulness of the simulation, which is directed to the knowledge of an object or system and, possibly, its improvement.

In Fig. 2.8, an unlimited set of possible objects is further divided into *real objects* and *projected* or *intended* (*proposed*) *objects*. All models, except the natural ones, enable the modeling of either real objects or only intended ones. However, there are a lot of objects and problems connected with them that cannot be solved nowadays. This can be due to difficulties in the realization of a physical model, or in the algorithmization of the solution, in the formulation of a mathematical or physical model, etc. The intersection of individual sets expresses the fact that many tasks can be solved with various models. The model choice depends on the aim of the solution, on the desired information, on the feasibility of the solution, and on the accessibility of modeling means, among other things. The growth of knowledge and development of natural and technical sciences, in connection with the mathematization and cybernetization of fields, has widened the limits of mathematical as well as cybernetic (functional) modeling, which is based on the similarity of the outer behavior of a system.

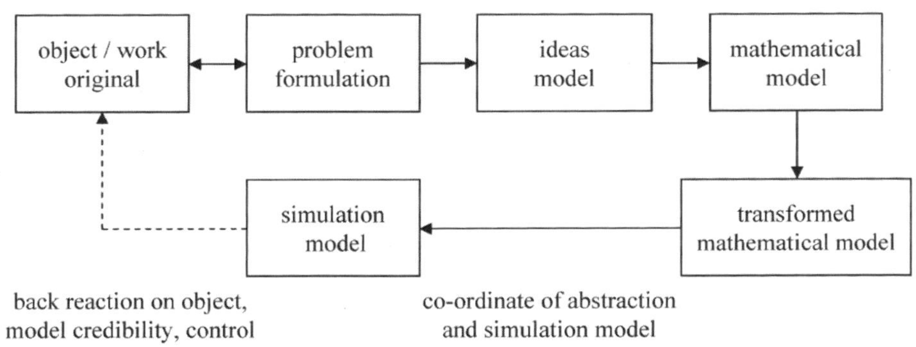

Fig. 2.7. Development of an abstract model in simulation model generation.

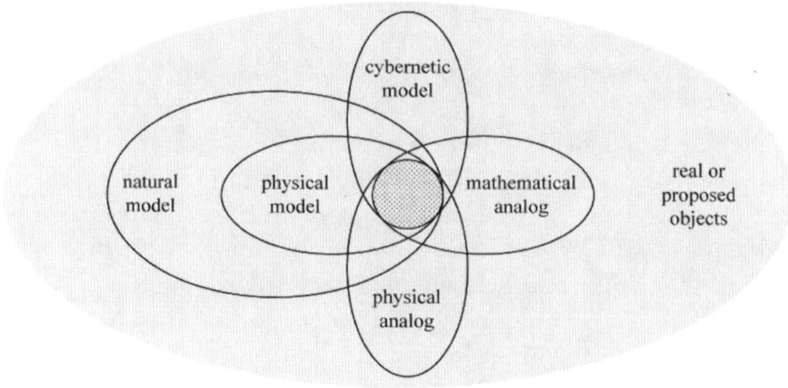

Fig. 2.8. Display of the sets of areas where various sorts of models are used.

2.2.2 Various Categorizations of Models

In Fig. 2.9, models are sorted according to several basic points of view. *Depending on the character of the modeling process* (A), models are divided into *deterministic* and *stochastic* ones. Deterministic models are distinguished by unambiguously determined causes and their consequences. In stochastic models, either the examined problem itself or the method of solution has a random character.

Similarity (B) of course relates to the similarity between an original and a model. The *way to treat the model information* (C) is among the fundamental points of view in dividing computer-aided models. *Model purpose* (D) concerns the function of a model as a means to obtain knowledge or to utilize it to control a process. *Modeling process control* (E) relates to whether the modeling process is passive from the point of view of external influence or is changing actively according to control conditions.

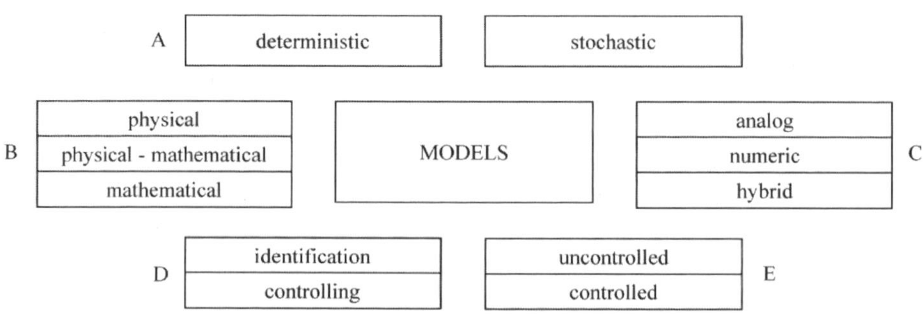

Fig. 2.9. Division of models according the character of the modeling process (A), similarity between a model and an object (B), way to treat the model information (C), purpose of a model (D) and modeling process control (E).

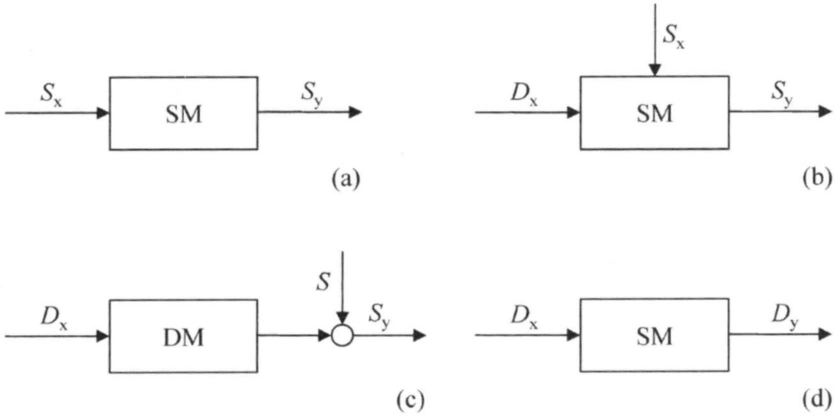

Fig. 2.10. Division of stochastic models (S, D – stochastic or deterministic quantity or process; SM, DM – stochastic or deterministic models).

In Fig. 2.10, four types of stochastic models are presented. If the examined process itself is stochastic, the model will be stochastic too (a). An example on this type of model can be found in nuclear physics where only the probabilities with which the original phenomena occur are known. Other cases involve a process which can be described in a deterministic way, but some parameters or disturbing influences on the object are random. If these influences cannot be neglected, the model will be stochastic (b). A mechanical system with random excitation force is an example. If the disturbing influences on the object can be neglected, a deterministic model (c) can be built. The resulting quantity is random due to the action of disturbing influences S on the model. In evaluating the results, the proceedings, like those for stochastic models, must be used even if the problem and the model are deterministic. A special group of stochastic models (d) is represented by those models which are used to solve deterministic problems with probability methods. These stochastic models and methods are discussed in Chapter 9.

In the majority of cases, models have been comprehended hitherto as a means of gaining new knowledge. Essentially, this is a passive role in which a model acts neither on an examined process nor on an object directly. Nowadays, the significance of models acting actively on a process or object increases. They are applied especially where the necessary information about process behavior cannot be obtained by direct measurement.

The technical point of view for the sorting of simulation models is the *means used for simulation.* All models, deterministic and stochastic,

Tab. 2.1 Sorting of models with respect to space and time

Models	Unsteady Continuous time	Unsteady Discrete time	Steady
Continuous space	CSCT	CSDT	CS
Discrete space	DSCT	DSDT	DS

analogue, numerical and hybrid, can be divided into *computer-aided,* built on programmable universal simulation means (namely computers), *single-purpose programmable computer-aided models* and *simulators*, and *single-purpose non-programmable models* and *simulators.*

Expressions of space and time in a model represent fundamental aspects of the division of models and simulation methods. It is displayed in Tab. 2.1. Among many other standpoints applied in distinguishing simulation models, the one for *model similarity preservation* is significant. In compliance with this point of view, simulation models can be sorted into complete, incomplete and approximate. *Complete models* ensure complete model similarity in space and time. *Incomplete models* (*partial, local*) are distinguished by partial similarity, e.g. in space and time. In practice, incomplete modeling often occurs when the similarity is preserved only in certain places (*O, A, B, C*) of a space or time system of coordinates, as can be seen in Fig. 2.11. In general, the function $U(X,\tau)$ is not similar on a model and an object. Modeling can be carried out when the investigation is limited to certain places (Fig. 2.11a) where the similarity conditions are fulfilled. Another example of incomplete modeling is that with similarity in partial sections *OA* and *OA'* of the characteristic, see Fig. 2.11b. This case is frequent, e.g. when the elastic stress of bodies is modeled. In *approximate models* the similarity criteria are usually not preserved. The modeling error depends on the approximate expression of the parameters and on its total influence on the course of the process as well.

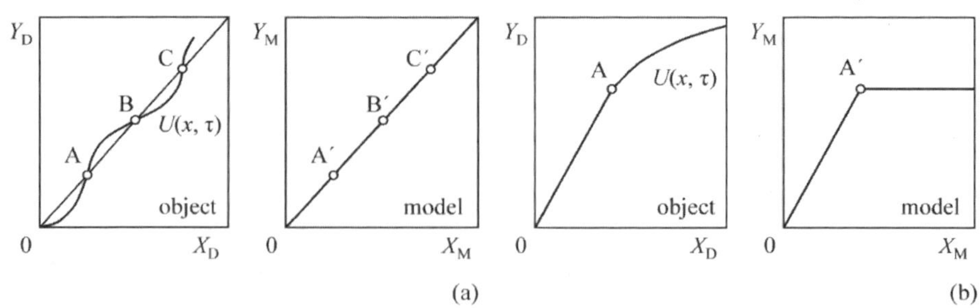

Fig. 2.11. Example of incomplete similarity in modeling – local (a) and initial (b) similarity.

2.2.3. Sorting Models According to Degree of Abstraction

The basic point of view applied in sorting models is the *level of abstraction* from a real object (original, work). The amount of information the model preserves compared to the object is decisive. According to the degree of abstraction, simulation models can be divided as follows.

Zero-abstraction-degree model
The model is the *original under working conditions* directly. The amount of information is one hundred percent.

First-abstraction-degree model (Fig. 2.12a)
The physical process is fully preserved. Some modeling scales are changed. Essentially, it is the object (original, prototype, machine) under various testing conditions. This is a *natural physical model.*

Second-abstraction-degree model (Fig. 2.12b)
The physical process is fully or partially preserved in the inner system structure in which another substance can act. For example, in liquid flow modeling the aggressive substance can be replaced by another non-aggressive one, etc. The laws of physical similarity are decisive. Some scales are not unit ones. The mathematical description is identical with that of the object but for the solution it need not be known. This is a *physical model.*

Third-abstraction-degree model (Fig. 2.12c)
The original physical process is replaced by an analogous one. The analogy between object and model inner structures is preserved. The physical similarity laws are fully or partially valid. This is based on the similarity of mathematical models of both analogue processes. The model is a *physical analogue.*

Fourth-abstraction-degree model
The similarity between the inner structures of the object and the model is not preserved. A system with distributed parameters is modeled by a system with concentrated parameters. The similarity exists in nodal points of the model only. The parameters distributed in the whole volume of an object cell are concentrated into one node in the model. The mathematical and, to a certain extent, even the physical similarity are preserved. The most widespread computer-aided models, working on the principle of numerical methods, and various discrete analogue models are among this group. If the concentration of model parameters must be emphasized, the model is called a passive network analogue.

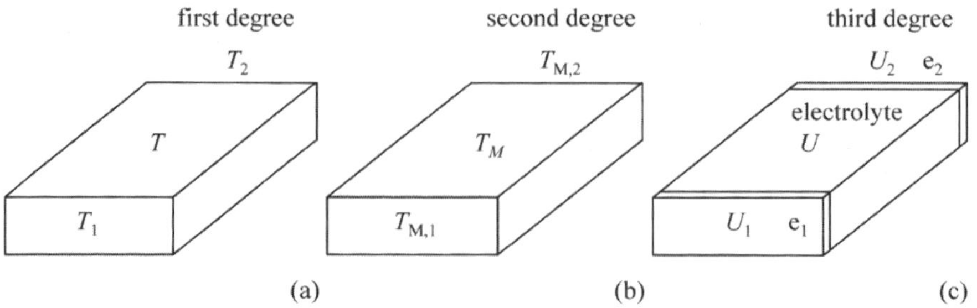

Fig. 2.12. Division of models according to the degree of abstraction – a natural physical model(a), physical model (b) and a physical analogue (c).

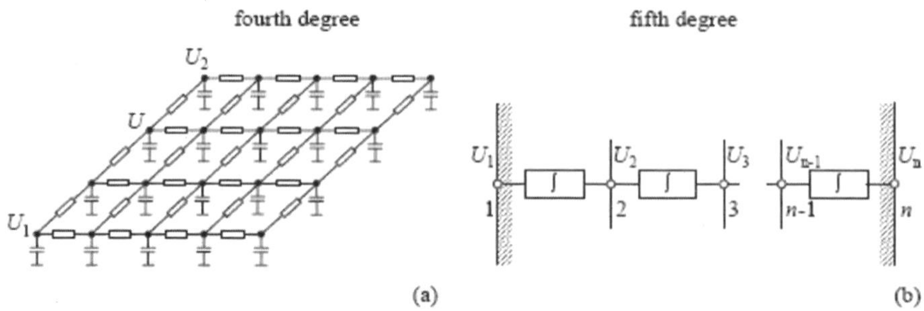

Fig. 2.13. Division of models according to the degree of abstraction – a passive network analogue (a) and active network analogue (b)

Fifth-abstraction-degree model

The original physical process in a system with distributed parameters is modeled by means of an active network analog. The mathematical process similarity is decisive at points where the inputs and outputs of elements of a mathematical model meet. The fundamental differential equation is converted to a set of ordinary differential or algebraic equations. It represents a *mathematical model.*

Sixth-abstraction-degree model

The process in a model is similar to that of an object in terms of external behavior only – functional similarity. Neither physical nor mathematical similarities in the inner structure are considered. It is a *functional model of a black box.* In Fig. 2.14, a functional system model is represented by a black box with the function $f_1(T)$ at the input and the function $f_2(T)$ at the output. The function F which is modeled depends, for example, on dimensions, physical material properties and external conditions.

Seventh-abstraction-degree model

This is a *functional model of two black boxes.* The similarity in the inner

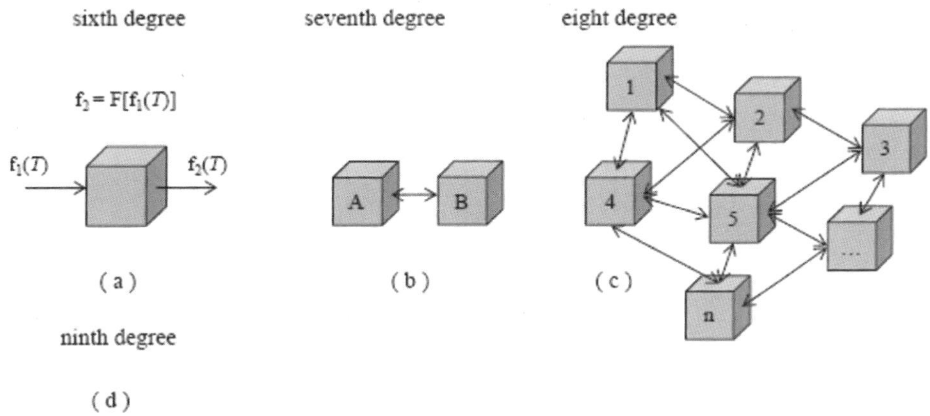

Fig. 2.14. Distribution of models according to the degree of abstraction – one-black-box functional model (a), two-black-box model (b) and multi-black-box model (c) hierar.fun.mod. (d).

structure: is not considered. Only the mathematical similarity in external mutual influence oil sub-systems A and B is considered. An example of a two-element functional model is in Fig. 2.14b, displaying the action of sub-system A on sub-system B.

Eighth-abstraction-degree model
This is a *functional model with many black boxes* and mutual influences among them. The mathematical similarity in external behavior of individual elements of a model set is decisive. Figure 2.14c shows a multi-element functional model of mutual influences among individual sub-systems.

Ninth-abstraction-degree model
This is a *two-set hierarchic functional model* in which there are inner mutual couplings in each of both sets, with resulting action of one black box (element) set on the second set.

Tenth-abstraction-degree model
These are the most complicated *functional multi-set models, organized in a hierarchy.* Such models are characteristically used for complicated biological systems especially.

Relation between model abstraction and similarity

Figure 2.15 shows a relationship between similarity and the degree of abstraction. To previously defined concepts of similarity and abstraction

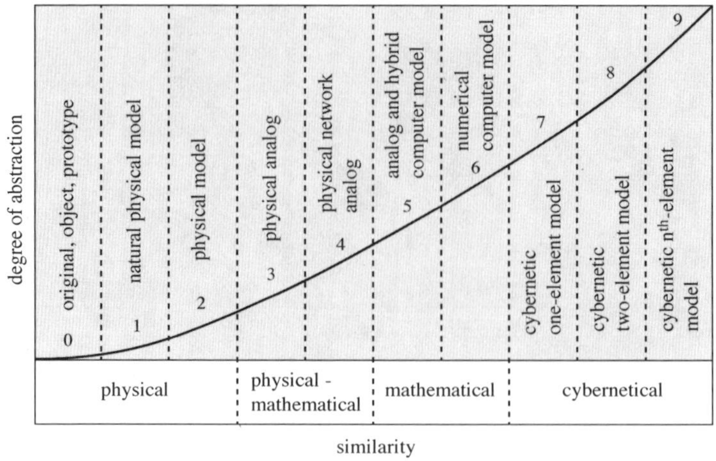

Fig. 2.15. Sorting models according to similarity and the degree of abstraction.

one can add that mathematical similarity concerns either an inner process (up to 5th degree of abstraction) or external system behavior (6th and higher degree of abstraction). According to this, all models can be divided into *models of inner and external behavior.* The inner behavior models can preserve similarity either at all points of a continuous model (up to 3rd degree of abstraction) or at selected nodal points only (models of 4th and 5th degree of abstraction).

2.2.4 Model Simplification and Credibility

Simplification is a phenomenon accompanying modeling. The more complicated the investigated problem is, the more simplification is expressed. Because there is no absolute model-original conformity even from the information standpoint, one of the fundamental questions in modeling is how to find the optimal relation between model simplification and credibility. This is valid for building both an abstract mathematical model and a physical simulation model.

There is no general rule to find this relationship. The character of the task to be solved, the aim of the solution and good knowledge of the problem's physical parameters are most significant. The designer's ability is very important to distinguish and sort out essential things from non-essential ones and to take the simplification to a limit where the model remains credible and the solution is low-cost but adequately accurate. Usually, simplification concerns the character of the process, dimensionality and shape of the modeled zone (region), physical parameters, variability of boundary and initial conditions, inner sources or troughs.

In achieving model credibility, the problem of its simplification arises especially if complicated technical and even non-technical systems, e.g. biological ones, are modeled. Figure 2.16 shows an example of human-body model simplification to parts in the shape of a multilayer plate or cylinder expressing the anatomical structure of the body. However, this simplification can be made especially if medium-integral quantities, e.g. blood temperature in a body, are being investigated. A practical case of thermal process modeling of the human body exposed to a thermal shock is discussed in Chapter 8.

The above-mentioned example of the simplification of a model of the human body, as a complex biological system, to a simple plate is an extreme case which has many analogies, even in other areas. On the other hand, there are a lot of cases in which the zone shape cannot be simplified. The fact that each mathematical and simulation model always has errors does not belittle the significance and results of modeling. In simulation, diverse differential diagrams of an approximate mathematical model can be defined very precisely, but one must consider whether higher modeling claims are satisfactorily counterbalanced by more precise results.

2.2.5. Tasks and Algorithmization of Solutions

Input analysis of the problem and determination of the task character are fundamentals requirements for success of modeling.

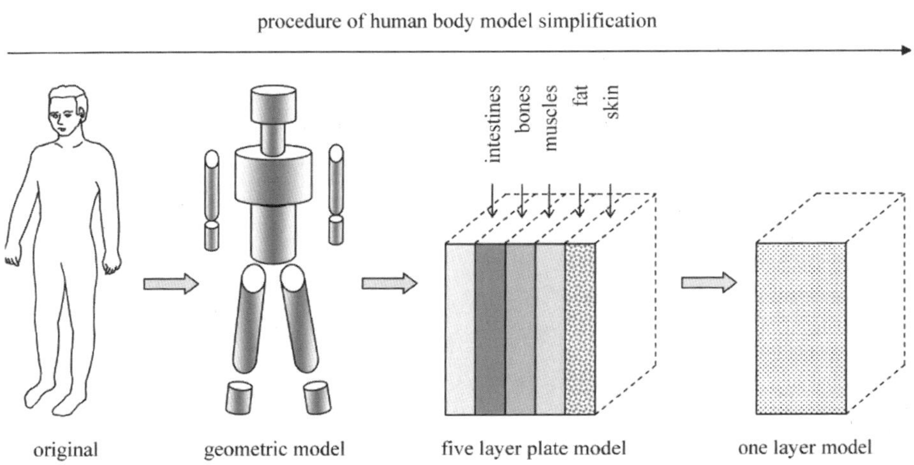

Fig. 2.16. Example of model simplification of a complex biological system – the human body.

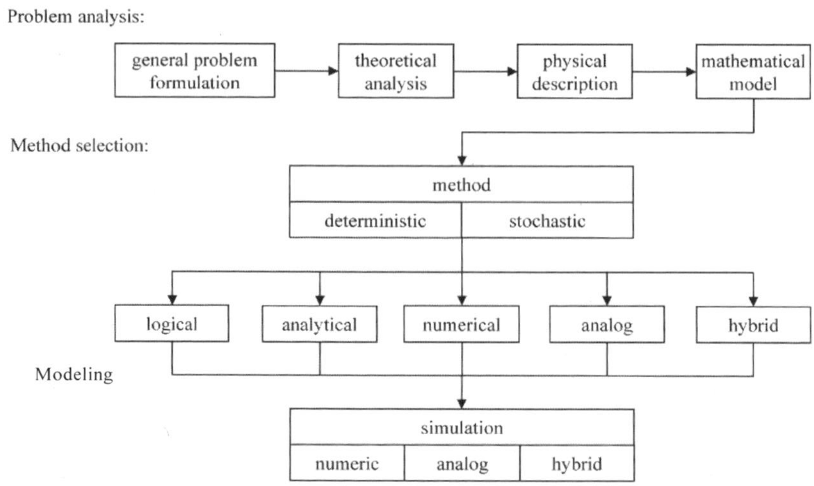

Fig. 2.17. Procedure of problem analysis, method selection, and model solution.

Problem analysis

The *input analysis* in which a complicated task is divided into partial physical tasks is an important part of every problem solution. Depending on the character of partial tasks, theoretical or experimental solutions or, possibly, both of them, are chosen. Figure 2.17 shows a procedure diagram for problem analysis, model selection and solution. The theoretical problem analysis, aimed at building physical and mathematical descriptions, is the most difficult part. The choice of the method, for which a deterministic or stochastic approach can be applied, is a further step. As will be described in further chapters, each of these approaches can give a completely different significance to the concept of the task complexity. Logical, analytical, numerical, analogous, and hybrid methods are available. After the method has been chosen, a *modeling solution* follows in which specific properties of numerical, analogue, and hybrid methods can be utilized.

Classification of tasks

In modeling, the tasks can be divided essentially into four basic groups, which differ in their characters and also the method of solution. There are often required *direct tasks,* and *indirect tasks, identification* and *optimization tasks* as well.

Figure 2.18 shows a diagram for classification of tasks in modeling. The task always involves a fundamental equation or a set of them and further conditions for an unambiguous solution, among which are boundary and initial conditions, and further restrictive and optimization conditions. In solving stationary tasks, a boundary condition is sufficient to obtain an unambiguous solution, whereas for non-stationary processes the initial condition must also be considered. In optimization, restrictive and optimization conditions are joined to the above-mentioned ones.

Direct tasks

In solving direct tasks, it is essential to determine the inner system response to external stimulation (boundary condition) and initial system state (initial condition), with given system parameters (zone shape, physical parameters, inner sources, movement of a boundary or a source) and known mathematical description. The solution results in a statement of a dependent variable as a function of place and time. Direct tasks can be sorted according various characteristics, e.g. into *steady* and *unsteady, linear* and *non-linear*, those with *simple* or *composed boundary conditions*, those with *movement of a boundary, phase change*, with *inner sources or sinks* of energy, and the like.

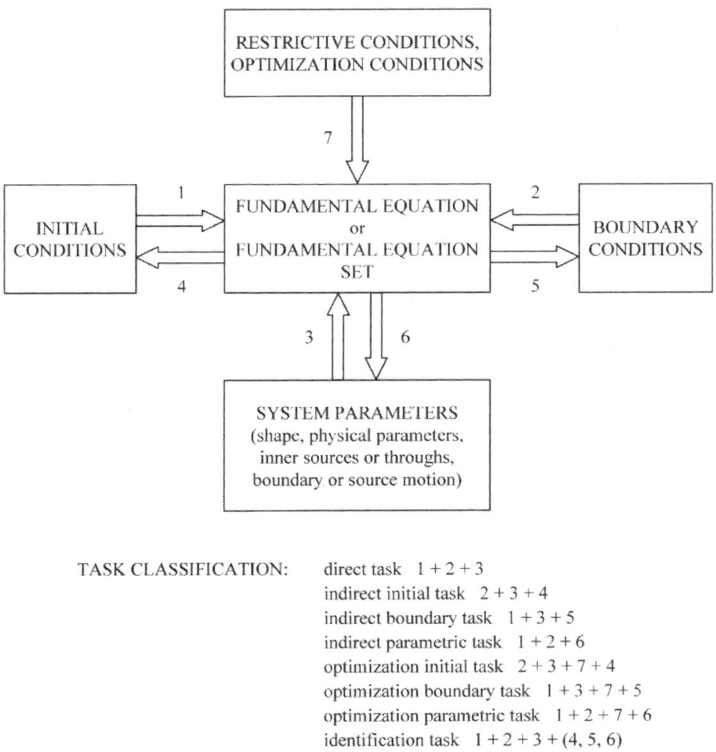

Fig. 2.18. Task classification in modeling.

The majority of direct tasks in the last century can be explained by the fact that the focus of problems was in analysis of results for investigated processes rather than in optimal control of these processes.

Indirect tasks

Indirect tasks are opposite to direct ones and are therefore called reversed or, sometimes, return tasks. The corresponding external influences, initial state or system parameters are determined for the known-in-advance or requested inner behavior of the system, expressed by a known dependent variable in a system. There are many indirect tasks which can be put into three basic groups. For a *boundary indirect task,* a boundary condition is sought with the known dependent variable values in the system, known system parameters, initial conditions and a mathematical description. For an *initial indirect task,* an initial condition is sought with given boundary conditions, system parameters, dependent variable values and a mathematical description. For a *parametric indirect task,* a certain system parameter (zone shape, physical parameters, inner sources, boundary or source motion) is sought with the given boundary and initial conditions, known dependent variable values and a mathematical description. A parametric task is often called an inner or inverse task. The solution of strongly non-linear parametric tasks is very complicated.

Identification tasks

Identification tasks are solved to state or, more often, to increase the accuracy of a system model, i.e. of a fundamental equation or a set of equations, and of equations corresponding to the initial conditions or system parameter distribution. Essentially, an input-to-output transmission of an investigated system is being sought. In solving, the system responses to external stimulations are being sought without knowing the inner system behavior, with the system represented in such a case by a black box for which the input-to-output transmission is being solved. In the case of linear systems, one can pass, by means of a reverse transformation, from a measured transmission to a fundamental mathematical model or to making it more precise. Sometimes, the identifications tasks are called inductive ones.

Optimization tasks

The above-mentioned direct, indirect and identification tasks deal with system behavior and a model of it. The optimization tasks are directed to optimize the system behavior in accordance with a given criterion, i.e. to control the process being underway therein, in order to fulfil the conditions for the optimization of the process with the given restrictive conditions (limiting the model variability during its optimization) and optimization conditions (expressing optimal values of a system dependent variable). Because the system behavior can be affected mainly by regulating the boundary or initial conditions or system parameters, the optimization tasks are closely related to indirect tasks. The restrictive and optimization conditions are given by the character of the task to be solved.

The optimization tasks can be divided into four groups. A *boundary optimization task* is most frequent. With the given initial condition and system parameters, the boundary condition distribution in time and space is being sought so that the dependent variable may fulfil the restrictive and optimization conditions in addition to a fundamental equation. Similarly, for an *initial optimization task,* the initial condition distribution is being sought, while for a *parametric optimization task* some system parameter distribution in time and space is being sought. A *mixed optimization task* is a combination of these tasks.

The solving optimization tasks involves a qualitative change in the approach to various technical and even non-technical task solutions. Indirect tasks as well as optimization ones can be solved as a large number of alternatives (approximations) of direct tasks too. One then speaks about an *alternative (approximation) indirect* or *optimization task.* Although the alternative task solution is simpler, it is usually much more time-consuming to solve the large number of alternative tasks than the indirect or optimization task.

"If an equation is dimensionally homogeneous, it can be reduced to relationship among a complete set of dimensionless products."

Edgar Buckingham *(8.7.1867–29.4.1940)*

3

Dimensional Analysis

Dimensional analysis involves a dimensional model analysis of acting quantities in the investigated process. It enables one to determine, in a simple algebraic way, dimensionless similarity criteria and functional relations, represented amongst them by a criterion equation. Further, it enables the conversion of physical quantities into other various fundamental sets of measuring units, the conversion of measuring units and other procedures. In modeling and experiment, its main function is to reduce the amount of independent variables, to simplify the solution and to generalize the results thereof. It can become an effective method, especially if a complete mathematical model of the investigated process is not known.

The first dimensionless combination of several diverse physical quantities, binding characteristic length L, dynamic viscosity η, density ρ and flow velocity w of a viscous fluid was introduced by Reynolds in 1883. The corresponding criterion is called the Reynolds number $\mathrm{Re} = \rho w L \eta^{-1}$.

Dimensional analysis uses unpretentious mathematical operations over units or dimensions of physical quantities, the aim of which is to obtain dimensionless criteria or functional relations among quantities. The dimensionality of the quantities expresses qualitative information about the physical quantity. In it, incomplete and various qualitative particularities of the physical quantity are reflected. Nevertheless, because functional couplings among physical quantities appear in dimensional expressions, even the nature of laws among the quantities of the investigated phenomenon can be determined from a dimensional analysis in some cases. Sometimes, even the structure of generalized

J. Kuneš, *Similarity and Modeling in Science and Engineering*,
DOI: 10.1007/978-1-907343-78-0_3,

variables can be defined. However, one cannot penetrate deeper into the substance of the physical problem based on knowledge of acting quantities and their dimensions only. Here, the realistic assessment of H.L. Langhaar is valid: ''the power and weakness of dimensional analysis is rooted in its generality''.

This is a method, simple from the practical point of view, which does not enable either solving a problem completely or revealing important inner couplings of an investigated phenomenon. However, it is an extraordinarily effective means of obtaining an idea about the behavior of a phenomenon if neither its complete mathematical nor physical descriptions are known. Usually, it is an important physical tool in every more complicated physical, scientific or industrial experiment. The main functions of dimensional analysis are the following:

• *determination of the number and form of dimensionless quantities* which represent the similarity criteria,

• *reduction of the number of independent variables in an experiment,* simplification of the solution and generalization of its results,

• *conversion of the basic set of units* of the measurement,

• *conversion of physical quantities* into another basic set of units of measurement,

• *determination of functional relations* in cases where the solver does not know more detailed information on the physical principle of the investigated phenomenon and no complete mathematical description of the phenomenon is known.

In application of dimensional analysis, the highest efficiency is reached in its combination with general physical ideas obtained by a solver directly from experiments. The depth of previous knowledge of the physical principles of the investigated phenomenon can influence and extend considerably the possibilities of the dimensional analysis.

3.1 Quantities, Dimensional Matrix and Similarity Criteria

In describing a process, dimensional as well as dimensionless quantities appear. Dimensional analysis utilizes dimensional quantities only.

3.1.1 Quantities and Units

Dimensional physical quantities and physical constants have fundamental significance in investigating various phenomena. The *physical quantity* is measurable and can be defined by means of a *measuring unit.* In such cases, based on physical quantities, the laws of Nature can be defined in the form of mathematical equations, representing

Tab. 3.1 Basic and complementary quantities and units in the SI system

Quantity	Quantity symbol	Unit name and symbol	Dimensional symbol
length	l	meter (m)	L
mass	m	kilogram (kg)	M
time	T	second (s)	T
electric current	I	ampere (A)	I
thermodynamic temperature	T	kelvin (K)	Θ
substance quantity	n	mole (mol)	N
luminous intensity	I	candela (cd)	J
planar angle	α, β	radian (rad)	α
space angle	Ω, w	steradian (sr)	Ω

mathematical models. This enables the solution of various tasks theoretically or experimentally.

For physical quantities, a certain mutual coupling is characteristic. With fundamental and complementary units of measurement chosen for a certain number of quantities, i.e. the so-called basic quantities, the measuring units of other (derived) quantities represent the functions of fundamental and complementary units. *The expression of an arbitrary measuring unit by means of fundamental and complementary units is called a dimension.* This is the formal product of all dimensional symbols with corresponding exponents. Table 3.1 shows the *fundamental and complementary quantities in the SI system* in compliance with the international standards, including the introduction of *dimensional symbols.*

3.1.2 Fundamental Theorems of Dimensional Analysis

Dimensional analysis is supported by two fundamental theorems. The *first theorem* expresses the *dimensional independence* of the selected set of fundamental units of measurement. The fundamental set of measuring units comprises the fundamental land complementary units of this set.

The relation between units is independent of the selected set of fundamental units of measurement.

Because arbitrary units can be expressed dimensionally in the form of a product of powers of the fundamental units, the relation between quantities does not change if the measuring units are changed. The theory of dimensionality was developed significantly in the work of J. Bertrand, who proved that *all physical laws,* including those not yet known, *must be subjugated to the dimensionality condition.*

The *second theorem* expresses the *dimensional homogeneity* of physical equations.

All quantities, used in the equations representing a physical model of a physical process must be expressed in the same set of units of measurement. With this condition observed, the dimensions of all additive members of the equation are the same and remain so even in an arbitrarily selected set of fundamental measuring units.

From the dimensionality point of view, an arbitrary equation describing an investigated physical process, and all members in the equation as well, can be expressed by means of a single set of measuring units. It such a case, all members of this equation have the same dimensions. The second theorem is extraordinarily significant in similarity and modeling theory and represents the base of *dimensionality theory.* It was formulated first by Fourier in 1822.

With the factors acting on the investigated process and the quantities involved in investigated relations determined, the type of the dependence sought can be determined from the principle of the homogeneity of equations. The significance of the equation homogeneity theorem is discussed in the following text.

3.1.3 Dimensional Matrix

The dimensional matrix is the cornerstone of dimensional analysis. From this matrix, the similarity criteria are obtained.

Let us use N dimensional quantities x_1, x_2, x_N to characterize an investigated system or process. It follows from the properties of physical values that, in a general case, an arbitrary variable can be expressed in a system of fundamental and complementary units SI $(L, M, T, I, Q, N, J, \alpha, \Omega)$ in the following form

$$\left[x_j\right] = L^{a_{1j}} M^{a_{2j}} T^{a_{3j}} I^{a_{4j}} \Theta^{a_{5j}} N^{a_{6j}} J^{a_{7j}} \alpha^{a_{8j}} \Omega^{a_{9j}}, \qquad (3.1)$$

$$j = 1, \ldots, N,$$

where $a_{1j}, \ldots, a_{9j}$ are unknown exponents. With fundamental and complementary units, with Y_i for $i = 1, \ldots r$, where r can reach the value of 9 in a general case. Equation (3.1) can be written in the form

$$\left[x_j\right] = \prod_{i=1}^{r} Y_i^{a_{ij}}, \qquad (3.2)$$

$$j = 1, \ldots, N.$$

The physical quantities describing the investigated system or process are determined after a careful physical analysis. Usually, a great many quantities are found, but it is useful not to consider those quantities which have a negligible influence on the investigated system or process, in order to simplify the solution. In this selection of physical quantities, the physical knowledge of the problem, the experience and the intuition of the solver are very important.

The dimensions of the above-mentioned variables x_1, x_2, x_N can be expressed by exponents a_{ij} arranged in the following *dimensional matrix*

$$\mathbf{A} = \begin{array}{c} \\ Y_1 \\ Y_2 \\ \vdots \\ Y_r \end{array} \begin{array}{c} \begin{array}{cccc} x_1 & x_2 & \cdots & x_N \end{array} \\ \begin{bmatrix} a_{11}, & a_{12}, & \cdots, & a_{1N} \\ a_{21} & a_{22}, & \cdots, & a_{2N} \\ \vdots & \vdots & \ddots & \vdots \\ a_{r1} & a_{r2} & \cdots, & a_{rN} \end{bmatrix} \end{array}, \tag{3.3}$$

where, e.g., a_{21} denotes the exponent at the fundamental dimensional symbol Y_2 of the variable x_1. Some of the exponents a_{ij} can equal zero.

3.1.4 Similarity Criteria

The dimensionless quantities are also often called similarity criteria. This name originated historically because these criteria are the same in similar systems and processes.

The *simple criterion P* (simplex) is a dimensionless quantity expressing the ratio of two physical quantities equal with respect to dimensions. The *parametric criterion* is a simple criterion expressed as the ratio of a quantity to its relative value. The *composed criterion* π (complex) is a dimensionless quantity expressing the coupling among several physical quantities different with respect to dimensions. The Π is used as the summary symbol for simple as well as composed similarity criteria.

Every criterion originates as a combination of quantities x_1, x_2, x_N, and, therefore, a common arbitrary criterion can be written in the following form

$$\prod_i = x_1^{e_1}\, x_2^{e_2} \cdots x_N^{e_N}, \tag{3.4}$$

$$i = 1, \ldots, N-r,$$

where e_1, e_N are unknown exponents, and $N-r$ is the number of all criteria. For the dimension of the criterion, the following holds

$$[\Pi_i] = 1, \tag{3.5}$$

$i = 1, \ldots, N - r$

After inserting the Eqs. (3.2) and (3.4) into (3.5), one obtains

$$\prod_{i=1}^{r} Y_i^{a_{i1} e_1 + a_{i2} e_2 + \cdots + a_{iN} e_N} = 1. \tag{3.6}$$

This equation is fulfilled if expressed by a system of linear algebraic equations

$$\sum_{j=1}^{N} a_{ij} e_j = 0, \tag{3.7}$$

$i = 1, \ldots, r.$

which can be written in the following matrix form

$$\mathbf{A e} = \mathbf{0}, \tag{3.8}$$

where **A** is the dimensional matrix and **e** is the column vector of exponents.

Basic properties of a dimensional matrix

With the matrix **A** having N columns, r rows, and the rank r, there are $N - r$ independent solutions. Thus, arbitrary $N - r + 1$ solutions will be linearly dependent on preceding ones. This means that $N - r$ independent solutions deliver $N - r$ independent similarity criteria and the $(N - r + 1)^{\text{th}}$ criterion is expressed by $N - r$ independent criteria, where N is the number of dimensional physical quantities x and r is the number of fundamental and complementary units Y.

If the rank of the matrix **A** is less than the number of rows or dimensions, i.e. in the matrix there is a row linearly dependent on other rows, as can happen now and again, some dimensions must be omitted so that the rank of the reduced matrix may equal the number of rows. Then the reduced dimensional matrix is processed as an ordinary dimensional one. Of course, the number of dimensions r must also be reduced by a corresponding number of omitted dimensions. Should the linearly dependent row not be omitted and the rank of the matrix **A** remains less than the number of rows, the sub-matrix $\mathbf{A}_Z$ could not be obtained with a non-zero determinant (by changing the columns in the dimensional matrix). However, in the majority of cases the matrix rank equals the row number, i.e. the number of basic and complementary dimensions.

If the dimensional quantity has one of the basic and complementary dimensions which no other dimensional quantity of the dimensional matrix has, then this quantity is dimensionally irrelevant. This means that it cannot be contained in similarity criteria to be determined, and this quantity must either be omitted from the dimensional matrix or, on the contrary, another quantity must be added which will contain this dimension as well.

3.2 Determination of Number and Form of Similarity Criteria

From the known number of acting characteristic quantities and chosen fundamental and complementary dimensional units, the Pi theorem determines how many dimensionless variables occur in an investigated physical process. The form of similarity criteria is determined using the matrix solution of the dimensional matrix **A.**

3.2.1 Pi theorem

The Pi theorem represents the fundamental similarity and modeling theorem which expresses substantially the process of condensing and generalizing the model information. Sometimes, but not quite accurately, it is called the *Buckingham π theorem.* This original π theorem was based on some inaccurate presumptions which did not enable its general formulation. From many other more accurate formulations of this theorem, mentioned, for example, by H.L. Langhaar and others, we are going to present the formulation of the extended Pi theorem by L.I. Sedov, as complemented by L.S. Eigenson, and which can be held as the most general one.

Edgar Buckingham (8.7.1867–29.4.1940), American physicist.
He dealt with the capillary theory and the energy potential in soil physics theory, and was the first to expound the dependence of soil hydraulic conductivity on the capillary potential. He is the originator of the Buckingham π theorem in the field of dimensional analysis.

In general, the investigated physical phenomenon can be expressed by N-dimensional physical quantities x in the form of the following equation

$$f\left(x_1,x_2\cdots,x_N\right)=0. \tag{3.9}$$

Further, n is defined as the number of quantities, different with respect to dimension, x_1, x_2, .., x_N and r as the number of fundamental and complementary dimensions of these quantities. The number of all dimensional physical quantities is supposed to be greater than the number of their fundamental and complementary dimensions.

Should $N = r$, the dimensional matrix would be square and either the submatrix $\mathbf{A}_Z$ would be a singular one (i.e. having zero-determinant) or no relation among acting quantities could be created[1]. With $N < r$, either the rank of the dimensional matrix would be $N - 1$ or no relation among acting quantities can be created[2].

Generally, dimensionless quantities can occur even in the description of an investigated physical phenomenon. The dimensionless quantities can be either constants, e.g. the Ludolph number π, the Euler number e, the stechiometric number α_i, etc., or variables, e.g. the Poisson number ν, emissivity ε, friction coefficient f, etc.

- The dimensionless quantities – the variables – are expressed, if possible, by dimensional quantities which then occur among N-dimensional physical quantities. Among the simple, possibly composed, dimensionless criteria found, the said dimensionless quantity – the variable – occurs. If it is not suitable to express these dimensionless quantities – the variables – by means of dimensional ones, e.g., due to the complexity of expressing them, one can process them as dimensionless quantities – the constants.
- The dimensionless quantities – the constants – are not quantities in the exact meaning of the word, but numbers only. Therefore, they cannot be expressed by dimensional quantities and are out of the Pi theorem and methods of generalized variables. After the simple and composed similarity criteria determination, these dimensionless quantities – the constants – can be assigned to simple criteria.
- There is one other possibility to work with dimensionless quantities which we neither wish to nor can express by dimensional quantities. We incorporate these dimensionless

[1]If the matrix $\mathbf{A}_Z$ were said not to be singular, the rank of the matrix $\mathbf{A}$ would have to be r and $N - r = 0$ similarity criteria would have to exist, which is a contradiction because the minimum number of similarity criteria is at least one, i.e. with $N = r$, the rank of dimensional matrices must be less than r and some dimension, e.g. of a row of a dimensional matrix, must be omitted.

[2]Because the number of searched similarity criteria $N - r$ must be greater than or equal to one, then $N - r \geq 1$ must be valid and therefore $r \leq N - 1$, i.e., the number of dimensions must be less than or equal to $N - 1$.

quantities into the number N of all physical quantities which describe the investigated process, where the number n of quantities, different as to dimension, and the number r of dimensions remain without change.

The formulation of the π theorem according to Buckingham:

The number of dimensionless composed criteria (k_k) *equals the difference between all quantities* (n), *different in dimension, acting in a process, and the number of fundamental and complementary dimensions* (r).

For the *number of composed criteria,* the following holds

$$k_k = n - r. \tag{3.10}$$

If in a group of acting quantities there are ones equal in dimensions, i.e. longitudinal coordinates x, y, z and characteristic length L, the process can be described not only by composed but by simple criteria as well.

The Buckingham π theorem does not define the number of criteria. This can be established by the more exact formulation of L.S. Eigenson as a supplement to the basic π theorem:

. . . *If there are quantities equal in dimension in a group of acting quantities, then the number* (k_s) *of simple criteria equals the difference between the total number* (N) *of acting quantities and the number* (n) *of quantities with different dimensions.*

Therefore, for the *total number of criteria* and the *number of simple criteria,* the following is valid

$$k_s = N - n, \tag{3.11}$$

$$k = k_k + k_s = N - r. \tag{3.12}$$

Therefore, the *Pi theorem* formulation is made by using the Buckingham π-theorem, enlarged by L.S. Eigenson's formulation.

Thus, the investigated process can be described, instead of using N variable dimensional quantities, by means of $n–r$ dimensionless composed criteria π and $N–n$ simple criteria P in the form of the following equation

$$\varphi\left(\pi_1, \pi_2, \ldots, \pi_{n-r}, P_1, P_2, \cdots, P_{N-n}\right) = 0. \tag{3.13}$$

The number of simple and composed similarity criteria can be determined by means of the Pi theorem. The form of criteria is obtained by one of three methods of generalized variables – by dimensional analysis, physical model analysis, or mathematical model analysis.

3.2.2 Matrix Solution of Similarity Criteria

The aim is to convert, by means of a certain method, the original dimensional matrix **A** to a solution matrix **B**, from which individual similarity criteria can be determined directly. The procedure of matrices transformation can be put into algorithmic form and processed advantageously by a computer.

1. The *dimensional matrix* **A** of (r, N) type is made and arranged so that the coefficients corresponding to decisive quantities should be in the first $N–r$ columns. These are dependent quantities and, in general, all those which ought to be involved in the similarity criteria only once and can be easily measured. In choosing the decisive quantities, the character of the experiment and the possibility of measuring the quantities are taken into account.

$$\mathbf{A} = \begin{array}{c} \\ Y_1 \\ Y_2 \\ \vdots \\ Y_r \end{array} \begin{array}{c} x_1 \quad x_2 \quad \cdots \quad x_{N-r} \quad x_{N-r+1} \quad \cdots \quad x_N \\ \left[\begin{array}{c|c} \quad \mathbf{A}_S \quad & \quad \mathbf{A}_Z \quad \end{array} \right] \end{array}, \tag{3.14}$$

where $\mathbf{A}_S$ is a *sub-matrix of decisive quantities* of $(r, N - r)$ type and $\mathbf{A}_Z$ is the *residual sub-matrix* of (r, r) type.

2. The quantities in the dimensional matrix should be arranged according to the following rules:

- In the sub-matrix $\mathbf{A}_Z$, every quantity must have different dimensions, otherwise the determinant of this sub-matrix would equal zero, i.e., if more dimensional quantities with equal dimensions are present in the dimensional matrix, then only one of them can be in the sub-matrix $\mathbf{A}_Z$.
- In the sub-matrix $\mathbf{A}_Z$, only independent dimensional quantities should be present, i.e. the input quantities of the investigated process.
- In the sub-matrix $\mathbf{A}_S$, dependent dimensional quantities, depending on other quantities, should be present and, in general, all quantities which have to occur only once in similarity criteria, can be isolated easily and can be measured without any difficulty.
- In the sub-matrix $\mathbf{A}_S$, possible dimensionless quantities should be present which are going through the Pi theorem and the dimensional analysis.

3. The regularity of the following residual sub-matrix is verified

$$\det \mathbf{A}_Z \neq 0. \tag{3.15}$$

With the determinant of the residual sub-matrix $\mathbf{A}_Z$ equalling zero, the rank of the **A** matrix must be determined. If its rank equals the number of rows, it is sufficient to shift the columns in the dimensional matrix **A** between the left and right sub-determinants so that the determinant of the residual sub-matrix $\mathbf{A}_Z$ would be non-zero.

With the rank of the matrix **A** less than the number of rows, i.e. the number of dimensions, a number of rows (dimensions) must be omitted so that the rank of the **A** matrix would equal the number of rows.

4. The *transposed matrices* $\mathbf{A}_S^T$ and $\mathbf{A}_Z^T$ are created, the *inverse matrix* $(\mathbf{A}_Z^T)^{-1}$ is calculated and then the matrix $\mathbf{B}_1$ of $(N-r,\ r)$ type is given by the following relation

$$\mathbf{B}_1 = -\mathbf{A}_S^T\left(\mathbf{A}_Z^T\right)^{-1}. \tag{3.16}$$

5. From the unity matrix **I** of $(N-r,\ N-r)$ type and the residual matrix of the solution $\mathbf{B}_1$, the *matrix of the solution* **B** of $(N-r,\ N)$ type is made as follows

$$\mathbf{B} = \begin{array}{c} \\ \prod_1 \\ \prod_2 \\ \vdots \\ \prod_{N-r} \end{array} \overset{\begin{array}{cccccc} x_1 & x_2 & \cdots & x_{N-r} & x_{N-r+1} & \cdots & x_N \end{array}}{\left[\begin{array}{c|c} \quad \mathbf{I} \quad & \quad \mathbf{B}_1 \quad \end{array}\right]}. \tag{3.17}$$

6. The individual similarity criteria Π_i are determined directly from the matrix **B**

$$\prod_i = \prod_{j=1}^{N} x_j^{b_{ij}}, \tag{3.18}$$

$$i = 1\ldots, N-r.$$

The practical procedure of transforming the dimensional matrix **A** to the solution matrix **B** and of determining the form of simple and composed criteria is illustrated in the Example 3.2. This example relates to the preceding one where the number of similarity criteria was determined.

3.2.3 Sets of Similarity Criteria

The set of similarity criteria which describes the investigated process is determined by the dimensional analysis procedure. However, if the columns in a dimensional matrix are suitably arranged over, other similarity criteria sets are obtained. All these similarity criteria sets describe the investigated process. The similarity criteria from one set can be expressed by means of the criteria of another arbitrary set.

If the similarity criteria describing the investigated process are being sought, it is appropriate to seek a similarity criteria set that contains as many established similarity criteria as possible, i.e. internationally utilized ones. From the physical point of view, nothing would prevent another criteria set from being used. However, in this case, the similarity criteria found and criterion equations which contain these criteria could not be compared with other authors' research works.

As mentioned above, another similarity criteria set can be obtained by appropriately rearranging columns in a dimensional matrix. Shifting columns in the sub-matrix $\mathbf{A}_S$ does not lead to a different set because it results in the interchanging of rows in the matrix $\mathbf{B}_1$ and the same similarity criteria are obtained. Neither does the shifting of columns in the sub-matrix $\mathbf{A}_Z$ lead to a different set, because it causes the interchanging of columns in the matrix $\mathbf{B}_1$, and the same similarity criteria are obtained. Only mutual shifts of columns between the sub-matrices $\mathbf{A}_S$ and $\mathbf{A}_Z$ can lead to different similarity criteria sets.

The *greatest possible number of different sets of similarity criteria* s is given by the number of column arrangements into sub-matrices $\mathbf{A}_S$ and $\mathbf{A}_Z$ where the order of columns in both sub-matrices is not important, i.e.

$$s = \binom{N}{N-r} = \binom{N}{r} = \frac{N!}{(N-r)!\,r!}. \qquad (3.19)$$

It corresponds to selecting r columns from N ones with the sequence not being important. However, the rank of the dimensional matrix is supposed to equal the number of dimensions, due to which a sub-matrix $\mathbf{A}_Z$ can be created with a non-zero determinant.

The *real number of different similarity criteria sets* equals the greatest possible number of sets s, reduced by the number of forbidden mutual interchanging of columns between the sub-matrices $\mathbf{A}_S$ and $\mathbf{A}_Z$ and by the number of possible mutual interchanges between the sub-matrices $\mathbf{A}_S$ and $\mathbf{A}_Z$, which lead to the same similarity criteria set. The forbidden mutual interchanges of columns between the sub-matrices $\mathbf{A}_S$ and $\mathbf{A}_Z$ are those which lead to a singular sub-matrix $\mathbf{A}_Z$, i.e. with a zero determinant.

In certain cases an automatic process is useful which tests all possible selections of input physical quantities for the residual sub-matrix, solves the corresponding dimensional matrix simultaneously and tests how many already applied similarity criteria are contained in the solution matrix found. Some of the found similarity criteria sets, containing the maximum number of the already applied criteria, is a set that is used for final arrangement of similarity criteria which is based on the solver's experience and intuition.

3.2.4 Examples

The procedure to determine the number of simple and composed similarity criteria using the Pi theorem is illustrated in the following example. In the next example, similarity criteria are obtained using the matrix solution procedure.

Example 3.1 *Aero-elasticity. Beam vibration in flowing air*

The number of aero-elastic similarity criteria for a beam vibrating due to flowing air is to be determined. The beam deflection y, air flow velocity w, modulus of elasticity E of the beam, beam length l, beam diameter d, beam material density ρ, air density ρ_v, and dynamic air viscosity η are the acting quantities. These quantities are summarized, including their dimensions.

Acting quantity	Quantity symbol	Dimensional exponent		
		L	M	T
beam deflection	y	1	0	0
flow velocity	w	1	0	–1
modulus of elasticity	E	–1	1	–2
beam length	l	1	0	0
beam diameter	d	1	0	0
beam density	ρ	–3	1	0
air density	ρ_v	–3	1	0
dynamic viscosity	η	–1	1	–1

In compliance with the above-mentioned list of quantities, the number of simple and composed similarity criteria can be determined by the Pi theorem.

Determination of the number of criteria by the Pi theorem:	
Total number of dimensional physical quantities:	$N = 8$
Number of dimensionally different quantities:	$n = 5$
Number of basic and complementary dimensions:	$r = 3$ (L, M, T)
Total number of dimensionless criteria:	$k = N - r = 5$
Number of simple criteria:	$k_s = N - n = 3$
Number of composed criteria:	$k_k = n - r = 2$

It follows according to the Pi theorem that the vibrating beam process in flowing air can be described by three simple criteria and two composed ones.

Example 3.2 *Aero-elasticity. Beam vibration in flowing air*

The form of criteria of aero-elastic similarity of a beam vibrating due to flowing air is to be determined. The quantities acting in the process have been summarized in the preceding Example 3.1. It has been found that the process can be described by three simple and two composed criteria. In this example, the similarity criteria will be determined by dimensional analysis.

The solution procedure is based on construction of a dimensional matrix, its conversion to a solution matrix and evaluation of the similarity criteria.

1. The dimensional matrix, arranged according to (3.14), is set up

$$\mathbf{A} = \begin{array}{c} \\ L \\ M \\ T \end{array} \begin{array}{c} \begin{array}{cccccccc} y & w & E & l & \rho & \rho_v & d & \eta \end{array} \\ \left[\begin{array}{ccccc|ccc} 1, & 1, & -1, & 1, & -3 & -3, & 1, & -1 \\ 0, & 0, & 1, & 0, & 1, & 1, & 0, & 1 \\ 0, & -1, & -2, & 0, & 0, & 0, & 0, & -1 \end{array}\right] \end{array}. \qquad (3.20)$$

2. The regularity of the residual sub-matrix $\mathbf{A}_Z$ is verified according to Eq. (3.15)

$$\det \mathbf{A}_Z = 1 \neq 0. \qquad (3.21)$$

3. The residual solution matrix is determined according from (3.16)

$$\mathbf{B}_1 = (-1)\begin{bmatrix} 1, & 0, & 0 \\ 1, & 0, & -1 \\ -1, & 1, & -2 \\ 1, & 0, & 0 \\ -3, & 1, & 0 \end{bmatrix} \begin{bmatrix} 0, & 1, & 0 \\ 1, & 3, & 0 \\ 1, & 2, & -1 \end{bmatrix} = \begin{bmatrix} 0, & -1, & 0 \\ 1, & 1, & -1 \\ 1, & 2, & -2 \\ 0, & -1, & 0 \\ -1, & 0, & 0 \end{bmatrix}. \quad (3.22)$$

4. The solution matrix **B** is set up according to (3.17)

$$\mathbf{B} = \begin{matrix} \\ \Pi_1 \\ \Pi_2 \\ \Pi_3 \\ \Pi_4 \\ \Pi_5 \end{matrix} \begin{matrix} \begin{matrix} y & w & E & l & \rho & \rho_v & d & \eta \end{matrix} \\ \begin{bmatrix} 1, & 0, & 0, & 0, & 0, & 0, & -1, & 0 \\ 0, & 1, & 0, & 0, & 0, & 1, & 1, & -1 \\ 0, & 0, & 1, & 0, & 0, & 1, & 2, & -2 \\ 0, & 0, & 0, & 1, & 0, & 0, & -1, & 0 \\ 0, & 0, & 0, & 0, & 1, & -1, & 0, & 0 \end{bmatrix} \end{matrix}. \quad (3.23)$$

The dimensionless similarity criteria can be determined directly from the solution matrix **B** by using the Eq. (3.18) where Π_1, Π_4 and Π_5 are simple criteria and Π_2 and Π_3 are composed ones.

$\Pi_1 = \dfrac{y}{d}$ – relative beam deflection,

$\Pi_2 = \dfrac{w\rho_v d}{\eta} = \dfrac{wd}{v}$ – Reynolds number,

$\Pi_3 = \dfrac{E\rho_v d^2}{\eta^2} = \dfrac{Ed^2}{v\eta}$ – Hedström number,

$\Pi_4 = \dfrac{l}{d}$ – relative beam length,

$\Pi_5 = \dfrac{\rho}{\rho_v}$ – relative density.

The aero-elastic process of beam vibration due to the air flow is described by the following dimensionless equation

$$\varphi\left(\frac{wd}{v}, \frac{Ed^2}{v\eta}, \frac{y}{d}, \frac{l}{d}, \frac{\rho}{\rho_v}\right) = 0. \quad (3.24)$$

The original process, expressed by eight dimensional quantities, is described, in a dimensionless form, by five dimensionless similarity criteria only.

3.3 Conversion of Units and Quantities

Every physical quantity can be expressed by its numerical value and a unit which can be written by means of the set of fundamental and complementary units, or the set of fundamental and complementary quantities and their dimensional symbols, respectively.

3.3.1 Conversion of Units

One set of fundamental and complementary units $X_1, X_2, \ldots, X_k$ can be converted to another set of fundamental and complementary units $Y_1, Y_2, \ldots, Y_m$, provided that $m > k$, in the *condition of dimensional independence of quantities,* expressed by a *conversion matrix* **K,** which is composed of exponents at dimensional symbols. The rows correspond to units Y_i and the columns to units X_j, and one can write

$$\mathbf{K} = \begin{array}{c} \\ Y_1 \\ Y_2 \\ \vdots \\ Y_m \end{array} \begin{array}{c} \begin{array}{cccc} X_1 & X_2 & \cdots & X_k \end{array} \\ \begin{bmatrix} k_{11}, & k_{12}, & \ldots, & k_{1k} \\ k_{21}, & k_{22}, & \ldots, & k_{2k} \\ \vdots & \vdots & \ddots & \vdots \\ k_{m1}, & k_{m2}, & \ldots, & k_{mk} \end{bmatrix} \end{array}. \tag{3.25}$$

Then, the fundamental and complementary units Y_i can be expressed in the following form

$$Y_i = K_i \prod_{j=1}^{k} X_j^{k_{ij}}, \tag{3.26}$$

$$i = 1, \ldots, m.$$

where K_i are dimensionless constants.

Because the rank of the conversion matrix is r, there is at least one non-zero r-order sub-determinant, with all sub-determinants of $r+1$ or higher order equalling zero. Obviously, the necessary condition to convert one fundamental measuring units set into another is that a k-order sub-determinant of the conversion matrix must differ from zero. In this case, the fundamental and complementary units are independent with respect to dimensions. This means that from m units Y_i, k units independent in dimension can be chosen, by means of which a set of units can be created which can replace fully the set based on X_j units.

3.3.2 Conversion of Quantities

Every physical quantity can be expressed by its numerical value and a unit composed of fundamental and complementary units. In the set of units $X_1, X_2, \ldots, X_k$, the quantity V_1 can be written in the following form

$$V_1 = r_1 \prod_{j=1}^{k} X_j^{ej}, \tag{3.27}$$

where r_1 is the numeric value of the quantity V_1 and e_j for $j = 1$, k is the vector of dimensional exponents of this quantity. In the dimensional matrix in the set of units $X_1, X_2, \ldots, X_k$, the vector of the exponents e_j conforms to the column corresponding to quantity V_1.

After being converted into a new set of units $Y_1, Y_2, \ldots, Y_k$, the quantity V_2 can be written as follows

$$V_2 = r_2 \prod_{l=1}^{k} Y_l^{f_l}, \tag{3.28}$$

where r_2 is the numeric value of quantity V_2 in the new set of units and f_l for $l = 1, \ldots, k$ is the vector of dimensional exponents of the quantity in the new set of units. In the dimensional matrix in the new set of units $Y_1, Y_2, \ldots, Y_k$, the vector of exponents f_l corresponds to the column corresponding to quantity V_2.

In the new set of units, the following holds for the numeric value of quantity V_2,

$$r_2 = r_1 \prod_{j=1}^{k} X_j^{e_j} \left(\prod_{l=1}^{k} Y_l^{f_l} \right)^{-1}, \tag{3.29}$$

where the units $Y_1, Y_2, \ldots, Y_k$, from the new set can be expressed by means of original units by using the conversion relation (3.26).

For the simplest case, with the unity conversion matrix, the simplified conversion relation (3.26) between the new and original sets of units is valid in the following form

$$Y_i = K_i X_i, \tag{3.30}$$

$$i = 1, \ldots, k.$$

Simultaneously, with the relation (3.30), the following is valid as well

$$f_l = e_l, \tag{3.31}$$

$$l = 1, \ldots, k.$$

After inserting the relations (3.30) and (3.31) into equation (3.29) and arranging it, one obtains

$$r_2 = r_1 \prod_{l=1}^{k} K_l^{-e_l}. \tag{3.32}$$

The relation (3.32) can be utilized advantageously, e.g., to convert the quantities between metric sets MKS (meter, kilogram, second) and CGS (centimetre, gram, second) and the non-metric set FPS (foot, pound, second) and, further, to convert the previously used units into the SI system.

3.3.3 Examples

In the following example, practical procedure to convert the sets of units is presented. The next example demonstrates how to proceed if converting quantities between MKS and FPS systems.

Example 3.3 *Conversion of a fundamental set of units of measurement*

The purpose is to find out whether a mechanical problem, expressed by fundamental quantities or units respectively, for example, length, mass and time (dimensional symbols L_1, M_1, T_1) in the SI-system, can be described by other fundamental quantities or units, such as, e.g., force, length and time (F_2, L_2, T_2), force, length and density (F_3, L_3, R_3), or force, velocity and power (F_4, W_4, P_4). If possible, the conversion relations between new and original fundamental quantities or units thereof will be expressed.

To convert one set of fundamental quantities or units into another one, the condition of a non-zero determinant of the conversion matrix must be fulfilled, in this case of a third-order determinant.

1. Conversion $(L_1, M_1, T_1) \rightarrow (F_2, L_2, T_2)$

	L_1	M_1	T_1
F_2			
L_2			
T_2			

$$\begin{vmatrix} 1, & 1, & -2 \\ 1, & 0, & 0 \\ 0, & 0, & 1 \end{vmatrix} = -1 \neq 0.$$

A mechanical system can be described by force, length and time (F_2, L_2, T_2). A new set of fundamental quantities or units can be expressed by conversion relations by means of the original set

$$F_2 = K_1 L_1 M_1 T_1^{-2}, \quad (3.33)$$

$$L_2 = K_2 L_1, \quad (3.34)$$

$$T_2 = K_3 T_1 . \quad (3.35)$$

2. Conversion $(L_1, M_1, T_1) \rightarrow (F_3, L_3, R_3)$

	L_1	M_1	T_1
F_3			
L_3			
R_3			

$$\begin{vmatrix} 1, & 1, & -2 \\ 1, & 0, & 0 \\ -3, & 1, & 0 \end{vmatrix} = -2 \neq 0.$$

A mechanical set can also be described by force, length and density (F_3, L_3, R_3). The new set of fundamental quantities or units can then be expressed by conversion relations by means of the original set

$$F_3 = K_1 L_1 M_1 T_1^{-2}, \quad (3.36)$$

$$L_3 = K_2 L_1, \quad (3.37)$$

$$R_3 = K_3 L_1^{-3} M_1. \quad (3.38)$$

3. Conversion $(L_1, M_1, T_1) \rightarrow (F_4, W_4, P_4)$

	L_1	M_1	T_1
F_4			
W_4			
P_4			

$$\begin{vmatrix} 1, & 1, & -2 \\ 1, & 0, & -1 \\ 2, & 1, & -3 \end{vmatrix} = 0.$$

A mechanical system cannot be described by force, velocity and power (F_4, W_4, P_4). In the determinant, the last row is a linear combination of two preceding rows. Besides, it is also obvious from physics laws that $P = FW$.

Example 3.4 *Conversion of a quantity into another set of measurement units*

Normal gravity acceleration is to be converted from the MKS-system into the FPS-system. Normal gravity acceleration can be written in the MKS-system as follows

$$g_{n1} = 9.80665 \text{ m s}^{-2}. \quad (3.39)$$

Both sets of units utilize the fundamental dimensions L, T and, therefore, the unity conversion matrix **K** between the sets L_1(m), T_1 (s) and L_2 (ft), T_2 (s) is obtained as follows

$$\mathbf{K} = \begin{matrix} & L_1 & T_1 \\ L_2 & 1, & 0 \\ T_2 & 0, & 1 \end{matrix}, \quad (3.40)$$

where the constants K_i for $i = 1,2$ are

$$K_1 = 0.3048, \qquad K_2 = 1. \tag{3.41}$$

The conversion relations between the fundamental units of these sets can be written in the following forms

$$L_2 = 0.3048\ L_1, \tag{3.42}$$

$$T_2 = T_1. \tag{3.43}$$

In the original set, the numerical value of the normal gravity acceleration is

$$r_1 = 9.80665, \tag{3.44}$$

and for the numerical value in the new set, the following holds

$$r_2 = r_1\ K_1^{-e_1} K_2^{-e_2}, \tag{3.45}$$

where the following is valid for the exponents e_1, e_2 at the quantity g_{n1} (the column at the quantity g_{n1} in the dimensional matrix in the L, M system)

$$e_1 = 1, \quad e_2 = -2 \tag{3.46}$$

For the numerical value of the quantity g_{n2}, one obtains

$$r_2 = 32.1740, \tag{3.47}$$

and the usual gravity acceleration in the FPS system can be written in the form

$$g_{n2} = 32.1740\ \text{ft s}^{-2}. \tag{3.48}$$

3.4 Determination of Functional Relations

In experimental techniques, dimensional analysis is applied mainly to determine the functional relations between dimensional quantities (or possible dimensionless similarity criteria) which describe the investigated process. These processes are applied extensively, especially when the mathematical equations describing this process are not known.

The basis consists of the *common procedure utilizing Rayleigh's algebraic method,* which deals with all dimensional quantities and basic and complementary dimensions thereof. However, any complete functional relation is not always found. In these cases, some of the *improved procedures to reduce the number of quantities in a functional relation* are used.

3.4.1 Functional Relation and Criterion Equation

Generally, the studied physical process is expressed by dimensional physical quantities in the form of a *functional relation among dimensional quantities*

$$f(x_1, x_2, \ldots, x_N) = 0. \tag{3.49}$$

After obtaining similarity criteria, this equation can be replaced by an equivalent relation, the so-called *functional relation of dimensionless quantities*

$$\varphi(\Pi_1, \Pi_2, \ldots, \Pi_{N-r}) = 0. \tag{3.50}$$

The criterion equation is a mathematical relation (3.50) in which the dimensionless similarity criteria and numerical parameters appear.

If possible, instead of the implicit forms of Eqs. (3.49) and (3.50), an explicit expression can be considered

$$x_N = \overline{f}(x_1, x_2, \ldots, x_{N-1}), \tag{3.51}$$

$$\Pi_{N-r} = \overline{\varphi}(\Pi_1, \Pi_2, \ldots, \Pi_{N-r-1}). \tag{3.52}$$

The general form of the functional relation of dimensional quantities, or dimensionless ones, is expressed by Eqs. (3.49) and (3.50). If the equation for the functional relation contains only products of quantities with various numerical powers, it is the *classic product form of the functional relation*

$$x_N = l_N \prod_{i=1}^{N-1} x_i^{l_i}, \tag{3.53}$$

$$\Pi_{N-r} = k_{N-r} \prod_{i=1}^{N-r-1} \Pi_i^{k_i}, \tag{3.54}$$

where l_i for $i = 1, \ldots, N$, are numerical parameters of the functional relation of dimensional quantities, and k_i for $i = 1, \ldots, N - r$ are numerical parameters of the functional relation of dimensionless quantities. To a certain extent, the numerical values of the functional relation of dimensional quantities can be established by common and improved dimensional analysis procedures. The residual values can then be determined, i.e. experimentally. The numerical parameter values of functional relation of dimensionless quantities, i.e. the criterion equation, are all established by, for example, experiments.

In the case of the classic product form of the criterion equation, $N - r$ unknown parameters must be defined, where $N - r$ is the number of dimensionless quantities in this criterion equation. In the simplest case, only $N - r$ experiments are satisfactory.

In the case of a general form of the criterion equation, it is far more complicated to determine or reconstruct the relation by dimensional analysis only. A process using the so-called polynomial development of functions can be useful. To ensure the required accuracy of function approximation, a satisfying-degree polynomial must be defined, the coefficients of which are determined by experiment. It is better to apply a higher-degree polynomial because an incorrect relation can be obtained if the polynomial degree is less than necessary. With the polynomial degree higher than necessary, the values of coefficients at superfluous polynomial members are several orders lower than at other coefficients. With the general form of the criterion equation, many more experiments are usually necessary to determine the required coefficients.

3.4.2 Common Procedure

In determining the functional relation by a *common procedure, the Rayleigh algebraic method* of dimensional expression of quantities is used. Sometimes, this very simple and fast method can lead to a complete solution (except for a dimensionless constant) and often avoids a very pretentious theoretical or experimental solution.

The principle of this procedure consists of assembling the functional relation in a classic product form (in cases where it is possible) in which the powers of dimensional quantities are sought. After replacing the dimensional quantities by dimensional symbols, an equation is obtained in which the fundamental and complementary dimensions appear. A set of algebraic equations for unknown parameters is obtained from exponents at corresponding dimensions.

3.4.3 Improved Procedures

These procedures are improvement of the common procedure and make use of a reduced number of quantities to obtain a complete functional relation describing the investigated process. These procedures are divided as follows:

- *procedure utilizing an increased number of dimensions,*
- *procedure utilizing a reduced number of dimensional quantities,*

• *procedure utilizing the connection of dimensionless quantities.*

The advantage of improved procedures is the reduced number of quantities in the functional relation to be determined.

Procedure Utilizing Increased Number of Dimensions

The *procedure based on utilizing increased number of dimensions* belongs to methods with reduction of the number of quantities in a functional relation whereas it utilizes the increased number of fundamental and complementary dimensions. This method is advantageous because of the reduced number of unknown parameters in the sought functional relation.

The idea to divide the dimensions and to obtain a larger number of dimensions was proposed by H.E. Huntley. His idea of physical quantities vectoriality consists of being able to extend the amount of fundamental and complementary dimensions. It reduces the number of unknown variables in solving the functional relation. Differentiated units of measurement of some basic physical values, especially length and mass, are used in this procedure. At the same time, this is supported by the fact that in spite of the same dimensional expression of two or more quantities, these quantities cannot be made identical from the physical and especially informational point of view.

For example, the longitudinal dimension L can be regarded as a dimension oriented in space (a vector), determined by components L_x, L_y, L_z in space generally with the meaning of a vector from the standpoint of the process investigated. Practically, this can be demonstrated with the example of measurement of a rectangular plate with a meter gauge. To measure three basic dimensions, i.e. thickness, width, and length, the gauge must be placed in three mutually perpendicular directions. In case of the length, this measurement can be extended further, namely by assigning the orientation to individual components. It will do to orient the dimensions of the considered plate, i.e. relative to its axes, etc. Then, positive and negative values of corresponding components will be obtained, due to which the number of fundamental independent quantities is doubled and, therefore, the dimensional matrix rank increases. In the first-mentioned example, this means to increase basic dimensions as much as by two, because the dimensions L_x, L_y, L_z appear instead of L. In another example, the number of fundamental dimensions can be extended as much as by five. The weight M can be distinguished in the amount M_m and in the sense of inertia M_i. In Table 3.2, the selected thermomechanical quantities are presented, including their fundamental as well as extended dimensions.

The application of this procedure depends on the physical nature of

Tab. 3.2 The basic and extended dimensions of thermomechanical quantities

Symbol	Quantity	Unit	Basic or extended dimensions	
			$L\ M\ T\ \Theta$	$L_x\ L_y\ L_z\ M_m\ M_i\ T\ \Theta$
a	thermal diffusivity[4]	$m^2 \cdot s^{-1}$	L^2T^{-1}	$L_x^2\ T^{-1}$
A, S	area[1]	m^2	L	L_yL_z
c	specific thermal capacity	$J \cdot kg^{-1} \cdot K^{-1}$	$L^2T^{-2}\ \Theta^{-1}$	$L_x^{2/3}\ L_y^{2/3}\ L_z^{2/3}M_m^{-1}M_iT^{-2}\ \Theta^{-1}$
C_T	thermal capacity	$J \cdot K^{-1}$	$L^2MT^{-2}\ \Theta^{-1}$	$L_x^{2/3}\ L_y^{2/3}\ L_z^{2/3}M_iT^{-2}\ \Theta^{-1}$
d	diameter[1]	m	L	$L_y^{1/2}\ L_x^{\ 1/2}$
F	inertial force[2]	N	LMT^{-2}	$L_xM_iT^{-2}$
p	gas pressure	Pa	$L^{-1}MT^{-2}$	$L_x^{-1/3}\ L_y^{-1/3}\ L_z^{-1/3}M_iT^{-2}$
Q	heat	J	L^2MT^{-2}	$L_x^{2/3}\ L_y^{2/3}\ L_z^{2/3}M_iT^{-2}$
V	volume	m^3	L^3	$L_x\ L_yL_z$
w_x	velocity[2]	$m \cdot s^{-1}$	LT^{-1}	L_xT^{-1}
α	heat transfer coefficient[4]	$W \cdot m^{-2} \cdot K^{-1}$	$MT^{-3}\ \Theta^{-1}$	$L_x^{2/3}\ L_y^{-1/3}\ L_z^{-1/3}M_iT^{-3}\ \Theta^{-1}$
η	dynamic viscosity[3]	$Pa \cdot s$	$L^{-1}MT^{-1}$	$L_x^{-1}M_iT^{-1}$
λ	specific thermal conductivity[4]	$W \cdot m^{-1} \cdot K^{-1}$	$LMT^{-3}\ \Theta^{-1}$	$L_x^{5/3}\ L_y^{-1/3}L_z^{\ -1/3}M_iT^{-3}\ \Theta^{-1}$
ν	kinematic viscosity	$m^2 \cdot s^{-1}$	L^2T^{-1}	$L_yL_zM_m^{-1}M_iT^{-1}$
ρ	density	$kg \cdot m^{-3}$	$L^{-3}M$	$L_x^{-1}L_y^{\ -1}L_z^{\ -1}M_m$
τ	shear stress[2]	Pa	$L^{-1}MT^{-2}$	$L_y^{-1/2}L_z^{-1/2}M_iT^{-2}$
Φ	heat flux	W	L^2MT^{-3}	$L_x^{2/3}\ L_y^{2/3}\ L_z^{2/3}M_iT^{-3}$

Note: 1 – perpendicular to i-axis direction; 2 – in the i-axis direction; 3 – in flow in the x-axis direction; 4 – in heat transfer in the i-axis direction

the investigated task and acting quantities. Care and consistent checking are very important to prevent erroneous division of dimensions.

Procedure Utilizing a Reduced Number of Dimensional Quantities

The *procedure utilizing a reduced number of dimensional quantities* is among the methods reducing the number of quantities in functional relation and utilizes the reduction of the total number of dimensional quantities.

The method consists of *connecting the dimensional quantities.* It is based on previous knowledge of the form in which two quantities, or possibly more, occur in the equation or in a set of equations connecting all dimensional quantities. The name and designation are attached to this group of quantities, and it becomes a new dimensional quantity which replaces all original quantities contained therein. In this way, the number of dimensional quantities is reduced at least by one.

With the number of dimensionless quantities already equalling one, the dimensional quantities are not connected, because the minimum number of dimensionless quantities is at least one.

The reduction of the amount of dimensional quantities is useful only if the original dimensional quantities are removed from the equation, or a set of them – regardless of this relation being known or not.

The advantage of this method consists in the reduced number of unknown parameters in a functional relation sought.

Procedure Utilizing the Connection of Dimensionless Quantities

The *process utilizing the connection of dimensionless quantities* is among the methods used to reduce the number of quantities in a functional relation.

If a resulting already-found functional relation contains several dimensionless quantities, it occurs sometimes that some of these quantities can be brought together into a group. Then this group forms a new similarity criterion and replaces fully the original dimensionless quantities.

Especially, this method simplifies the graphic presentation of resulting relations advantageously, i.e. there are fewer necessary diagrams, and further reduces the amount of independent variables in experiments. Thus, it simplifies and generalizes the results.

3.4.4 Examples

The following examples of functional relation determination are in the area of solid body mechanics and fluid mechanics. In the first two examples, a common procedure is used, and the next two examples utilize the procedure to increase the number of dimensions by dividing them. The procedure utilizing the reduction of the number of dimensional quantities is applied in further examples. Two examples are purposely the same, with the aim of comparing various methods of solving one task.

Example 3.5 *Physics. String vibration*

The eigenfrequency f of a stretched string with length l, longitudinal density ρ_l and stretching force F has to be determined. The following equation is valid for the eigenfrequency of a string

$$f = \varphi(F, l, \rho_l) = k F^a l^b \rho_l^c, \tag{3.55}$$

where k is a dimensionless constant. Equation (3.55) can be written in the following form

$$T^{-1} = \left(M L T^{-2}\right)^a L^b \left(L^{-1} M\right)^c. \tag{3.56}$$

After assembling the equations from exponents at corresponding fundamental dimensions, one obtains

$$\left.\begin{array}{llll} L: & 0 & = & a+b-c, \\ M: & 0 & = & a+c, \\ T: & -1 & = & -2a, \end{array}\right\} a = \frac{1}{2}, b = -1, c = -\frac{1}{2}. \tag{3.57}$$

By inserting this into Eq. (3.55), the expression for the string eigenfrequency is determined to be

$$f = \frac{k}{l}\sqrt{\frac{F}{\rho_l}}, \tag{3.58}$$

in which the constant $k = 1/2$ is determined by another method, e.g. experimentally.

Example 3.6 *Mechanics of bodies. Deflection of a conductor*

The relation between the stretching force F and deflection s of a horizontally free stretched l-long wire conductor is to be determined. The wire mass m is firmly held on both ends and subject to gravitational acceleration g only.

The expression for the stretching force can be written in the following form

$$F = \varphi(m, l, s, g) = k m^a l^b s^c g^d. \tag{3.59}$$

The corresponding dimensional equation is

$$L M T^{-2} = M^a L^b L^c \left(L T^{-2}\right)^d. \tag{3.60}$$

Unknown exponents can be determined from the following equations

$$\left.\begin{aligned} L:&\quad 0 = b+c+d,\\ M:&\quad 1 = a,\\ T:&\quad -2 = -2d, \end{aligned}\right\} a=1,\ b=-c,\ d=1. \tag{3.61}$$

Because there are only three equations available for four unknown exponents, one exponent remains undetermined. Therefore, Eq. (3.59) can be written as follows

$$F = k\,m\,g\left(\frac{s}{l}\right)^{c}. \tag{3.62}$$

The constant k and the unknown coefficient c must be determined experimentally or by theoretical solution. As the example shows, a common procedure does not always lead to a complete solution. However, there are ways to obtain a complete solution even without carrying out experiments in these cases.

Example 3.7 *Mechanics of bodies. Dynamics*

To determine the energy of a stretched string oscillating at its eigenfrequency.

Acting quantity	Quantity symbol	Dimensional exponent		
		L	M	T
frequency	f	0	0	−1
string length	l	1	0	0
longitudinal density	ρ_l	−1	1	0
string deflection	z	1	0	0
energy	E	2	1	−2

The string energy is described by the following equation

$$E = \varphi(f, l, \rho_l, z, E) = k\, f^{a}\, l^{b} \rho_l^{c}\, z^{d}. \tag{3.63}$$

Solution utilizing common procedure

Equation (3.63) can be transformed to the following dimensional form

$$L^2 M\, T^{-2} = \left(T^{-1}\right)^{a} L^{b} \left(L^{-1} M\right)^{c} L^{d}. \tag{3.64}$$

By comparing the exponents at corresponding dimensions, one

obtains $a = 2$, $c = 1$, $b = 3-d$. Therefore, the resulting string energy equation is as follows

$$E = k\,f^2 l^3 \rho_l \left(\frac{z}{l}\right)^d. \tag{3.65}$$

The solution is incomplete because the remaining parameter d has not been determined. Further, the constant k must be determined.

Solution utilizing improved procedure

Let us use an L-long vector. L_x is the length oriented in the along-the-string direction and it is important to state its weight. L_z characterizes the string deflection and energy. Therefore, Eq. (3.64) can be given in the following form

$$L_z^2\, M\, T^{-2} = \left(T^{-1}\right)^a L_x^b \left(L_x^{-1} M\right)^c L_z^d. \tag{3.66}$$

If comparing the exponents at fundamental dimensions, one obtains

$$\left.\begin{array}{lrcl} L_x: & 0 & = & b-c, \\ L_z: & 2 & = & d, \\ M: & 1 & = & c, \\ T: & -2 & = & -a, \end{array}\right\} a=2,\ b=1,\ c=1,\ d=2. \tag{3.67}$$

By inserting this into Eq. (3.63), the result of the solution is obtained in the following form

$$E = k f^2\, l\, \rho_l\, z^2. \tag{3.68}$$

where the constant k is determined, e.g., experimentally.

Example 3.8 *Fluid mechanics. Passage of a viscous fluid through a pipeline*

The relation for mass flow of a viscous incompressible fluid flow through a round-cross-section pipeline has to be determined by making use of the common and improved methods.

For the mass flow, the following holds

$$Q_m = f\left(p, \rho, \eta, r\right) = k\, p^a \rho^b \eta^c r^d. \tag{3.69}$$

Acting quantity	Quantity symbol	Dimensional exponent		
		L	M	T
mass flow	Q_m	0	1	–1
fluid pressure	P	–1	1	–2
density	ρ	–3	1	0
dynamic viscosity	η	–1	1	–1
pipeline radius	r	1	0	0

Solution utilizing common procedure

The preceding equation can be expressed in the shape of dimensional one

$$M\,T^{-1} = \left(L^{-1}M\,T^{-2}\right)^a \left(L^{-3}M\right)^b \left(L^{-1}M\,T^{-1}\right)^c L^d \,. \tag{3.70}$$

By comparing the exponents at corresponding dimensional symbols, one obtains

$$\left.\begin{aligned} L:&\quad 0 = -a-3b-c+d,\\ M:&\quad 1 = a+b+c,\\ T:&\quad -1 = -2a-c, \end{aligned}\right\} \begin{aligned} b &= a,\\ c &= 1-2a,\\ d &= 2a+1. \end{aligned} \tag{3.71}$$

Therefore, the mass flow equation can be written as follows

$$Q_m = k\,p^a\,\rho^a\,\eta^{1-2a}\,r^{2a+1}. \tag{3.72}$$

The solution by this procedure is not complete because the exponent a has not been determined. Further, the constant k must be determined.

Solution utilizing improved procedure

In the case of mass, the application of the Huntley theorem leads to diversifying this concept as a measure of the substance amount (M_m) and as a measure of inertia (M_s). Due to this, the number of basic dimensions is increased by one and the number of unknown variables drops by the same value in the corresponding dimensional equation. Instead of Eq. (3.70), one can write

$$M_m\,T^{-1} = \left(L^{-1}M_sT^{-2}\right)^a \left(L^{-3}M_m\right)^b \left(L^{-1}M_sT^{-1}\right)^c L^d, \tag{3.73}$$

and further it follows that

$$\left.\begin{array}{llll} L: & 0 & = & -a-3b-c+d, \\ M_m: & 1 & = & b, \\ M_s: & 0 & = & a+c, \\ T: & -1 & = & -2a-c, \end{array}\right\} \begin{array}{lll} a & = & 1, \\ b & = & 1, \\ c & = & -1, \\ d & = & 3. \end{array} \tag{3.74}$$

The resulting equation has the following form

$$Q_m = k\frac{p\rho r^3}{\eta}, \tag{3.75}$$

where the constant k is determined by, for example, experiments. For this procedure, the physical substance of particular quantities and their actions in the process studied must be known.

Example 3.9 *Fluid mechanics. Free fall of a drop*

By using the common and improved methods of dimensional analyses, the relation for maximum velocity of a free-falling drop will be determined. The drop radius is considered constant and a constant air density value and a constant gravity acceleration value are also assumed.

Acting quantity	Quantity symbol	Dimensional exponent		
		L	M	T
maximum velocity	v	1	0	–1
dynamic viscosity	η	–1	1	–1
drop radius	r	1	0	0
drop density	ρ	–3	1	0
gravitational acceleration	g	1	0	–2

For maximum free-fall velocity, the following holds

$$v = f(\eta, r, \rho, g) = k\eta^a r^b \rho^c g^d. \tag{3.76}$$

Solution utilizing common procedure

The preceding equation can be expressed in the form of a dimensional matrix

$$LT^{-1} = \left(L^{-1}M\,T^{-1}\right)^a L^b \left(L^{-3}M\right)^c \left(LT^{-2}\right)^d. \tag{3.77}$$

By comparing the exponents at corresponding dimensional symbols, one obtains

$$\left.\begin{aligned} L: &\quad 1 = -a+b-3c+d, \\ M: &\quad 0 = a+c \\ T: &\quad -1 = -a-2d \end{aligned}\right\} \begin{aligned} a &= -c \\ 2b &= 3c+1, \\ 2d &= 1+c. \end{aligned} \tag{3.78}$$

Therefore, the equation for the maximum free-fall velocity can be written in the following form

$$v = k\eta^{-c} r^{\frac{3c+1}{2}} \rho^c g^{\frac{1+c}{2}}. \tag{3.79}$$

The solution is incomplete, with the parameter c not determined. Further, the constant k must be determined.

Solution utilizing improved procedure

We know that at the maximum fall velocity, the resulting resistance force F, which determines the fall velocity, equals the gravity force G acting on the drop. Therefore, at the maximum drop fall velocity, the following holds

$$F = G = \frac{4}{3}\pi r^3 \rho g. \tag{3.80}$$

The product ρg can be assumed to occur even in the relation sought, i.e. these quantities can be replaced by a new quantity G_r as follows

$$G_r = \rho g. \tag{3.81}$$

In this way, the number of variables in Eq. (3.76) decreases by one, due to which this equation can be rewritten in the following form

$$v = f\left(\eta,\, r\, G_r\right) = k\, \eta^a r^b G_r^c. \tag{3.82}$$

Instead of Eq. (3.77), one can write

$$LT^{-1} = \left(L^{-1}M\,T^{-1}\right)^a L^b \left(L^{-2}M\,T^{-2}\right)^c, \tag{3.83}$$

and further, the following holds

$$\left.\begin{aligned} L: &\quad 1 = -a+b-2c, \\ M: &\quad 0 = a+c, \\ T: &\quad -1 = -a-2c, \end{aligned}\right\} \begin{aligned} a &= -1, \\ b &= 2, \\ c &= 1, \end{aligned} \tag{3.84}$$

The resulting equation has the following form

$$v = k\frac{r^2 G_r}{\eta} = k\frac{r^2 g \rho}{\eta}, \tag{3.85}$$

where the constant k can be determined by experiments, for example. This procedure requires some knowledge of the physical substance of the process.

Example 3.10 *Fluid mechanics. Passage of a viscous fluid through a pipeline*

Using the common and improved methods, the relation for the mass flow of viscous incompressible fluid flowing through a round-cross-section pipeline is to be determined.

Acting quantity	Quantity symbol	Dimensional exponent		
		L	M	T
mass flow	Q_m	0	1	–1
fluid pressure	p	–1	1	–2
density	Q	–3	1	0
dynamic viscosity	η	–1	1	–1
pipeline radius	r	1	0	0

For the mass flow, the following holds

$$Q_m = f(p, \rho, \eta, r) = k\, p^a \rho^b \eta^c r^d. \tag{3.86}$$

Solution utilizing common procedure

The common procedure gives the following result (see Example 3.8)

$$Q_m = k\, p^a \rho^a \eta^{1-2a} r^{2a+1}. \tag{3.87}$$

This solution is not complete because the parameter a is not determined. Further, the constant k must be determined.

Solution utilizing improved procedure

Instead of the mass flow Q_m and the fluid density Q, the volume flow Q_V can be considered and the following holds

$$Q_V = \frac{Q_m}{\rho}. \tag{3.88}$$

The original quantities Q_m and ρ are replaced by a new quantity Q_V, due to which the number of variables in Eq. (3.86) drops by one and this equation can be written in the following form

$$Q_V = k\, p^a \eta^b r^c. \tag{3.89}$$

This equation can be expressed by the following dimensional equation

$$L^3\, T^{-1} = \left(L^{-1} M\, T^{-2}\right)^a \left(L^{-1} M\, T^{-1}\right)^b L^c. \tag{3.90}$$

Comparison of the exponents at the corresponding dimensional symbols results in

$$\left.\begin{array}{llcl} L: & 3 & = & -a-b+c, \\ M: & 0 & = & a+b, \\ T: & -1 & = & -2a-b \end{array}\right\} \begin{array}{lcr} a & = & 1, \\ b & = & -1, \\ c & = & 3. \end{array} \tag{3.91}$$

After having been transferred back to the mass flow, the resulting equation becomes

$$Q_m = k\frac{p \rho r^3}{\eta}, \tag{3.92}$$

where the constant k is determined, e.g., by experiment. The relation obtained is the same as that in Example 3.8.

3.5 Applications

Following up the preceding examples, the following examples show some more complicated practical applications of dimensional analysis in various areas: heat conduction and thermal stress in solid bodies, heat transfer in fluids, electricity and magnetism, and machinery.

3.5.1 Thermomechanics of Solid Bodies

In the following example, the functional relation for relative deformation of a steel ball thrown into hot fluid is determined by a common dimensional analysis procedure. The number and form of similarity criteria, for deterministic and stochastic description of a heat field, is then determined in the following two examples.

Example 3.11 *Mechanics of bodies. Thermal deformation*

The similarity criteria and dimensionless functional relation describing the relative deformation of a steel ball thrown into hot fluid with a constant temperature is to be determined.

Acting quantity	Quantity symbol	Dimensional exponent			
		L	M	T	Θ
relative deformation	ε	0	0	0	0
ball diameter	$D = L$	1	0	0	0
thermal conductivity	λ	1	1	–3	–1
specific volume heat	cg	–1	1	–2	–1
temperature difference	$\Delta T = T_l - T_S$	0	0	0	1
heat transfer coefficient	α	0	1	–3	–1
volume thermal expansion coefficient	β	0	0	0	–1
time	τ	0	0	1	0

According to the quantities chosen, the number of simple and composed similarity criteria can be determined by means of the Pi theorem.

Determination of the number of criteria by the Pi theorem:

Total number of dimensional physical quantities:	$N = 7$
Number of dimensionally different quantities:	$n = 7$
Number of basic and complementary dimensions:	$r = 4$ (L, M, T, Θ)
Total number of dimensionless criteria:	$k = N - r = 3$
Number of simple criteria:	$k_s = N - n = 0$
Number of composed criteria:	$k_k = n - r = 3$

Essentially, the right side of the list of physical quantities acting in the process represents a dimensional matrix with four rows expressing fundamental dimensional symbols, and eight columns expressing physical quantities. The functional relation can be written

$$\varepsilon = k D^a \lambda^b (c\rho)^c \Delta T^d \alpha^e \beta^f \tau^g, \tag{3.93}$$

where k is the numerical constant. By inserting the dimensions of individual physical quantities, one obtains

$$1 = L^a \left(L M T^{-3} \Theta^{-1}\right)^b \left(L^{-1} M T^{-2} \Theta^{-1}\right)^c \Theta^d \left(M T^{-3} \Theta^{-1}\right)^e \Theta^{-f} T^g. \tag{3.94}$$

After comparing the exponents at corresponding dimensional symbols, one can write

$$\left.\begin{array}{llll} L: & 0 & = & a+b-c, \\ M: & 0 & = & b+c+e, \\ T: & 0 & = & -3b-2c-3e+g, \\ \Theta: & 0 & = & -b-c+d-e-f, \end{array}\right\} \begin{array}{lll} e & = & -b-c, \\ a & = & -b+c, \\ f & = & d, \\ g & = & -c. \end{array} \tag{3.95}$$

After inserting this into Eq. (3.93), one obtains

$$\varepsilon = k D^{-b+c} \lambda^{b} (c\rho)^{c} \Delta T^{d} \alpha^{-b-c} \beta^{d} \tau^{-c}, \tag{3.96}$$

and further,

$$\varepsilon = k\left(\frac{\lambda}{D\alpha}\right)^{b}\left(\frac{c\,\rho\,D}{\alpha\,\tau}\right)^{c}(\beta\Delta T)^{d}. \tag{3.97}$$

After modifying for a 3rd-kind boundary condition, the following holds

$$\varepsilon = k\left(\frac{\lambda}{\alpha L}\right)^{b+c}\left(\frac{L_2}{\alpha\tau}\right)^{c}(\beta\Delta T)^{d} = k\,Bi^{-b-c}\,Fo^{-c}\,Gc^{d}. \tag{3.98}$$

In general, the following relation can be written

$$\varepsilon = \varphi(Bi, Fo, Gc), \tag{3.99}$$

in which there are three composed similarity criteria

$$Bi = \frac{\alpha L}{\lambda} \quad \text{– Biot number.}$$

$$Fo = \frac{a\tau}{L_2}, \quad \text{– Fourier number.}$$

$$Gc = \beta\Delta T \quad \text{– Gay-Lussac number.}$$

However, the above-mentioned criteria can be determined also directly by suitable selection of exponents from the expressions (3.95). First, the criterion $\pi_3 = Gc$ has to be determined from the condition $d = f$. Next, two constants are chosen, e.g. $b = 1$, $c = -1$, so that

$e = 0$, $g = 1$, $a = 2$ is derived, and after inserting these into (3.98), the Fourier criterion $\pi_2 = Fo$ is determined, where $a = \lambda(c\rho)^{-1}$ is valid. By choosing $b = -1$, $c = 0$, $d = 0$, one obtains $a = 1$, $e = 1$ and the Biot criterion $\pi_1 = Bi$.

Example 3.12 *Heat conduction. Deterministic temperature field*

The number of simple and composed similarity criteria describing a three-dimensional linear unsteady temperature field in a body in a deterministic manner, with inner heat sources, has to be determined by the Pi theorem. The body has a characteristic longitudinal dimension L, thermal diffusivity a, thermal conductivity λ, and temperature in arbitrary place and time $T(x,y,z)$. Inside the body, the inner heat sources with volume density of the heat flux $q_V(x,y,z)$ are acting. The 3rd-kind boundary condition is expressed by the temperature of outer surroundings T_P = const and the heat transfer coefficient α. The initial condition is given by the temperature $T_{ini}(x,y,z,0)$.

The temperature field is described by dimensional quantities in the following form

$$f\left(x, y, z, L, T, T_P, T_{ini}, a, \tau, q_V, \lambda, \alpha\right) = 0. \qquad (3.100)$$

Determination of the number of criteria by the Pi theorem:	
Total number ofdimensional physical quantities:	$N = 12$
Number of dimensionally different quantities:	$n = 7$
Number of basic and complementary dimensions:	$r = 4$ (L, M, T, Θ)
Total number of dimensionless criteria:	$k = N - r = 8$
Number of simple criteria:	$k_s = N - n = 5$
Number of composed criteria:	$k_k = n - r = 3$

The *simple criteria* can be determined from Eq. (3.100) directly and they are dimensionless coordinates and temperatures

$$X = \frac{x}{L},\ Y = \frac{y}{L},\ Z = \frac{z}{L},$$

$$\Theta = \frac{T}{T_{ini}},\ \Theta_P = \frac{T_P}{T_{ini}}.$$

The *composed criteria* can be determined by dimensional analysis or some of the similarity analysis methods. Analysis of mathematical heat conduction model is especially suitable. The composed criteria are

$$Fo = \frac{a\tau}{L^2} \quad \text{– Fourier number.}$$

$$Po = \frac{q_V L^2}{\lambda T_{ref}} \quad - \text{ Pomerantsev number.}$$

$$Bi = \frac{\alpha L}{\lambda} \quad - \text{ Biot number,}$$

where the Fourier number expresses the dimensionless process time, the Pomerantsev number expresses the inner heat source, and the Biot number expresses convective heat transfer on the surface.

The unsteady heat transfer can be expressed in the dimensionless form

$$\rho(X, Y, Z, \Theta, \Theta_P, Fo, Po, Bi) = 0, \tag{3.101}$$

due to which the number of variables in the thermal process is reduced by four quantities

Example 3.13 *Heat conduction. Stochastic temperature field*

The number of simple and composed thermal similarity criteria describing stochastic three-dimensional unstable movement temperature field, with a 1st-kind boundary condition, has to be determined The local, surface, and reference temperatures T, T_S, T_{ref}, coordinates x, y, z, reference length L, noise N_i (m) with the spectral power density S (m^2 s), velocity components w_x, w_y, w_z, reference velocity w_{ref}, thermal conductivity λ, thermal diffusivity a, volume density of the heat flux q_V, time τ, random walk time τ_b and reference time τ_{ref} are the acting quantities. The temperature field is described by the following equation

$$f(x, y, z, L, N_i, S, T, T_S, T_{ref}, w_x, w_y, w_z, w_{ref}, \lambda, a, q_V, \tau, \tau_B, \tau_{ref}) = 0. \tag{3.102}$$

Determination of the number of criteria by the Pi theorem:	
Total number of dimensional physical quantities:	$N = 19$
Number of dimensionally different quantities:	$n = 8$
Number of basic and complementary dimensions:	$r = 4$ (L, M, T, Θ)
Total number of dimensionless criteria:	$k = N - r = 15$
Number of simple criteria:	$k_s = N - n = 11$
Number of composed criteria:	$k_k = n - r = 4$

The *simple criteria* can be determined directly from the ratio of quantities, equal in dimensions, and these are dimensionless coordinates, temperatures, times, and velocities

$$X=\frac{x}{L},\ Y=\frac{y}{L},\ Z=\frac{z}{L},\ \overline{N}_i=\frac{N_i}{L},$$

$$\Theta=\frac{T}{T_{ref}},\ \Theta_s=\frac{T_s}{T_{ref}},\ \tau_{rel}=\frac{\tau}{\tau_{ref}},\ \tau_{B,rel}=\frac{\tau_B}{\tau_{ref}},$$

$$W_x=\frac{w_x}{w_{ref}},\ W_y=\frac{w_y}{w_{ref}},\ W_z=\frac{w_z}{w_{ref}}.$$

The *composed criteria* can be determined by dimensional analysis or some of the similarity analysis methods. These are

$$Fo=\frac{a\tau}{L^2}\quad \text{– Fourier number,}$$

$$Pe=\frac{w_{ref}L}{a}\quad \text{– Péclet number,}$$

$$Po=\frac{q_V L^2}{\lambda T_{ref}}\quad \text{– Pomerantsev number,}$$

$$No_i=N_i\sqrt{\frac{\tau}{S}}\quad \text{– dimensionless noise number,}$$

where the Fourier number expresses the dimensionless process time, the Péclet number expresses the velocity, and the Pomerantsev number expresses the inner heat source. The index i at the dimensionless noise number denotes the independence of individual processes only.

The dimensionless relation of the stochastically described temperature field is expressed by the following equation

$$\rho\left(X,\ Y,\ Z, \overline{N}_i,\ \Theta,\ \Theta_s\ \tau_{rel}, \tau_{B,rel}, W_x, W_y, W_z,\ Fo,\ Pe,\ Po, No_i\right)=0, \quad (3.103)$$

which comprises eleven simple and four composed similarity criteria.

3.5.2 Thermomechanics of Fluids

In the following example, the functional relation for thermal conductivity of gas is determined by the common and improved procedures based on increasing the number of dimensions. The functional relation for heat transfer by forced convection is deduced by the common procedure in a further example. The heat transfer similarity criteria, in saturated steam film condensation, are formulated with the same procedure as in the previous example.

Example 3.14 *Physics. Thermal conductivity of gas*

The relation between thermal conductivity and molecular properties of the gas will be determined.

Acting quantity	Quantity symbol	Dimensional exponent			
		L	M	T	Θ
thermal conductivity	λ	1	1	–3	–1
specific thermal capacity	c_p	2	0	–2	–1
weight	m	0	1	0	0
numerical density of elements	n	3	0	0	0
average velocity of molecules	u	1	0	–1	0
average free path of molecules	l	1	0	0	0
gas pressure	p	–1	1	–2	0

Quantity symbol	Dimensional exponent						
	L_x	L_y	L_z	M_m	M_i	T	Θ
λ	5/3	–1/3	–1/3	0	1	–3	–1
c_p	2/3	2/3	2/3	–1	1	–2	–1
m	0	0	0	1	0	0	0
n	–1	–1	–1	0	0	0	0
u	1	0	0	0	0	–1	0
l	1	0	0	0	0	0	0
p	–1/3	–1/3	–1/3	0	1	–2	0

The upper part of the list of acting quantities represents a dimensional matrix for a common procedure with four basic dimensions (L, M, T, Θ), whereas in the bottom part there is a case of the improved procedure with seven basic dimensions $(L_x, L_y, L_z, M_m, M_i, T, \Theta)$. In the second case, the matrix has the same number of rows and columns, and the improved procedure enables a complete solution. However, by using the common dimensional analysis procedure, one obtains an incomplete solution only.

Dimensional expression of thermal conductivity

For the thermal conductivity dimension, the following holds

$$[\lambda]=\left[\frac{Q}{A\alpha\nabla T}\right]=\frac{L_1^2\,MT^{-2}}{L_2^2\,T\Theta L_3^{-1}}=L\,MT^{-3}\,\Theta^{-1}, \tag{3.104}$$

where the indexes 1, 2, and 3 at the length have to highlight its different physical significance only. With some physical quantities, the different vector character of the length and mass, represented by the dimensional symbols L and M is obvious. For example, in the dimensional expression of the heat quantity Q, the length L_1 occurs in the significance of the volume, i.e. of the product $L_x\ L_y\ L_z$. As the resulting L_1^2 in the denominator of the expression (3.104) must be valid, the product $L_x^{2/3}L_y^{2/3}L_z^{2/3}$ will be obtained for the volume. The surface A is perpendicular to the heat flux in the x-axis direction and therefore $A = L_yL_z$. At the temperature gradient in the x-axis direction, the length L_x must appear. From this physical view, the vector meaning of the length is obvious. A similar case holds for the mass which is acting as inertial mass M_i of molecules from the physical point of view in the case considered, but does not act as a measure for the amount M_m of the substance. With the length L considered as a vector and the mass M having two physical significances, the gas thermal conductivity dimension can be expressed in the form

$$[\lambda]=\left[\frac{Q}{A\alpha\nabla T}\right]=\frac{L_x^{2/3}\,L_y^{2/3}\,L_z^{2/3}\,M_i\,T^{-2}}{L_y\,L_z\,T\,\Theta\,L_x^{-1}}=$$

$$=L_x^{5/3}\,L_y^{-1/3}\,L_z^{-1/3}\,M_i\,T^{-3}\,\Theta^{-1}. \tag{3.105}$$

In general, thermal conductivity can be expressed in the following form of the product functional relation

$$\lambda = k\,m^a\,n^b\,u^c\,l^d\,p^e\,c_p^f, \tag{3.106}$$

in which k is a numerical constant and the exponents a, b, c, d, e, f represent the unknown variables sought.

Solution utilizing common procedure

The previous functional relation (3.106) is expressed by means of dimensional symbols

$$LM\,T^{-3}\,\Theta^{-1}=M^aL^{-3b}\left(LT^{-1}\right)^c L^d\left(L^{-1}M\,T^{-2}\right)^e\left(L^2\,T^{-2}\,\Theta^{-1}\right)^f. \tag{3.107}$$

Hence, it follows that

$$
\begin{array}{lrcl}
L: & 1 & = & -3b+c+d-e+2f, \\
M: & 1 & = & a+e, \\
T: & -3 & = & -c-2e-2f, \\
\Theta: & -1 & = & -f,
\end{array}
\left.\vphantom{\begin{array}{l}1\\1\\1\\1\end{array}}\right\}
\begin{array}{lcl}
f & = & 1, \\
a & = & 1-e, \\
c & = & 1-2e, \\
d & = & 3b+3e-2.
\end{array}
\tag{3.108}
$$

Inserting to Eq. (3.106) one can obtain incomplete solution in the form

$$\lambda = k m^{1-e} n^{b} u^{1-2e} l^{3b+3e-2} p^{e} c_p. \tag{3.109}$$

Usually, the unknown exponents b, e and the dimensionless constant k must be determined experimentally.

Solution utilizing improved procedure

By inserting the dimensions into the relation (3.106), one obtains

$$
\begin{aligned}
L_x^{5/3} L_y^{-1/3} L_z^{-1/3} M_i T^{-3} \Theta^{-1} = {} & M_m^a \left(L_x L_y L_z\right)^{-b} \left(L_x T^{-1}\right)^c \\
& L_x^d \left(L_x^{-1/3} L_y^{-1/3} L_z^{-1/3} M_i T^{-2}\right)^e \\
& \left(L_x^{2/3} L_y^{2/3} L_z^{2/3} M_m^{-1} M_i T^{-2} \Theta^{-1}\right)^f .
\end{aligned}
\tag{3.110}
$$

By solving this relation, the following formulation is obtained

$$
\begin{array}{lrcl}
L_x: & 5/3 & = & -b+c+d-1/3e+2/3f, \\
L_y, L_z: & -1/3 & = & -b-1/3e+2/3f, \\
M_m: & 0 & = & a-f, \\
M_i: & 1 & = & e+f, \\
T: & -3 & = & -c-2e-2f, \\
\Theta: & -1 & = & -f,
\end{array}
\left.\vphantom{\begin{array}{l}1\\1\\1\\1\\1\\1\end{array}}\right\}
\begin{array}{lcl}
a & = & 1, \\
b & = & 1, \\
c & = & 1, \\
d & = & 1, \\
e & = & 0, \\
f & = & 1.
\end{array}
\tag{3.111}
$$

By inserting this into Eq. (3.106), the complete solution is obtained in the following form

$$\lambda = k\, m n u l c_p = k_1 \rho T l c_p, \tag{3.112}$$

except for the dimensionless constant k or k_1. Equation (3.112) implies that for gas the average movement velocity of molecules is proportional to absolute temperature, with product mn corresponding to gas density ρ.

Example 3.15 *Forced convection. Heat transfer*

The dimensionless functional relation which describes the heat transfer by forced convection with the fluid flow in a round-cross-section pipeline will be determined.

Acting quantity	**Quantity symbol**	**Dimensional exponent**			
		L	M	T	Θ
flow velocity	w	1	0	–1	0
fluid density	g	–3	1	0	0
thermal conductivity	λ	1	1	–3	–1
specific thermal capacity	c_p	2	0	–2	–1
dynamic viscosity	η	–1	1	–1	0
pipeline diameter	D	1	0	0	0
heat transfer coefficient	α	0	1	–3	–1

According to the summary of quantities, the number of simple and composed similarity criteria can be determined in compliance with the Pi theorem.

Determination of the number of criteria by the Pi theorem:	
Total number of dimensional physical quantities:	$N = 7$
Number of dimensionally different quantities:	$n = 7$
Number of basic and complementary dimensions:	$r = 4$ (L, M, T, Θ)
Total number of dimensionless criteria:	$k = N - r = 3$
Number of simple criteria:	$k_s = N - n = 0$
Number of composed criteria:	$k_k = n - r = 3$

The heat transfer will be described by the criterion equation containing three composed similarity criteria. By using dimensional analysis and the dimensional independence condition following from it, the heat transfer coefficient can be expressed by the following equation

$$\alpha = k\, w^a \rho^b \lambda^c \eta^d c_p^e D^f, \tag{3.113}$$

in which k is a dimensionless constant. By inserting the corresponding dimensions into the preceding equation, one obtains

$$M T^{-3} \Theta^{-1} = \left(L T^{-1}\right)^{a} \left(M L^{-3}\right)^{b} \left(M L T^{-3} \Theta^{-1}\right)^{c} \left(M L^{-1} T^{-1}\right)^{d} \left(L^{2} T^{-2} \Theta^{-1}\right)^{e} L^{f}. \tag{3.114}$$

By comparison of exponents at each dimension on both sides of the equation, one obtains

$$\left.\begin{array}{lcl} L: & 0 = & a-3b+c-d+2e+f, \\ M: & 1 = & b+c+d \\ T: & 3 = & a+3c+d+2e, \\ \Theta: & 1 = & c+e, \end{array}\right\} \begin{array}{lcl} b & = & a, \\ d & = & 1-a-c, \\ e & = & 1-c, \\ f & = & a-1. \end{array} \tag{3.115}$$

Substitution into Eq. (3.113) gives

$$\alpha = k\, w^{a} \rho^{a} \lambda^{c} \eta^{1-a-c} c_{p}^{1-c} D^{a-1}, \tag{3.116}$$

and by further rearranging one obtains

$$\frac{\alpha D}{\lambda} = k \left(\frac{\rho w D}{\eta} \right)^{a} \left(\frac{c_p \eta}{\lambda} \right)^{1-c}. \tag{3.117}$$

The *composed criteria* can be determined by dimensional analysis or any method of similarity analysis

$$Nu = \frac{\alpha D}{\lambda} \quad \text{– Nusselt number,}$$

$$Re = \frac{\rho w D}{\eta} \quad \text{– Reynolds number,}$$

$$Pr = \frac{c_p \eta}{\lambda} \quad \text{– Prandtl number.}$$

The relation describing the heat transfer by forced convection with the fluid flowing in a pipeline can be written in criterion form

$$Nu = k\, Re^{a} Pr^{1-c}. \tag{3.118}$$

Equation (3.118) is Nusselt's equation, in which the three unknown constants must be determined by three different measurements.

Example 3.16 *Condensation. Heat transfer*

The similarity criteria describing heat transfer at film condensation of saturated steam flowing through a vertical bare pipe will be determined.

Acting quantity	Quantity symbol	Dimensional exponent			
		L	M	T	Θ
heat transfer coefficient	α	0	1	–3	–1
temperature difference	$\Delta T = T_n - T_s$	0	0	0	1
specific latent heat of conden–sation	$l_{v,l}$	2	0	–2	0
condensate thermal conducti–vity	λ	1	1	–3	–1
condensate dynamic viscosity	η	–1	1	–1	0
gravity acceleration	g	1	0	–2	0
characteristic length	L	1	0	0	0

Depending on the quantities used, the number of simple and composed similarity criteria can be determined in compliance with the Pi theorem.

Determination of the number of criteria by the Pi theorem:	
Total number of dimensional physical quantities:	$N = 7$
Number of dimensionally different quantities:	$n = 7$
Number of basic and complementary dimensions:	$r = 4$ (L, M, T, Θ)
Total number of dimensionless criteria:	$k = N - r = 3$
Number of simple criteria:	$k_s = N - n = 0$
Number of composed criteria:	$k_k = n - r = 3$

With steam condensation, the heat transfer can be expressed by the following functional relation

$$\alpha = k \Delta T^a l_{v,l}^b \lambda^c \eta^d g^e L^f , \tag{3.119}$$

in which k is a dimensionless constant. After inserting the dimensions, one obtain

$$MT^{-3}\Theta^{-1} = \Theta^a (L^2T^{-2})^b (LMT^{-3}\Theta^{-1})^c (L^{-1}MT^{-1})^d (LT^{-2})^e L^f , \tag{3.120}$$

From comparison of the exponents, it follows that

$$\begin{array}{llll} L: & 0 & = & 2b+c-d+e+f, \\ M: & 1 & = & c+d, \\ T: & -3 & = & -2b-3c-d-2e, \\ \Theta: & -1 & = & a-c, \end{array} \left\} \begin{array}{lll} a & = & c-1, \\ b & = & 1-c-e, \\ d & = & 1-c, \\ f & = & e-1. \end{array} \right. \tag{3.121}$$

By choosing $c = -1$, $e = 2$, the values $a = -2$, $d = 2$, $f = 1$, and $b = 0$ result. After inserting them into Eq (3.119), one can write

$$\alpha = k\,\Delta T^{-2}\,\lambda^{-1}\,\eta^2\,g^2\,L, \tag{3.122}$$

and the following criterion is obtained

$$\pi_1 = \frac{\eta^2\,g^2 L}{\alpha\lambda\,\Delta T^2}.$$

By choosing $c = 1$, $e = 1$, the results are $a = d = f = 0$, and $b = -1$. After inserting them into Eq. (3.119), it can be written

$$\alpha = k\,l_{v,l}^{-1}\,\lambda g, \tag{3.123}$$

and the following criterion is derived

$$\pi_2 = \frac{\lambda g}{\alpha l_{v,l}}$$

By choosing $c = 0$, $e = 0$, the results are $a = -1$, $d = 1$, $f = -1$, and $b = 1$. Equation (3.119) is rewritten in the following form

$$\alpha = k\,\Delta T^{-1}\,l_{v,l}\,\eta\,L^{-1}, \tag{3.124}$$

and the following criterion is derived

$$\pi_3 = \frac{\alpha\,\Delta T\,L}{l_{v,l}\,\eta} Re_{bk1}.$$

3.5.3 Electricity and Magnetism

In the following example, the similarity criteria describing electric currents in real dielectrics and semiconductors are determined by the common dimensional analysis procedure. In a further example, the similarity criteria are derived by the same procedure for the three-dimensional electromagnetic field.

Example 3.17 *Electronics. Current conduction in semiconductors*

The physical similarity criteria describing electric currents limited by a space charge in real dielectric materials and semiconductors will be determined. The electrodes on the element material represent a planar parallel set.

Acting quantity	Quantity symbol	Dimensional exponent			
		L	M	T	I
planar current density	J	2	0	0	1
voltage	U	2	1	–3	–1
material thickness	h	1	0	0	0
magnetic induction	B	0	1	–2	–1
volume charge density	ρ	–3	0	1	1
permittivity	ε	–3	–1	4	2

According to this survey of quantities, the number of simple and composed similarity criteria can be determined in conformity with the Pi theorem.

Determination of the number of criteria by the Pi theorem:	
Total number of dimensional physical quantities:	$N = 6$
Number of dimensionally different quantities:	$n = 6$
Number of basic and complementary dimensions:	$r = 4$ (L, M, T, I)
Total number of dimensionless criteria:	$k = N - r = 2$
Number of simple criteria:	$k_s = N - n = 0$
Number of composed criteria:	$k_k = n - r = 2$

For the composed criteria sought, the following equation is valid

$$\pi_{1,2} = J^a U^b h^c B^d \rho^e \varepsilon^f . \tag{3.125}$$

Further, the dimensional equation can be written in the following form

$$1 = \left(L^{-2}I\right)^a \left(L^2MT^{-3}I\right)^b L^c \left(MT^{-2}I^{-1}\right)^d \left(L^{-3}TI\right)^e \left(L^{-3}M^{-1}T^4I^2\right)^f . \tag{3.126}$$

After arranging, the following set is valid

$$\left.\begin{array}{llll} L: & 0 & = & 2a+2b+c-3e-3f, \\ M: & 0 & = & b+d-f, \\ T: & 0 & = & -3b-2d+e+4f, \\ I: & 0 & = & a-b-d+e+2f. \end{array}\right\} \quad (3.127)$$

For chosen $a = 1$, $b = 0$, one obtains $c = -1$, $d = 1$, $e = -2$, and $f = 1$, and after inserting these values into Eq. (3.125), the required criterion is obtained

$$\pi_1 = \frac{J \varepsilon B}{h \rho^2}.$$

The criterion π_1 expresses the similarity relation among the current density, material thickness, permittivity, magnetic induction and volume charge density.

Similarly, for the chosen $a = 0$, $b = 1$, one obtains $c = -2$, $d = 0$, $e = -1$, and $f = 1$, and the second criterion is

$$\pi_2 = \frac{U \varepsilon}{h^2 \rho}.$$

The criterion π_2 expresses the similarity relation among voltage, material thickness, permittivity and volume charge density.

By considering the diffusion of charge carriers, expressed by diffusion coefficient D, the original number of acting quantities is extended by one. Using the same procedure as in the case of the previous criteria, the following criterion is obtained

$$\pi_3 = \frac{Jh}{D \rho}.$$

If the diffusion component of current dominates over the drift component of current, the solution is derived in the form

$$\varphi\left(\pi_2, \pi_3\right) = 0. \quad (3.128)$$

With both current components having the same role, the following new criterion must be introduced

$$\pi_4 = \frac{D \varepsilon B}{h^2 \rho}.$$

In the case considered for semiconductors and dielectrics, the use of the deduced similarity criteria can be limited by various disturbing influences. This fact must be taken into consideration in transferring the results from the model sample to the original.

Example 3.18 *Electrical engineering. Electromagnetic field*

The similarity criteria for a three-dimensional electromagnetic field will be determined.

Acting quantity	Quantity symbol	Dimensional exponent			
		L	M	T	I
space coordinates	x, y, z	1	0	0	0
characteristic relative length	L	1	0	0	0
electric voltage	U	2	1	–3	–1
electric current	I	0	0	0	1
permeability	μ	1	1	–2	–2
specific electric conductivity	γ	–3	–1	3	2
permittivity	ε	–3	–1	4	2
frequency	f	0	0	–1	0
time	τ	0	0	1	0

Depending on the quantities used, the number of simple and composed similarity criteria can be determined in compliance with the Pi theorem.

Determination of the number of criteria by the Pi theorem:	
Total number of dimensional physical quantities:	$N = 11$
Number of dimensionally different quantities:	$n = 8$
Number of basic and complementary dimensions:	$r = 4$ (L, M, T, I)
Total number of dimensionless criteria:	$k = N - r = 7$
Number of simple criteria:	$k_s = N - n = 3$
Number of composed criteria:	$k_k = n - r = 4$

The *simple criteria* can be determined directly and there are three dimensionless coordinates

$$X = \frac{x}{L},\ Y = \frac{y}{L},\ Z = \frac{z}{L}.$$

The *composed criteria* can be determined by the following procedure. First, the decisive quantities (μ, γ, ε, f), representing the fundamental parameters of an electromagnetic field, are chosen. Then, the individual criteria can be written as follows

$$\begin{aligned}\pi_1 &= [U]^{a_1}[I]^{b_1}[\tau]^{c_1}[L]^{d_1}[\mu]= \\ &= \left(L^2MT^{-3}I^{-1}\right)^{a_1}(I)^{b_1}(T)^{c_1}(L)^{d_1}\left(LMT^{-2}I^{-2}\right),\end{aligned} \tag{3.129}$$

$$\begin{aligned}\pi_2 &= [U]^{a_2}[I]^{b_2}[\tau]^{c_2}[L]^{d_2}[\gamma]= \\ &= \left(L^2MT^{-3}I^{-1}\right)^{a_2}(I)^{b_2}(T)^{c_2}(L)^{d_2}\left(L^{-3}M^{-1}T^3I^2\right),\end{aligned} \tag{3.130}$$

$$\begin{aligned}\pi_3 &= [U]^{a_3}[I]^{b_3}[\tau]^{c_3}[L]^{d_3}[\varepsilon]= \\ &= \left(L^2MT^{-3}I^{-1}\right)^{a_3}(I)^{b_3}(T)^{c_3}(L)^{d_3}\left(L^{-3}M^{-1}T^4I^2\right),\end{aligned} \tag{3.131}$$

$$\begin{aligned}\pi_4 &= [U]^{a_4}[I]^{b_4}[\tau]^{c_4}[L]^{d_4}[f]= \\ &= \left(L^2MT^{-3}I^{-1}\right)^{a_4}(I)^{b_4}(T)^{c_4}(L)^{d_4}\left(T^{-1}\right).\end{aligned} \tag{3.132}$$

In the case of criterion π_1, by comparing the exponents at the corresponding dimensions one obtains

$$\left.\begin{array}{llll} L: & 0 & = & 2a_1+d_1+1, \\ M: & 0 & = & a_1+1, \\ T: & 0 & = & -3a_1+c_1-2, \\ I: & 0 & = & -a_1+b_1-2, \end{array}\right\} \begin{array}{lll} a_1 & = & -1, \\ b_1 & = & 1 \\ c_1 & = & -1, \\ d_1 & = & 1, \end{array} \tag{3.133}$$

and hence,

$$\pi_1 = \frac{IL\mu}{U\tau}.$$

Similarly, the further three criteria are determined

$$\pi_2 = \frac{UL\gamma}{I},\ \pi_3 = \frac{UL\varepsilon}{I\tau},\ \pi_4 = \tau f.$$

After modification by which the variables U and I are eliminated from two criteria, one obtains

$$\pi_1 = L^2 f^2 \mu\varepsilon,\ \pi_2 = L^2 f\mu\gamma,\ \pi_3 = \frac{U}{I}\sqrt{\frac{\varepsilon}{\mu}},\ \pi_4 = \tau f.$$

Therefore, the space electromagnetic field is described in the dimensionless form by the following equation

$$\varphi(X, Y, Z, \pi_1, \pi_2, \pi_3, \pi_4) = 0, \tag{3.134}$$

which contains three simple and four composed similarity criteria. From the similarity criteria it is obvious that if the longitudinal and time scales are changed, the environmental parameters in the model must be changed as well.

3.5.4 Mechanical Engineering

In the following example, the similarity criteria describing the mechanical reliability of a loaded shaft are being determined with a common procedure of dimensional analysis.

Example 3.19 *Machinery. Mechanical reliability of machine parts*

A dimensionless mechanical reliability model of a rotating torque-loaded spline shaft has to be determined.

Acting quantity	Quantity symbol	Dimensional exponent		
		L	M	T
torque	M_k	2	1	–2
revolutions	n	0	0	–1
average time to failure	τ	0	0	1
length	l	1	0	0
diameter	D	1	0	0
material hardness	B	–1	1	–2
out-of-axis angle of the shaft	α	0	0	0

Depending on the quantities used, the number of simple and composed similarity criteria can be determined in compliance with the Pi theorem.

Determination of the number of criteria by the Pi theorem:	
Total number of dimensional physical quantities:	$N = 7$
Number of dimensionally different quantities:	$n = 4$
Number of basic and complementary dimensions:	$r = 3$ (L, M, T)
Total number of dimensionless criteria:	$k = N - r = 4$
Number of simple criteria:	$k_s = N - n = 3$
Number of composed criteria:	$k_k = n - r = 1$

One parametric criterion, e.g. P_3, is represented by the dimensionless out-of-axis angle a of the shaft and can be written as

$$\Pi_3 \equiv P_3 \equiv \alpha.$$

For three other criteria, the following holds

$$\Pi_i = M_k^a n^b \tau^c l^d D^e B^f, \tag{3.135}$$

$i = 1, 2, 4.$

The dimensional equation can be written in the following form

$$1 = \left(L^2 M T^{-2}\right)^a \left(T^{-1}\right)^b T^c L^d L^e \left(L^{-1} M T^{-2}\right)^f. \tag{3.136}$$

In solving the dimensional equation, the following set must be valid L

$$\left.\begin{aligned} L:&\quad 0 = 2a+d+e-f, \\ M:&\quad 0 = a+f, \\ T:&\quad 0 = -2a-b+c-2f. \end{aligned}\right\} \tag{3.137}$$

Because there are six unknown variables in three equations of the set, three of them must be chosen. When choosing, one must know the problem to be solved and the aims of solving it. In a given case, it is suitable to choose $c = 1$, $a = d = 0$, which means that the criterion will contain the average lifetime τ. By solving this, $b = 1$, $e = f = 0$ is obtained. After inserting the values into Eq. (3.135), the following parametric criterion is derived

$$\Pi \equiv P_1 = \tau n.$$

To eliminate the independent variable τ from other criteria, one chooses $c = 0$ and then, for example, $a = 0$, $d = 1$. In a similar manner, $e = -1$, $b = f = 0$ is obtained and a further criterion will be

$$\Pi_2 \equiv P_2 = \frac{l}{D}.$$

The remaining criterion can be determined for $c = d = 0$, $f = 1$. By solving this, one obtains $a = -1$, $e = 3$, $b = 0$, and the corresponding composed criterion will have the following form

$$\Pi_4 \equiv \pi_1 = \frac{BD^3}{M_k}.$$

The empirical dimensionless mathematical model of mechanical reliability of the shaft is expressed by the functional relation of one composed criterion and three simple parametric ones

$$\varphi\left(\Pi_1, \Pi_2, \Pi_3, \Pi_4\right) \equiv \varphi\left(P_1, P_2, P_3, \pi_1\right) = 0. \tag{3.138}$$

"Science is set towards generalization: and generalization means simplification."

Albert Szent-Györgyi *(16.9.1893–22.10.1986)*

4

Methods of Similarity Analysis

These methods are applied to determine similarity criteria and proceed from two formal models of an investigated process: a physical phenomenological and a mathematical model. However, the physical model is understood here to be an abstract physical model represented by a formal description of the structure of an investigated process, specifically from the standpoint of acting forces or energies. On the other hand, the mathematical model represents a complete mathematical description of the investigated physical process. Usually, it is a fundamental equation and corresponding conditions of unambiguity involving boundary and initial conditions or optimization and restrictive ones in controlled processes.

Sometimes, the method of analysis for the formal physical model is called, not quite precisely, the similarity method. Usually, it starts from a descriptive phenomenological physical model of an investigated phenomenon in which acting forces, energy, momenta and, possibly, even other physical effects appear. The significance of this method is limited mainly to the sphere of clarification of the physical problem. In any case, its application to determine the similarity criteria is infrequent. With increasing knowledge of mathematical models, the method of mathematical model analysis is becoming more significant all the time.

4.1 Physical Phenomenological Model Analysis

This method belongs the oldest methods for determining similarity

J. Kuneš, *Similarity and Modeling in Science and Engineering*,
DOI: 10.1007/978-1-907343-78-0_4,

criteria. It was already utilized in the 19th century prior to the discovery of the Pi theorem and of dimensional analysis. At present, it is rarely utilized for this purpose, but more frequently to check the similarity criteria physically.

The principle of this method is relatively simple, but a good physics intuition is needed for it. It proceeds from a phenomenological physical model of an investigated phenomenon. The basis consists of dimensional homogeneity of mathematical equations, proceeding from the physical significance of individual members in equations. These members represent acting forces, energies, momenta, and other physical effects. The phenomenological expression of some forces and energies in mechanics and electrical engineering is presented in the following subsection 4.1.1.

When using the physical analysis method to determine the similarity criteria, the procedure consists of two stages. In the first one, all characteristic acting forces or energies are determined. In the next stage, the dimensionless criteria are determined so that the ratios among individual forces or energies may be derived. The number of criteria is given by the number of independent forces or energies which appear in the investigated phenomenon. For the practical procedure, good knowledge of the physical phenomenon is required. Usually, it is sufficient to know the structure and nature of acting forces but not the complete mathematical model of the process.

4.1.1 Phenomenological Expression of Forces and Energies

A survey of forces and energies in mechanics and electrical engineering is in the Tables 4.1 and 4.2. A survey of symbols for the quantities is given in the Appendix A.

4.1.2 Physical Significance of Similarity Criteria

The physical significance of the similarity criteria expressing the ratio between acting forces or energies enables one to get an idea about the dynamics and changing nature of an investigated process. This is well known with the Reynolds number *Re,* which characterizes the viscous fluid flow. Depending on its size, the flow can be divided into three fundamental characteristic regions – laminar, transient, and turbulent. Other complex processes can be similarly characterized.

To understand the physical meaning of the similarity criterion, a more complicated case from the mechanics of polyphase sets can be presented as an example, specifically, steam condensation and fluid boiling, the dynamics of coating deposition, the process of formation

Tab. 4.1 Phenomenological expression of forces and energies in mechanics

Forces in mechanics:				
F_p	$\propto$	$\Delta p L^2$	...	pressure force
F_{i1}	$\propto$	$\rho w^2 L^2$	...	inertial force
F_{i2}	$\propto$	$\rho w \tau^{-1} L^3$	...	inertial force
F_v	$\propto$	$\eta L^2 \tau^{-1}$	...	viscous–elastic force
F_g	$\propto$	$g \rho L^3$	...	gravity force
F_σ	$\propto$	σL	...	surface stress force
F_{ela1}	$\propto$	EL^2	...	elastic force of solid substances
F_{com}	$\propto$	KL^2	...	fluid compressibility force
F_f	$\propto$	$\eta w L$	...	friction force
F_h	$\propto$	$\rho g h L^2$	...	hydrostatic force
F_{cap}	$\propto$	$p_{cap} L^2$	...	capillary force
F_{fil}	$\propto$	$Q \eta k^{-1}$	...	filtration force
F_{cen}	$\propto$	$\rho L^3 r \omega^2$	...	centrifugal force
F_{Cor}	$\propto$	$\rho L^3 \omega v_r$	...	Coriolis force
Energies in mechanics:				
E_{cond}	$\propto$	$\lambda L T \tau$	...	heat conduction
E_{conv}	$\propto$	$\alpha L^2 T \tau$	...	heat transfer by convection
E_{rad}	$\propto$	$\sigma L^2 T^4 \tau$	...	heat transfer by radiation
E_{mob}	$\propto$	$\rho c w T L^2 \tau$	...	movement heat
E_{com}	$\propto$	$L^3 \Delta p$	...	compression energy
E_H	$\propto$	$\rho c T L^3$	...	enthalpy
F_{fric}	$\propto$	$L \eta w^2 \tau$	...	friction energy
E_{chem}	$\propto$	$L^3 \rho e$	...	chemical–thermal energy
E_{kin1}	$\propto$	$m w^2$	...	kinetic energy
E_{kin2}	$\propto$	$L^2 m \tau^{-2}$	...	kinetic energy
E_{kin3}	$\propto$	$F w \tau$	...	kinetic energy
E_{pot}	$\propto$	mgh	...	potential energy

and disintegration of bubbles in the moving or non-moving fluid. The surface stress force F_a, the gravity (Archimedes) force F_g, and the inertia force F_i are acting forces. The Weber numbers, expressed by the acting forces ratio phenomenologically, are characteristic criteria of physical similarity:

- *Weber number for a single-phase fluid*

$$We = \frac{F_i}{F_\sigma} = \frac{\rho w^2 L}{\sigma}, \tag{4.1}$$

- *Weber condensation number for a two-phase stationary fluid (steam)*

Tab. 4.2 Phenomenological expression of forces and energies in electrical engineering

		Forces in electrical engineering		
F_{E1}	$\propto$	εU^2	...	electric force
F_{E2}	$\propto$	$\varepsilon E^2 L^2$	...	electric force
F_{E3}	$\propto$	EDL^2	...	electric force
F_{E4}	$\propto$	WL^{-1}	...	electric force
F_{M1}	$\propto$	$\gamma w B^2 L^3$	...	electrodynamic force
F_{M2}	$\propto$	γEBL^3	...	electrodynamic force
F_{M3}	$\propto$	$B^2 L^2\ \mu^{-1}$	...	magnetic force
F_{M4}	$\propto$	BHL^2	...	magnetic force
F_{M5}	$\propto$	BLI	...	magnetic force
F_{M6}	$\propto$	mBL^{-1}	...	magnetic force
		Energies in electrotechnics:		
F_{E1}	$\propto$	UQ	...	electric energy
F_{E2}	$\propto$	$Q^2 C^{-1}$	...	electric energy
F_{E3}	$\propto$	DEL^3	...	electric energy
F_{E4}	$\propto$	CU^2	...	electric energy
F_{E5}	$\propto$	$\varepsilon E^2 L^3$	...	electric energy
F_{M1}	$\propto$	BHL^3	...	magnetic energy
F_{M2}	$\propto$	$B^2 \mu^{-1} L^3$	...	magnetic energy
F_{M3}	$\propto$	MI^2	...	magnetic energy
F_{M4}	$\propto$	$\mu H^2 L^3$	...	magnetic energy
F_{M5}	$\propto$	$I\Psi$	...	magnetic energy
F_{M6}	$\propto$	$L_i I^2$	...	magnetic energy

$$We = \frac{F_g}{F_\sigma} = \frac{(\rho_l - \rho_v) g L^2}{\sigma}, \tag{4.2}$$

- *Weber wet-steam number for a two-phase moving fluid* (*wet steam*)

$$We = \frac{F_i}{F_\sigma} = \frac{\rho_v (w_v - w)^2 L}{\sigma}. \tag{4.3}$$

The *Weber boiling number for the two-phase moving fluid* has the same form as the Weber wet-steam number (4.3). It differs only in the significance of velocity. Velocity w_v is the steam bubble velocity and w is the flow velocity of the boiling fluid.

The *Weber boiling number for the two-phase non-moving fluid* (*Bond number*) has the same form as the Weber condensation number (4.2) but different physical significance.

In the example of the Weber number, its physical meaning can be clarified, specifically for the dynamics of drop movement in steam condensation and for steam bubble movement with the fluid boiling. The complexity of the condensation process implies, in addition to other things, various kinds of steam condensation. However, in all condensation cases the Weber number is the characteristic criterion which expresses the relation between the surface stress force and gravity or inertia force.

The way the size of Weber number influences the nature of the condensation and boiling process is obvious from Fig. 4.1. In case (A), it involves steam condensation on a cold wall, in case (B), it involves the fluid boiling over a hot wall or possibly evaporation or dissipation of a moving drop over a hot wall.

For case (A) there are four different steam condensation states presented in Fig. 4.1. In an immobile two-phase environment, drop condensation is characterized by a low value of the Weber number *We*, where $We \to 0$ for a limited case of a spherical drop (a). Due to the gravity force influence increasing over the surface stress force, the Weber number increases and the original spherical drop shape is deformed to the shapes (b) and (c) afterwards. A further increase of the gravity force influence over the surface stress and, therefore, rise of the Weber number as well, results in a loss of the condensate shape cohesion which is characteristic for membrane condensation (d) with which a continuous condensate layer forms. The nature of a condensation surface has a significant role in influencing the surface

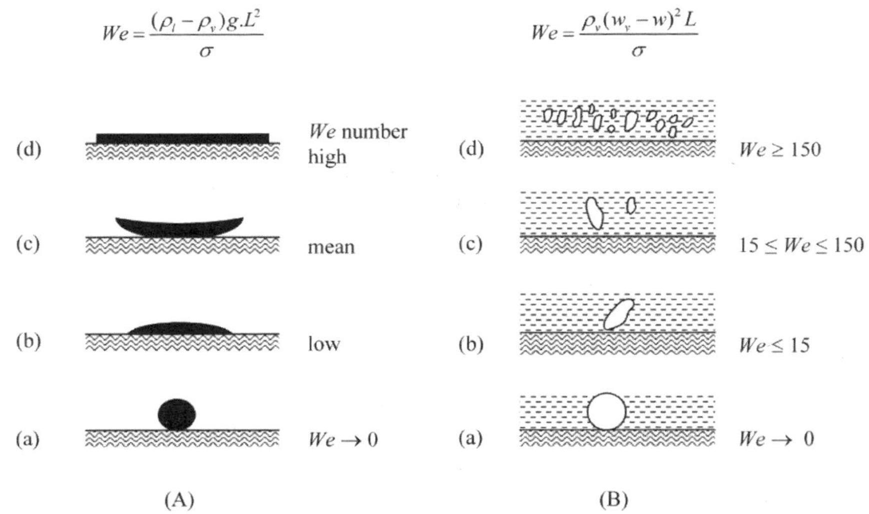

Fig. 4.1. Physical significance of the Weber number in the process of condensation and originating a coating on a cold wall (A), and with fluid boiling near a hot surface (B).

stress force and hence even the way condensation and heat transfer occur. A similar role is played by the Weber number in processes of deposition of thin layers and coatings.

In case (B) in Fig. 4.1, four different states of particle interaction (gas or steam bubbles, drops) with the body surface are expressed, and zones of the Weber boiling number are given. With the fluid boiling, this can involve the origin and development of a steam bubble, for example. With limitation on observing the influence of the surface stress and inertia forces only, then at the beginning of the steam bubble origin, the growth of which is given by Jacob number Ja_1, the bubble (a), adhering to the surface, has a spherical shape and the Weber steam number $We \rightarrow 0$ because the surface-stress force influence prevails over the inertia force ($w_v \rightarrow w$). With the steam bubble detaching from the wall surface, the significance of the inertia force rises, the bubble becomes deformed (b), and with a further increase of the Weber number, divides (c) and finally disintegrates (d). In fact, this process is more complicated, and besides inertia and surface stress forces, even the uplift force (Froude boiling number Fr_b, steam pressure criterion K_p), the viscosity force (Reynolds boiling number Re_b), and the heat transfer intensity (Jacob number of evaporation velocity Ja_2) occur.

The suggested physical significance of the Weber number for the condensation and boiling cases is found in other areas, e.g. in the sphere of micro- and nanotechnologies in forming layers and coatings.

4.1.3 Examples

The application of the physical phenomenological model method is obvious in the following two examples. The first one represents a hydrodynamic task, described by the Navier–Stokes equation generally, in which each member expresses an acting force. The second example represents an aero-thermo-mechanical task and involves deduction of the similarity criteria not only from the forces ratio but also from the ratio of energies, even for boundary conditions. It is also obvious that the same criterion, e.g. the Strouhal criterion, can be expressed either from the forces ratio (first example) or from that of energies (second example).

Example 4.1 *Hydrodynamics. Viscous flow of incompressible fluid*

Fundamental hydrodynamic similarity criteria for viscous incompressible fluid flow have to be determined. The flow process can be expressed in phenomenological form as a result of influences of the pressure force F_p, friction force F_f, gravity force F_g, inertial convection force F_{i1}, and local force F_{i2}, so that it can be written

$$F_p + F_f + F_g = F_{i1} + F_{i2}. \tag{4.4}$$

The composed hydrodynamic similarity criteria can be obtained directly from the ratio of individual forces if these can be expressed by corresponding dimensional quantities

$$\frac{F_p}{F_{i1}} \rightarrow Eu = \frac{p}{\rho w^2} \quad \text{– Euler number.}$$

$$\frac{F_{i1}}{F_f} \rightarrow Re = \frac{wL}{\nu} \quad \text{– Reynolds number.}$$

$$\frac{F_{i1}}{F_g} \rightarrow Fr = \frac{w^2}{gL} \quad \text{– Froude number.}$$

$$\frac{F_{i1}}{F_{i2}} \rightarrow Sh = \frac{w\tau}{L} \quad \text{– Strouhal number.}$$

The viscous incompressible fluid flow is described by the dimensionless equation

$$\varphi(Re, Eu, Fr, Sh, W, X, \ldots) = 0, \tag{4.5}$$

containing four composed similarity and other parametric criteria.

Example 4.2 *Space aero-thermo-mechanics*

The dimensionless aero-thermo-mechanical similarity criteria for movement and heating of a spherical body in fluid have to be determined without considering the compressibility influence. The body is affected by gravity force F_g, aerodynamic resistance F_R, hydrostatic uplift force F_h, and inertia force F_i. The body is subjected to aerodynamic heating and radiation.

The *dynamic balance of the body* can be expressed as a result of forces affecting the body

$$F_g = F_i + F_h + F_R. \tag{4.6}$$

Then, by dividing individual members of Eq. (4.6) by, for example, force $|F_j|$, it follows that

$$-\frac{F_g}{F_i} + 1 + \frac{F_h}{F_i} + \frac{F_R}{F_i} = 0. \tag{4.7}$$

Because the total aerodynamic resistance F_R can be presumed to equal the sum of shape (pressure) and friction resistances, the following holds

$$F_R = F_p + F_f, \tag{4.8}$$

and Eq. (4.7) is transformed to

$$\underset{Fr^{-1}}{\underset{\downarrow}{-\frac{F_g}{F_i}}}+1+\underset{R}{\underset{\downarrow}{\frac{F_h}{F_g}}}\,\underset{Fr^{-1}}{\underset{\downarrow}{\frac{F_g}{F_i}}}+\underset{Re^{-1}}{\underset{\downarrow}{\frac{F_f}{F_i}}}+\underset{Eu}{\underset{\downarrow}{\frac{F_p}{F_i}}}=0, \tag{4.9}$$

where $R = \rho_l \rho_s^{-1}$ is the ratio of the fluid density and the sphere material. In the fluid, the body movement is described by three composed criteria and simple criteria for the density and coordinates

$$\varphi_1 = (Re, Eu, Fr, R, X, \ldots) = 0. \tag{4.10}$$

Similarly, the thermal similarity criteria, describing the *heating of a moving body,* can be determined. An unsteady heat flux through the body is considered. For the energy balance, the inner energy increment ΔU of a body equals the heat sum from the aerodynamic heating E_{con}[4] and radiation E_{rad} on the body surface

$$\Delta U = E_{con} + E_{rad}. \tag{4.11}$$

With the aerodynamic heating neglected, the heat equality transferred by radiation E_{rad} and conduction E_{cond} is valid on the body surface as follows

$$E_{rad} = E_{cond}. \tag{4.12}$$

By forming the ratios between heats and describing them with corresponding dimensional expressions, the composed dimensionless thermal similarity criteria are obtained

$$\frac{E_{conv}}{E_{cond}} \rightarrow Pe = \frac{wL}{a} \quad \text{– Péclet number,}$$

$$\frac{E_{cond}}{U} \rightarrow Fo = \frac{a\tau}{L^2} \quad \text{– Fourier number,}$$

where U is the inner body energy. The dimensionless movement velocity can be expressed in the following form

$$\frac{E_{conv}}{U} \rightarrow Sh = \frac{w\tau}{L} = Pe\,Fo \quad - \text{ Strouhal number.}$$

In a similar way, the criterion for a nonlinear boundary condition, determining the thermal energy transfer on the body surface by radiation, can be determined

$$\frac{E_{rad}}{E_{cond}} \rightarrow Sf = \frac{\sigma}{\lambda} T^3 L \quad - \text{ Stefan (Stark) number.}$$

Sometimes, the inverse value is used

$$\frac{E_{cond}}{E_{rad}} \rightarrow Bo = \frac{\lambda}{\sigma T^3 L} \quad - \text{ Boltzmann number.}$$

The body heating is described by the following dimensionless equation

$$\varphi_2\left(Pe, Fo, Sf, \Theta, X, \ldots\right) = 0 . \qquad (4.13)$$

where neither Strouhal nor Boltzmann numbers appear in the equation, as they can be expressed by the criteria mentioned already in the equation.

The body movement in the incompressible viscous fluid when considering the influence of gravitation, dynamic heating, and radiation is expressed in the following dimensionless form

$$\varphi\left(Re, Eu, Fr, Pe, Fo, Sf, R, \Theta, X, \ldots\right) = 0 , \qquad (4.14)$$

which is composed of six composed criteria and other simple ones, of which Θ is the dimensionless temperature and X the dimensionless coordinate.

4.2 Mathematical Model Analysis

The considerable comprehensiveness of fundamental physical equations, their unambiguity, speed and simplicity in obtaining similarity criteria made the mathematical model analysis of a physical process the most effective method of similarity theory. Steadily increasing importance of mathematics in various technical and non-technical fields contributes to this. The mathematical model and its degree of credibility have become a measure of reality understanding as well as the level of development of the field

Essentially, the *similarity analysis method of a mathematical model* proceeds from the already-mentioned Fourier theorem of dimensional homogeneity of physical equations, in which every additive member is equal with respect to dimension. Usually, the method utilizes three different procedures based on the formal arrangement of the equations. However, a complete mathematical process description is required, namely a basic equation or set of equations and corresponding equations for boundary, initial, optimization and restrictive conditions, i.e. the unambiguity conditions.

4.2.1 Procedure Utilizing Scale Similarity Indicators

This is based on *scale expressions.* With them, in a basic equation and unambiguity condition equations, all variables are extended without changing the original significance of equations and their individual members. At every additive equation member, the group of scale expressions are equal to one. These groups of scale expressions are called *similarity indicators.* From them, the similarity criteria sought can be determined directly.

Let us use dimensional quantities x_i in the mathematical process description. Then, the scale expressions μ_i for all quantities x_i can be introduced

$$\mu_{x_i} = \frac{x_{i,M}}{x_{i,D}}, \qquad i = 1,\ldots,k, \tag{4.15}$$

where $x_{i,M}$ expresses the quantity on the model (M) and $x_{i,D}$ the quantity on the object (D).

4.2.2 Procedure Utilizing Reference Quantities

In converting fundamental process equations and unambiguity condition equations to dimensionless ones, the *reference quantities* are used, by which selected variables in the original equations are extended. As a result, composed expressions are obtained, usually containing one process quantity, equation parameters, and reference quantities. The dimensionless composed expressions obtained are similarity criteria, and the resulting equations are already converted into a dimensionless form.

Let us use dimensional quantities x_i in the mathematical process description. Then, for selected quantities, the reference quantities $x_{i,\,ref}$ will be introduced, by means of which the following simple parametric criteria can be expressed

$$P_i = \frac{x_i}{x_{i,ref}}, \qquad i = 1, \dots, m. \tag{4.16}$$

4.2.3 Procedure Utilizing Integral Analogues

This method is based on the possibility to replace the mathematical operators and other symbols in the basic equation and in the unambiguity condition equations by a ratio of mutually corresponding quantities called *integral analogues.* The signs of derivation and integrals are omitted because they do not express the dimensionality in equations. At the same time, all vector components in coordinate axes are replaced by absolute vector values and the coordinates are replaced by characteristic length dimensions. All equality, addition and division signs are replaced by similarity or proportionality ones. The dimensionless similarity criteria sought are equal to mutual ratios of obtained dimensional members.

Integral analogues of some mathematical symbols
Space derivation

$$[\nabla] = [\text{div}] = [\text{grad}] = [\text{rot}] = \left[\frac{\partial}{\partial x}\right] = \left[\frac{\text{d}}{\text{d}x}\right] = \left[L^{-1}\right], \tag{4.17}$$

$$[\nabla u] = [\text{grad}\, u] = \left[\frac{\partial u}{\partial x}\right] = \left[\frac{\text{d}u}{\text{d}x}\right] = \left[\frac{\Delta u}{\Delta x}\right] = \left[\frac{u}{x}\right] = \left[u\, L^{-1}\right], \tag{4.18}$$

$$[\nabla H] = [\text{div}\, H] = \left[\frac{\partial H}{\partial x}\right] = \left[H\, L^{-1}\right], \tag{4.19}$$

$$[\nabla \times H] = [\text{rot}\, H] = \left[H\, L^{-1}\right], \tag{4.20}$$

$$\left[\nabla^2\right] = [\nabla\nabla] = [\nabla] = \left[\frac{\partial^2}{\partial x^2}\right] = \left[L^{-2}\right], \tag{4.21}$$

$$\left[\nabla^n\right] = \left[\frac{\partial^n}{\partial x^n}\right] = \left[L^{-n}\right]. \tag{4.22}$$

Time derivation

$$\left[\nabla_\tau^n u\right] = \left[\frac{\partial^n u}{\partial \tau^n}\right] = \left[\frac{\text{d}^n u}{\text{d}\tau^n}\right] = \left[u\, \tau^{-n}\right]. \tag{4.23}$$

Integrals

$$\left[\int_b^a y\,\mathrm{d}x\right]=[y\,x]=\left[L^2\right], \tag{4.24}$$

$$\left[\int\int k\,\mathrm{d}y\,\mathrm{d}z\right]=[k\,y\,z]=\left[k\,L^2\right]. \tag{4.25}$$

4.2.4 Example

In the following example, the three mentioned procedures of a mathematical model analysis are compared, specifically, for the case of unsteady heat conduction with various boundary conditions. The simplicity as well as speed of procedures, leading to unambiguous determination of thermal similarity criteria, are characteristic properties of this method.

Example 4.3 *Unsteady linear heat conduction. Motion temperature field with various boundary conditions*

The procedure of using three different methods for the mathematical model analysis in determining the similarity criteria and conversion of equations into the dimensionless state has to be demonstrated.

The fundamental *Fourier–Kirchhoff equation* for an unsteady temperature motion field has the following form

$$a\nabla^2 T + w\,\nabla T = \frac{\partial T}{\partial \tau}. \tag{4.26}$$

The considered *boundary conditions:*

1st-kind time-variable condition expresses the linear change of surface temperature

$$T_s = T_{S,ini}\left(1+b\tau\right), \tag{4.27}$$

2nd-kind condition expresses the density of the surface heat flux

$$q_A = -\lambda\frac{\partial T}{\partial n}, \tag{4.28}$$

3rd-kind condition expresses the convection heat transfer on the area surface

$$-\lambda \frac{\partial T}{\partial n} = \alpha \Delta T, \tag{4.29}$$

with the following references: T designates temperature, ΔT is the thermal difference, T_S, $T_{S,ini}$ are temperatures of the wall surface and its initial value, q_A is the heat flux surface density, τ is time, a is thermal diffusivity, λ is thermal conductivity, a is the heat transfer coefficient, w is velocity, b is a constant, x, y, z are the coordinates, n is the normal perpendicular to the surface; the reference quantities: L is the reference length, T_{ref} is the reference temperature, and w_{ref} is the reference velocity are further acting quantities.

Determination of the number of criteria by the Pi theorem:	
Total number of dimensional physical quantities:	$N = 18$
Number of dimensionally different quantities:	$n = 9$
Number of basic and complementary dimensions:	$r = 4$ (L, M, T, Θ)
Total number of dimensionless criteria:	$k = N - r = 14$
Number of simple criteria:	$k_s = N - n = 9$
Number of composed criteria:	$k_k = n - r = 5$

The simple similarity criteria are used only in a procedure utilizing reference quantities where the composed criteria for the fundamental equation and all other boundary conditions are being determined. The composed similarity criteria obtained by the procedure, using the scale similarity indicators, are represented by the fundamental equation and the 1st-kind boundary condition. The procedure using integral analogues is shown in the fundamental equation and 3rd-kind boundary condition.

Procedure utilizing scale similarity indicators

The scales for temperatures μ_T, lengths μ_L, time μ_τ, thermal diffusivity μ_a, velocity μ_w and velocity constant μ_b are introduced as follows

$$\mu_T = \frac{T_M}{T_D}, \ \mu_L = \frac{L_M}{L_D}, \ \mu_\tau = \frac{T_M}{T_D}, \tag{4.30}$$

$$\mu_a = \frac{a_M}{a_D}, \ \mu_w = \frac{w_M}{w_D}, \ \mu_b = \frac{b_M}{b_D}, \tag{4.31}$$

and express the ratio of a corresponding quantity on a model (M) to that on an object (D). Equation (4.26) for a model has the following form

$$a_M \nabla^2 T_M + w_M \nabla T_M = \frac{\partial T_M}{\partial \tau_M}. \tag{4.32}$$

After inserting the scales, one obtains the equation for the object

$$\frac{\mu_a \mu_T}{\mu_L^2} a_D \nabla^2 T_D + \frac{\mu_w \mu_T}{\mu_L} w_D \nabla T_D = \frac{\mu_T}{\mu_\tau} \frac{\partial T_D}{\partial \tau_D}, \tag{4.33}$$

without changing the equation's significance. With Eq. (4.33) multiplied by $\mu_L^2 \mu_a^{-1} \mu_T^{-1}$, one obtains

$$a_D \nabla^2 T_D + \frac{\mu_w \mu_L}{\mu_a} w_D \nabla T_D = \frac{\mu_L^2}{\mu_a \mu_\tau} \frac{\partial T_D}{\partial \tau_D}. \tag{4.34}$$

From Eq. (4.34), the conditions for the thermal similarity, expressed by so-called similarity indicators, imply the following

$$\frac{\mu_w \mu L}{\mu_a} = 1, \quad \frac{\mu_L^2}{\mu_a \mu_\tau} = 1. \tag{4.35}$$

After reconverting it from the scales to physical quantities, the following composed criteria are derived

$$P_e = \frac{wL}{a} \quad \text{– Péclet number,}$$

$$Fo = \frac{a\tau}{L^2} \quad \text{– Fourier number.}$$

The Péclet number expresses the ratio of the fluid heat flux by flow to that by conduction, whereas the Fourier number expresses the dimensionless time.

Similarly, the equation for the 1st-kind boundary variable time condition for the model can be rewritten into the following form

$$T_{S,M} = T_{S,ini,M}\left(1 + b_M \tau_M\right), \tag{4.36}$$

and after application of the scales

$$\mu_T T_{S,D} = \mu_T T_{S,ini,D}\left(1 + \mu_b \mu_\tau b_D \tau_D\right). \tag{4.37}$$

With Eq. (4.37) multiplied by the expression μ_T^{-1}, one obtains

$$T_{S,D} = T_{S,ini,D}\left(1 + \mu b \ \mu\tau \ b_D \tau_D\right), \tag{4.38}$$

from which the similarity indicator results

$$\mu_b \, \mu_\tau = 1. \tag{4.39}$$

After inserting expression for μ_τ from Eq. (4.35)

$$\mu_b \frac{\mu_L^2}{\mu_a} = 1. \tag{4.40}$$

After reconverting from the scales to physical quantities, one obtains

$$Pd = \frac{bL^2}{a} \quad - \text{ Predvoditel number,}$$

which expresses the time change of the surface temperature.

Procedure utilizing reference quantities

With reference length L, temperature T_{ref}, and velocity w_{ref} introduced, simple similarity criteria can be defined

$$X = \frac{x}{L}, \; Y = \frac{y}{L}, \; Z = \frac{z}{L}, \; N = \frac{n}{L}, \; W = \frac{w}{w_{ref}},$$

$$\Theta = \frac{T}{T_{ref}}, \; \Delta\Theta = \frac{\Delta T}{T_{ref}}, \; \Theta_S = \frac{T_S}{T_{ref}}, \; \Theta_{S,ini} = \frac{T_{S,ini}}{T_{ref}}.$$

After inserting these relations into the original Eq. (4.26), the following dimensionless form is derived

$$\frac{aT_{ref}}{L^2} \nabla^2 \Theta + \frac{w_{ref} T_{ref}}{L} W \, \nabla\Theta = T_{ref} \frac{\partial \Theta}{\partial \tau}, \tag{4.41}$$

and similarly for the 1st-, 2nd-, and 3rd-kind boundary conditions

$$T_{ref} \, \Theta_S = T_{ref} \, \Theta_{S,ini} \left(1 + b\tau\right), \tag{4.42}$$

$$q_A = -\frac{\lambda T_{ref}}{L} \frac{\partial \Theta}{\partial N}, \tag{4.43}$$

$$-\frac{\lambda T_{ref}}{L} \frac{\partial \Theta}{\partial N} = \alpha \, T_{ref} \Delta\Theta. \tag{4.44}$$

With Eqs. (4.41) to (4.44) modified, one obtains

$$\nabla^2\Theta + \underbrace{\frac{W_{ref}L}{a}}_{Pe} W\,\nabla\Theta = \frac{\partial\Theta}{\partial\underbrace{\left(\frac{a\tau}{L^2}\right)}_{Fo}}, \tag{4.45}$$

$$\Theta_S = \Theta_{S,ini}\left(1 + \underbrace{\frac{bL^2}{a}}_{Pd}\underbrace{\frac{\tau a}{L^2}}_{Fo}\right), \tag{4.46}$$

$$\underbrace{\frac{q\mathrm{A}L}{\lambda T_{ref}}}_{Ki} = -\frac{\partial\Theta}{\partial N}, \tag{4.47}$$

$$-\frac{\partial\Theta}{\partial N} = \underbrace{\frac{\alpha L}{\lambda}}_{Bi}\Delta\Theta, \tag{4.48}$$

where *Pe, Fo, Pd, Ki* and *Bi* are Péclet, Fourier, Predvoditel, Kirpitchev, and Biot numbers. The Kirpitchev number expresses the ratio of the surface heat flux density to the heat conduction flux density in a body. The Biot number expresses the ratio of the heat flux transferred by convection on a body surface to the heat flux transferred by conduction in a body.

Then, the original Eqs. (4.26) to (4.29) can be written in the following forms

$$\nabla^2\Theta + PeW\nabla\Theta = \frac{\partial\Theta}{\partial Fo}, \tag{4.49}$$

$$\Theta_S = \Theta_{S,ini}\left(1 + Pd\,Fo\right), \tag{4.50}$$

$$Ki = -\frac{\partial\Theta}{\partial N}, \tag{4.51}$$

$$-\frac{\partial\Theta}{\partial N} = Bi\,\Delta\Theta. \tag{4.52}$$

Procedure utilizing integral analogues

By expressing the dimensionality of quantities and mathematical operators, dimensional members *A, B* and *C* can be derived from the fundamental equation as follows

$$a\nabla^2 T + w\nabla T = \frac{\partial T}{\partial \tau}, \qquad (4.53)$$

$$\begin{array}{ccc} \downarrow & \downarrow & \downarrow \\ \underbrace{\frac{aT}{L^2}}_{A} & \underbrace{\frac{wT}{L}}_{B} & \underbrace{\frac{T}{\tau}}_{C} \end{array}$$

by the mutual ratio from which the following composed similarity criteria can be obtained

$$\frac{B}{A} \rightarrow Pe = \frac{wL}{a} \quad - \text{ Péclet number,} \qquad (4.54)$$

$$\frac{A}{C} \rightarrow Fo = \frac{a\tau}{L^2} \quad - \text{ Fourier number.}$$

Similarly, e.g., the 3rd-kind boundary condition (4.29) can be expressed in the form

$$-\lambda \frac{\partial T}{\partial n} = \alpha \Delta T,$$

$$\begin{array}{cc} \downarrow & \downarrow \\ \underbrace{\lambda \frac{T}{L}}_{D} & \underbrace{\alpha T}_{E} \end{array}$$

from which the following number can be obtained

$$\frac{E}{D} \rightarrow Bi = \frac{\alpha L}{\lambda} \quad - \text{ Biot number.}$$

4.3 Applications

In the following examples, three procedures for the mathematical model analysis in various fields are explained: heat conduction, mechanics and thermal stress in solid bodies, heat transfer in fluids, electricity and magneto-hydrodynamics, physical chemistry, rheology and tribology, technology, ecology and medicine.

4.3.1 Thermomechanics of Solid Bodies

In the following example, a mathematical temperature field model in an anisotropic environment is transformed to an isotropic environment by making use of scale similarity indicators. Subsequent examples concern body mechanics. The first of them describes damped beam vibration, whereas the second describes the dynamics of musical string instruments. Both examples represent the procedure utilizing integral analogues.

The next two examples relate to the heat stress zone. In the first of them, the heat stress similarity criteria are deduced in a linear system with inner heat sources, by the procedure utilizing reference quantities, whereas in the second one, modeling scales for the general heat-stressed body state are determined by means of combining the procedures utilizing integral analogues and similarity scale indicators.

Example 4.4 *Physics. Unsteady heat conduction in an anisotropic environment. Mathematical model transformation*

The transformation scale of a mathematical model of a temperature field with inner heat sources in an anisotropic cylinder, to be transformed to a model in an isotropic cylinder, has to be determined. The anisotropy is given by different thermal conductivity in radial λ_r, peripheral λ_v and axial λ_z directions of the cylinder and is presumed to be constant. In the coordinate system of *r, p, z,* the *thermal process in an anisotropic cylinder* can be expressed by the following equation

$$\lambda_r \frac{\partial^2 T}{\partial r^2} + \frac{\lambda_r}{r}\frac{\partial T}{\partial r} + \frac{\lambda_\varphi}{r^2}\frac{\partial^2 T}{\partial \varphi^2} + \lambda_z \frac{\partial^2 T}{\partial z^2} + q_V = c\rho \frac{\partial T}{\partial \tau}, \tag{4.55}$$

where T denotes temperature, q_v is the volume heat flux density (inner source), $c\rho$ is the specific volume heat, and τ is time.

Similarly, for the *transformed isotropised temperature field,* the following holds

$$\lambda_{rt} \frac{\partial^2 T_t}{\partial r_t^2} + \frac{\lambda_{rt}}{r_t}\frac{\partial T_t}{\partial r_t} + \frac{\lambda_{\varphi t}}{r_t^2}\frac{\partial^2 T_t}{\partial \varphi_t^2} + \lambda_{zt} \frac{\partial^2 T_t}{\partial z_t^2} + qv_t = (c\rho)t \frac{\partial T_t}{\partial \tau_t}, \tag{4.56}$$

but where $\lambda_{rt} = \lambda_{\varphi t} = \lambda_{zt} = \lambda_t$.

With the introduced scales

$$\mu_r = \frac{r}{r_t},\ \mu_\varphi = \frac{\varphi}{\varphi_t},\ \mu_z = \frac{z}{z_t},\ \mu_T = \frac{T}{T_t},\ \mu_{\lambda r} = \frac{\lambda_r}{\lambda_t}, \tag{4.57}$$

$$\mu\lambda_\varphi = \frac{\lambda_\varphi}{\lambda_t},\ \mu\lambda_z = \frac{\lambda_z}{\lambda_t},\ \mu_q = \frac{q_V}{q_{Vt}},\ \mu_c = \frac{c\rho}{(c\rho)_t},\ \mu_\tau = \frac{\tau}{\tau_t}, \tag{4.58}$$

Eq. (4.55) changes to

$$\frac{\mu\lambda_r\,\mu_T}{\mu_r^2}\lambda_{rt}\frac{\partial^2 T_t}{\partial r_t^2}+\frac{\mu\lambda_r\,\mu_T}{\mu_r^2}\frac{\lambda_{rt}}{r_t}\frac{\partial T_t}{\partial r_t}+\frac{\mu\lambda_\varphi\,\mu_T}{\mu_r^2\,\mu_\varphi^2}\frac{\lambda_{\varphi t}}{r_t^2}\frac{\partial^2 T_t}{\partial \varphi_t^2}+$$

$$\frac{\mu\lambda_z\,\mu_T}{\mu_z^2}\lambda_{zt}\frac{\partial^2 T_t}{\partial z_t^2}+\mu_q\,q_{Vt}=\frac{\mu_c\,\mu_T}{\mu_\tau}(c\rho)_t\frac{\partial T_t}{\partial \tau_t}. \tag{4.59}$$

From comparison of Eqs. (4.56) and (4.59), the *similarity conditions* imply

$$\frac{\mu\lambda_r\,\mu_T}{\mu_r^2}=\frac{\mu\lambda_\varphi\,\mu_T}{\mu_r^2\,\mu_\varphi^2}=\frac{\mu\lambda_z\,\mu_T}{\mu_z^2}=\mu_q=\frac{\mu_c\,\mu_T}{\mu_\tau}. \tag{4.60}$$

By modifying Eq. (4.60), the following *coordinates transformation scales* are obtained

$$\mu_r=\sqrt{\frac{\lambda_r\mu_T}{\lambda_t\mu_q}},\ \mu_\varphi=\sqrt{\frac{\lambda_\varphi}{\lambda_r}},\ \mu_z=\sqrt{\frac{\lambda_z\mu_T}{\lambda_t\mu_q}}. \tag{4.61}$$

The scales for sources μ_q and temperatures μ_T are usually chosen and, for other scales, the following *similarity indicator* holds according to (4.56), (4.59) and (4.60)

$$\frac{\mu_c\mu_T}{\mu_q\mu_\tau}=1. \tag{4.62}$$

Example 4.5 *Mechanics of bodies. Damped vibrations*

The number and form of the dynamic mechanical similarity criteria for a damped vibrating beam should be determined. The acting quantities are: longitudinal density of the beam ρ_l, specific transverse beam load q, environment dynamic viscosity η, modulus of elasticity E, polar moment J, coordinate x, coordinate y perpendicular to the beam, time τ, frequency f and reference length L.

Determination of the number of criteria by the Pi theorem:	
Total number of dimensional physical quantities:	$N = 10$
Number of dimensionally different quantities:	$n = 8$
Number of basic and complementary dimensions:	$r = 3$ (L, M, T)
Total number of dimensionless criteria:	$k = N - r = 7$
Number of simple criteria:	$k_s = N - n = 2$
Number of composed criteria:	$k_k = n - r = 5$

The simple criteria are the dimensionless coordinates

$$X=\frac{x}{L},\ Y=\frac{y}{L}.$$

The mathematical model of the vibrating beam is expressed by the following equation

$$\underbrace{EJ\frac{\mathrm{d}^4y}{\mathrm{d}x^4}}_{\downarrow\ EJ\frac{y}{L^4} = A}+\underbrace{\rho l\frac{\mathrm{d}^2y}{\mathrm{d}\tau^2}}_{\downarrow\ \frac{\rho l\,y}{\tau^2} = B}+\underbrace{\eta\frac{\mathrm{d}y}{\mathrm{d}\tau}}_{\downarrow\ \frac{\eta y}{\tau} = C}=\underbrace{q(x,\tau)}_{\downarrow\ q = D}. \tag{4.63}$$

The beam deflection results from the effect of inertia force *B*, friction force *C* and external load *D*. Individual mechanical similarity criteria can be obtained by creating various ratios of acting forces, e.g.,

$$\frac{A}{D}\rightarrow\pi_1=\frac{EJy}{qL^4}\quad \text{– beam deflection criterion.}$$

$$\frac{B}{D}\rightarrow\pi_2=\frac{\rho_l y}{q\tau^2}=\frac{\rho_l L^4}{EJ\tau^2}=\frac{\rho_l L^4 f^2}{EJ}\quad \text{– vibration frequency criterion,}$$

$$\frac{C}{D}\rightarrow\frac{\eta y}{q\,\tau}=\frac{\eta L^4}{EJ\tau},$$

$$\frac{B}{C}\rightarrow\frac{\rho_l}{\eta\tau}\Rightarrow\pi_3=\frac{\eta^2L^4}{EJ\rho_l}\quad \text{– viscous damping criterion.}$$

The damped beam vibration is described by three composed criteria to which the other two composed criteria following from boundary and initial conditions are attached.

Example 4.6 *Mechanics of bodies. Dynamics of musical string instruments*

The number and form of mechanical dynamic similarity criteria for string instruments – violin, viola, violoncello, and double bass – are to be determined. Acting quantities are: momentary string deviation *y*, longitudinal along-the-string direction coordinate *x*, time τ, frequency *f*, mechanical strain σ in the string, string density ρ, and reference length

L. In all instruments mentioned, the same string strain and density are assumed

Determination of the number of criteria by the Pi theorem:	
Total number of dimensional physical quantities:	$N = 7$
Number of dimensionally different quantities:	$n = 5$
Number of basic and complementary dimensions:	$r = 3$ (L, M, T)
Total number of dimensionless criteria:	$k = N - r = 4$
Number of simple criteria:	$k_s = N - n = 2$
Number of composed criteria:	$k_k = n - r = 2$

The simple criteria are the following dimensionless coordinates

$$X = \frac{x}{L}, \; Y = \frac{y}{L}.$$

The mathematical model of the vibrating string in the above-mentioned instruments is expressed by the d'Alambert wave equation

$$\frac{\partial^2 y}{\partial \tau^2} = \frac{\sigma}{\rho}\frac{\partial^2 y}{\partial x^2}. \tag{4.64}$$

$$\downarrow \qquad \downarrow$$

$$\underbrace{\frac{y}{\tau^2}}_{A} \qquad \underbrace{\frac{\sigma}{\rho}\frac{y}{L^2}}_{B}$$

By establishing the ratio AB^{-1}, the dynamic mechanical similarity criterion for string instruments is obtained in the following form

$$\frac{A}{B} \rightarrow \frac{\rho}{\sigma}\frac{L^2}{\tau^2} \Rightarrow \pi = \frac{\rho}{\sigma} L^2 f^2.$$

The second composed criterion can be obtained from conditions of ambiguity. In building string instruments, the value of this criterion is often identical, in consequence of which the dynamic similarity condition of instruments is secured – it usually equals 0.2.

The deduced criterion is remarkable due to its physical and musical significance. First, the value is deduced from an instrument prototype that is constructed using previous experience. The criterion is the result of a complicated abstraction process in which a mathematical model can or need not be used. Knowledge of the criterion enables one to make instruments of similar properties. Here, the criterion asserts itself in the synthesis process. Finally, the role of the considered criterion is very significant in characterizing the musical properties of instruments made. A wide range of properties is concentrated remarkably into one number, which is the dynamic mechanical similarity criterion. It is similar to the case with other similarity criteria, those depth of the physical significance are not usually fully appreciated.

It is also true that for relative lengths of string instruments, as mentioned in the literature, e.g. the value of 0.9 for the violin, 1 for the viola, 2 for the violoncello, and 2.8 for the double bass, according to Stradivari, corresponding frequencies are obtained for certain musical tones. For example, for the viola, the a^1 tone appears at the frequency of 440 Hz; however, for a twice as big violoncello, the same tone has a frequency twice as deep (220 Hz). This complies with the criterion deduced, which in turn corresponds precisely with reality.

Example 4.7 *Thermal stress in a linear system with internal heat sources*

The number and form of the thermoelastic similarity criteria will be determined for coupled processes of linear unsteady heat conduction and thermal stress in an unloaded system with an inner thermal source. This relates to the thermal stress process which originates in heated or cooled material. The inner source or sink corresponds to transcrystalization heat in heating (–) or cooling (+). To determine the criteria, it is sufficient to consider one-dimensional heat conduction.

Temperature T, coordinate x, displacement u, Lamé functions λ, μ (Pa), thermal expansion coefficient a, thermal conductivity λ_t, specific heat capacity c, time τ, volume heat flux density q_V, material density ρ, reference temperature T_{ref}, and reference length L are the acting quantities.

Determination of the number of criteria by the Pi theorem:	
Total number of dimensional physical quantities:	$N = 13$
Number of dimensionally different quantities:	$n = 9$
Number of basic and complementary dimensions:	$r = 4$ (L, M, T, Θ)
Total number of dimensionless criteria:	$k = N - r = 9$
Number of simple criteria:	$k_s = N - n = 4$
Number of composed criteria:	$k_k = n - r = 5$

The coupled heat conduction and thermal stress processes are described by the following set

$$\underbrace{\frac{\partial^2 T}{\partial x^2}}_{A} - \underbrace{(3\lambda + 2\mu)\frac{\alpha T}{\lambda_t}\frac{\partial^2 u}{\partial x \partial \tau}}_{B} \pm \underbrace{\frac{q_V}{\lambda_t}}_{C} = \underbrace{\frac{c\rho}{\lambda_t}\frac{\partial T}{\partial_\tau}}_{D}, \tag{4.65}$$

$$\underbrace{\frac{\partial^2 u}{\partial x^2}}_{E} - \underbrace{\frac{3\lambda + 2\mu}{\lambda + 2\mu}\alpha\frac{\partial T}{\partial x}}_{F} = \underbrace{\frac{\rho}{\lambda + 2\mu}\frac{\partial^2 u}{\partial \tau^2}}_{G}. \tag{4.66}$$

In the theory of thermoelasticity the set of Eqs. (4.65) and (4.66) represents a mathematical process model of the displacement caused by the one-dimensional temperature field with internal heat sources and with consideration of the influence of temperature and thermal stress distributions on the process. The first member A expresses the temperature distribution on the x-axis, the member B the mutual coupling of space and time change of the displacement, the member C the influence of the inner heat source, the member D the time variation of temperature, the member E the displacement distribution in the x-axis, the member F the temperature gradient and thermal volume change influences, the member G the thermal-wave distribution and thermal stress influences.

The simple criteria are

$$X = \frac{x}{L},\ U = \frac{u}{L},\ \Theta = \frac{T}{T_{ref}},\ \frac{\lambda}{\mu}.$$

After inserting the reference quantities into Eqs. (4.65) and (4.66) and modifying them, one obtains

$$\frac{\partial^2 \Theta}{\partial X^2} - \underbrace{\frac{(3\lambda + 2\mu)\alpha a}{\lambda_t}}_{K_1}\Theta\frac{\partial^2 U}{\partial X\, \partial \underbrace{\left(\frac{a\tau}{L^2}\right)}_{Fo}} \pm \underbrace{\frac{q_V L^2}{\lambda_t T_{ref}}}_{Po} = \underbrace{\frac{c\rho}{\lambda_t}}_{1} a \frac{\partial \Theta}{\partial \underbrace{\left(\frac{a\tau}{L^2}\right)}_{Fo}}, \tag{4.67}$$

$$\frac{\partial^2 U}{\partial X^2} - \underbrace{\frac{3\lambda + 2\mu}{\lambda + 2\mu}\alpha T_{ref}}_{Gc_{te}}\frac{\partial \Theta}{\partial X} = \frac{\partial^2 U}{\partial \underbrace{\left(\frac{a\tau}{L^2}\right)^2}_{Fo}} \underbrace{\frac{a^2}{L^2}\frac{\rho}{\lambda + 2\mu}}_{Ma_{te}^2}. \tag{4.68}$$

In the set of Eqs. (4.67) and (4.68), the composed criteria are

$K_1 = (3\lambda + 2\,\mu)\dfrac{\alpha}{c\rho}$ – thermoelastic coupling criterion,

$$Fo = \frac{a\tau}{L^2} \quad \text{– Fourier number,}$$

$$Po = \frac{q_v L^2}{\lambda_t T_{ref}} \quad \text{– Pomerantsev heat number,}$$

$$Gc_{te} = \frac{3\lambda + 2\mu}{\lambda + 2\mu}\alpha T_{ref} = \beta T_{ref} \quad \text{– Gay-Lussac thermoelastic number,}$$

$$Ma_{te} = \frac{w_t}{w_{te}} \quad \text{– Mach thermoelastic number,}$$

where

$$a = \frac{\lambda_t}{c\rho} \quad \text{– thermal diffusivity,}$$

$$w_t = \frac{a}{L} \text{– heat diffusion velocity,}$$

$$w_{te} = \sqrt{\frac{\lambda + 2\mu}{\rho}} \quad \text{– thermoelastic stress propagation velocity.}$$

The starting set of Eqs. (4.65) and (4.66) can be expressed in the dimensionless form

$$\frac{\partial^2\Theta}{\partial X^2} - K_1\Theta\frac{\partial^2 U}{\partial X\,\partial Fo} \pm Po = \frac{\partial\Theta}{\partial Fo}, \tag{4.69}$$

$$\frac{\partial^2 U}{\partial X^2} - Gc_{te}\frac{\partial\Theta}{\partial X} = Ma_{te}^2\frac{\partial^2 U}{\partial Fo^2}. \tag{4.70}$$

Example 4.8 *Linear thermal stress. Modeling scales determination*

To determine the modeling scales for thermal stress of a body not subjected to external forces. On the one hand, the thermal stress is expressed in displacements and, on the other hand, by the thermoelastic displacement potential. In the process, displacement quantities *u*, Poisson number ν, coefficient of linear thermal expansion α, temperature *T*, coordinates *x*, *y*, *z*, reference length *L*, Young modulus *E*, shear elasticity modulus *G*, volume modulus of elasticity *K*, coefficient of volume thermal expansion β, deformation ε and tension σ are used.

Thermal stress expressed in displacements

For thermal stress the following is valid

$$\underbrace{\nabla^2 u}_{} + \underbrace{\frac{1}{1-2v}\nabla\left[\nabla u - 2(1+v)\alpha T\right]}_{} = 0. \tag{4.71}$$

$$\downarrow \qquad\qquad \downarrow \qquad\qquad \downarrow$$

$$\underbrace{\frac{u}{L^2}}_{A} \qquad \underbrace{\frac{u}{(1-2v)L^2}}_{B} \qquad \underbrace{\frac{2(1+v)\alpha T}{(1-2v)L}}_{C}$$

If using the integral analogue method, one obtains the expressions A, B and C and by creating their ratios one gets the *thermoelastic similarity criteria*

$\frac{A}{B} \to K_p = 1 - 2\nu = \frac{E}{3K}$ – parametric criterion of mechanical properties of a body,

$\frac{A}{C} \to K_u = \frac{u(1-2\nu)}{2L(1+\nu)\alpha T}$ – displacement criterion.

The *modeling scales* are specified from the equality of mutually corresponding similarity criteria of an object and a model, so that, e.g., for the displacement scale it is

$$K_{u,D} = K_{u,M}, \tag{4.72}$$

and hence

$$\mu_u = \frac{uM}{uD} = \frac{(1-2\nu_D)(1+\nu_M)}{(1+\nu_D)(1-2\nu_M)} \frac{\alpha_M}{\alpha_D} \frac{T_M}{T_D} \frac{L_M}{L_D} = \mu_\beta \mu_T \mu_L = \mu_L, \tag{4.73}$$

because

$$\mu_\beta \mu_T = 1, \tag{4.74}$$

where μ_L, μ_β and μ_T are the scales of length, thermal volume expansion, and temperatures. Other scales can be determined from the following displacement-to-deformation relation

$$\varepsilon = \frac{1}{2}\nabla u, \tag{4.75}$$

so that the deformation scale is

$$\mu_\varepsilon = \frac{\varepsilon_M}{\varepsilon_D} = \frac{u_M}{L_M}\left(\frac{u_D}{L_D}\right)^{-1} = \frac{(1-2\nu_D)(1+\nu_M)}{(1+\nu_D)(1-2\nu_M)} \frac{\alpha_M}{\alpha_D} \frac{T_M}{T_D} = \mu_\beta \mu_T = 1, \tag{4.76}$$

which means that the deformation ε is the criterion by itself. For the displacement-to-stress relation, the following holds

$$\sigma = \frac{E}{1+\nu}\left[\frac{1}{2}\nabla u - \delta \nabla u\right] = \frac{E}{2(1+\nu)}\nabla u = G\nabla u \quad \text{for } \delta = 0, \tag{4.77}$$

where δ is the Kronecker symbol (being 0 or 1) and G is the shear elasticity modulus, so that the stress scale for $\delta = 0$ is

$$\mu_\sigma = \frac{\sigma_M}{\sigma_D} = \frac{E_M}{2(1+\nu_M)}\frac{u_M}{L_M}\left(\frac{E_D}{2(1+\nu_D)}\frac{u_D}{L_D}\right)^{-1} =$$

$$= \frac{1-2\nu_D}{1-2\nu_M}\frac{E_M}{E_D}\frac{\alpha_M}{\alpha_D}\frac{T_M}{T_D} = \mu_K\,\mu_\alpha\,\mu_T, \tag{4.78}$$

where μ_K is the scale of the volume modulus of elasticity.

With the thermal stress given in displacements, the modeling scales can be expressed by the relations (4.73), (4.76), and (4.78).

Thermal stress expressed by thermoelastic potential

For thermal stress, the following holds

$$\nabla^2\Phi = \frac{1+\nu}{1-\nu}\alpha T = \beta T. \tag{4.79}$$

$$\downarrow \qquad\qquad \downarrow$$

$$\underbrace{\frac{\Phi}{L^2}}_{A} \qquad \underbrace{\frac{1+\nu}{1-\nu}\alpha T}_{B}$$

The expressions A and B can be determined by the integral analogue method and from their ratio the displacement thermoelastic potential criterion can be defined as follows

$$\frac{A}{B} \rightarrow K_{\Phi u} = \frac{\Phi(1-\nu)}{(1+\nu)\alpha T L^2} = \frac{\Phi}{L^2 \beta T}.$$

From the equality of the criterion $K_{\Phi u,D}$ on the object and the criterion $K_{\Phi u,M}$ on the model, the scale of the displacement thermoelastic potential can be determined as follows

$$\mu_\Phi = \frac{\Phi_M}{\Phi_D} = \frac{(1+\nu_M)(1-\nu_D)}{(1-\nu_M)(1+\nu_D)}\frac{L_M^2}{L_D^2}\frac{\alpha_M}{\alpha_D}\frac{T_M}{T_D} = \mu_\beta\,\mu_L^2\,\mu_T. \tag{4.80}$$

Because the relations for displacement, deformation and stress, deduced from the thermoelastic displacement potential, are valid, the following modeling scales, deduced from the fundamental equation for thermal stress expressed in displacements, can be written

$$u = \nabla\Phi \rightarrow \mu_u = \mu_\beta \, \mu_L \, \mu_T = \mu_L, \tag{4.81}$$

$$\varepsilon = \nabla^2\Phi \rightarrow \mu_\varepsilon = \mu_\beta \, \mu_T = 1, \tag{4.82}$$

$$\sigma = 2G\nabla^2\Phi \left(\delta = 0\right) \rightarrow \mu_\sigma = \mu_K \mu_\alpha \, \mu_T. \tag{4.83}$$

With the thermal stress expressed by the thermoelastic potential, the modeling scales are expressed by the relations (4.80) to (4.83).

4.3.2 Thermomechanics of Fluids

The following examples describe unsteady non-linear heat conduction in a body with a combined non-linear boundary condition and the heat transfer in the viscous fluid flow. Both examples proceed by using reference quantities.

Example 4.9 *Unsteady non-linear heat conduction with a combined non-linear boundary condition*

The number and form of thermal similarity criteria for non-linear heat transfer during rapid cooling of a moving body by convection and radiation will be determined. The thermal process is described by the functional relation of 19 dimensional physical quantities

$$\varphi = \left(x, y, z, n, L, T, T_p, T_s, T_i, T_{ref}, \lambda, \lambda_{ref}, c\rho, \left(c\rho\right)_{ref}, w, q_V, \alpha, \sigma, \tau\right) = 0, \tag{4.84}$$

where x, y, z and L are the coordinates and relative length, n is the normal perpendicular to the surface, T, T_P, T_S, T_i, T_{ref} are the temperatures of the body, surroundings, and surface, initial and reference temperatures, λ, λ_{ref} are the thermal conductivity and its reference value, $c\rho$, $(c\rho)_{ref}$ is the specific volume heat and its reference value, w is the velocity of movement, q_V is the volume density of the heat flux, α is the heat transfer coefficient, σ is absorbance, radiating capacity, and τ is time.

Determination of the number of criteria by the Pi theorem:	
Total number of dimensional physical quantities:	$N = 19$
Number of dimensionally different quantities:	$n = 9$
Number of basic and complementary dimensions:	$r = 4$ (L, M, T, Θ)
Total number of dimensionless criteria:	$k = N - r = 15$
Number of simple criteria:	$k_s = N - n = 10$
Number of composed criteria:	$k_k = n - r = 5$

The mathematical model of unsteady non-linear heat conduction in a moving body with inner heat sources can be described by the following equation

$$\mathrm{div}\left[\lambda(T)\nabla T\right]+c(T)\,\rho(T)\,w\nabla T+q_V=c(T)\,\rho(T)\frac{\partial T}{\partial \tau}. \quad (4.85)$$

For the boundary condition, the following holds

$$-\lambda\frac{\partial T}{\partial n}=\alpha\left(T_S-T_P\right)+\sigma\left(T_S^4-T_P^4\right), \quad (4.86)$$

and the boundary condition is

$$T(x,y,z,0)=T_i. \quad (4.87)$$

The simple dimensionless quantities are

$$X=\frac{x}{L},\ Y=\frac{y}{L} \ \text{– coordinates,}$$

$$Z=\frac{z}{L},\ N=\frac{n}{L} \ \text{– the coordinate and the normal to surface,}$$

$$\Theta=\frac{T}{T_{ref}},\ \Theta_P=\frac{T_P}{T_{ref}} \ \text{– temperature of the body and that of the environment,}$$

$$\Theta_S=\frac{T_S}{T_{ref}},\ \Theta_i=\frac{T_i}{T_{ref}} \ \text{– surface and initial temperatures,}$$

$$\Lambda=\frac{\lambda}{\lambda_{ref}},\ C=\frac{c\rho}{(c\rho)_{ref}} \ \text{– thermal conductivity and specific volume heat capacity.}$$

After inserting the reference quantities in Eqs. (4.85) to (4.87) and modifying them, one obtains

$$\operatorname{div}\left[\lambda(\Theta)\nabla\Theta\right]+C(\Theta)\underbrace{\frac{(c\rho)_{ref}\,wL}{\lambda_{ref}}}_{Pe}\nabla\Theta+\underbrace{\frac{q_V\,L^2}{\lambda_{ref}\,T_{ref}}}_{Po}=C(\Theta)\frac{\partial\Theta}{\partial\underbrace{\left(\frac{a\tau}{L^2}\right)}_{Fo}},\tag{4.88}$$

$$-\frac{\partial\Theta}{\partial N}=\underbrace{\frac{\alpha L}{\lambda}}_{Bi}(\Theta_S-\Theta_P)+\underbrace{\frac{\sigma T_{ref}^3\,L}{\lambda}}_{Sf}\left(\Theta_S^4-\Theta_P^4\right),\tag{4.89}$$

$$\Theta(X,Y,Z,0)=\Theta_i.\tag{4.90}$$

The composed criteria, in the set of Eqs. (4.88) to (4.90), are

$$Pe=\frac{wL}{a}\quad\text{– Péclet number,}$$

$$Po=\frac{q_V\,L^2}{\lambda\,T_{ref}}\quad\text{– Pomerantsev number,}$$

$$Fo=\frac{a\tau}{L^2}\quad\text{- Fourier number,}$$

$$Bi=\frac{\alpha L}{\lambda}\quad\text{– Biot number,}$$

$$Sf=\frac{\sigma\left(T_S^3-T_P^3\right)L}{\lambda}\quad\text{– Stefan number.}$$

In a cooled body, the thermal process can be expressed by the functional relation of fifteen dimensionless quantities

$$\Phi\left(Pe, Po, Fo, Bi, Sf, X, Y, Z, N, \Theta, \Theta_P, \Theta_S, \Theta_i, \Lambda, C\right)=0\,,\tag{4.91}$$

where five quantities are composed criteria and ten are simple parameters.

The starting set of Eqs. (4.85) to (4.87) can be rewritten into dimensionless form

$$\operatorname{div}\left[\Lambda(\Theta)\nabla\Theta\right]+C(\Theta)Pe\nabla\Theta+Po=C(\Theta)\frac{\partial\Theta}{\partial Fo},\tag{4.92}$$

$$-\frac{\partial\Theta}{\partial N}=Bi\left(\Theta_S-\Theta_P\right)+Sf\left(\Theta_S^4-\Theta_P^4\right),\tag{4.93}$$

$$\Theta(X,Y,Z,0)=1.\tag{4.94}$$

Example 4.10 *Forced convection. Conduction heat transfer in viscous fluid flow*

The number and form of thermo-hydrodynamic similarity criteria for conduction heat transfer in a flowing viscous fluid with internal heat sources will be determined and a mathematical process model will be expressed in dimensionless form with a 2nd-kind boundary condition. An incompressible fluid with invariable thermophysical parameters is considered.

In the flowing fluid, the thermal process is described by a functional relation consisting of 16 variable physical quantities

$$\varphi = \left(T,\, T_i,\, T_{ref},\, x,\, y,\, z,\, n,\, L,\, w,\, w_{ref},\, q_A,\, q_V,\, \eta,\, \lambda,\, a,\, \tau\right) = 0, \tag{4.95}$$

where T, T_i and T_{ref} are the fluid temperature, and initial and reference temperatures, x, y, z, L are coordinates and reference length, n is a normal perpendicular to the surface, w, w_{ref} are velocity and reference velocity, q_A, q_V are the surface and volume densities of the heat flux, η is dynamic viscosity, λ is thermal conductivity, a is thermal diffusivity, and τ is time.

Determination of the number of criteria by the Pi theorem:	
Total number of dimensional physical quantities::	$N = 16$
Number of dimensionally different quantities:	$n = 9$
Number of basic and complementary dimensions:	$r = 4$ (L, M, T, Θ)
Total number of dimensionless criteria:	$k = N - r = 12$
Number of simple criteria:	$k_s = N - n = 7$
Number of composed criteria:	$k_k = n - r = 5$

In expressing the similarity criteria, one can start from the mathematical model for heat transfer in a flowing viscous fluid, which can be deduced from the fluid entropy change in the following form

$$\underbrace{\nabla^2 T}_{A} + \underbrace{\frac{\eta}{\lambda}(\nabla w)^2}_{B} - \underbrace{\frac{w}{a}\nabla T}_{C} + \underbrace{\frac{q_V}{\lambda}}_{D} = \underbrace{\frac{1}{a}\frac{\partial T}{\partial \tau}}_{E}. \tag{4.96}$$

The equation is fulfilled by the 2nd-kind boundary condition

$$q_A = -\lambda \frac{\partial T}{\partial n}, \tag{4.97}$$

and by the initial condition

$$T(x, y, z) = T_i. \tag{4.98}$$

For example, Eq. (4.96) represents the mathematical thermal process model of material quenching in a flowing viscous fluid. At the same time, the influence of the quenched body velocity of submerging into the fluid and the flow velocity there of, the influence of inner heat sources originating, e.g., in transcrystalizing changes in the material in the quenching process, are considered. The physical significance of individual additive members in Eq. (4.96) is as follows: A is the volume temperature distribution in the system, B is the thermal effect of viscous friction of the flowing fluid, C is the movement velocity in the fluid, D is an inner heat source and E represents the time temperature change in the system.

Simple dimensionless quantities are

$$X = \frac{x}{L},\ Y = \frac{y}{L} \quad \text{– coordinates,}$$

$$Z = \frac{z}{L},\ N = \frac{n}{L} \quad \text{– the coordinate and the normal to the surface,}$$

$$\Theta = \frac{T}{T_{ref}},\ \Theta_i = \frac{T_i}{T_{ref}} \quad \text{– fluid temperature and initial value thereof,}$$

$$W = \frac{w}{w_{ref}} \quad \text{– flow velocity.}$$

By inserting the reference quantities into Eqs. (4.96) to (4.98) and modifying them, one obtains

$$\nabla^2\Theta + \underbrace{\frac{\eta w_{ref}^2}{\lambda T_{ref}}}_{Br}(\nabla W)^2 - \underbrace{\frac{wL}{a}}_{Pe}\nabla\Theta + \underbrace{\frac{q_V L^2}{\lambda T_{ref}}}_{Po} = \frac{\partial\Theta}{\partial\underbrace{\left(\frac{a\tau}{L^2}\right)}_{Fo}}, \tag{4.99}$$

$$\underbrace{\frac{q_A L}{\lambda T_{ref}}}_{Ki} = -\frac{\partial\Theta}{\partial N}, \tag{4.100}$$

$$\Theta(X,Y,Z,0) = \Theta_i. \tag{4.101}$$

In the Eqs. (4.99) to (4.101), the composed criteria are

$$Br = \frac{\eta w_{ref}^2}{\lambda T_{ref}} \quad \text{– Brinkmann number,}$$

$$Pe = \frac{wL}{a} \quad \text{– Péclet number,}$$

$$Po = \frac{q_V L^2}{\lambda T_{ref}} \quad \text{– Pomerantsev number,}$$

$$Fo = \frac{a\tau}{L^2} \quad \text{– Fourier number,}$$

$$Ki = \frac{q_A L}{\lambda T_{ref}} \quad \text{– Kirpitchev number.}$$

In a general three-dimensional case, the thermal process in a viscous fluid flow can be described by the following equation

$$\Phi\left(Br, Pe, Po, Fo, Ki, X, Y, Z, N, \Theta, \Theta_i, W\right) = 0, \tag{4.102}$$

consisting of five composed criteria and seven simple ones.

The starting set of Eqs. (4.96) to (4.98) can be rewritten into the following dimensionless forms

$$\nabla^2\Theta\left(X, Y, Z\right) + Br\,\left(\nabla W\right)^2 - Pe\nabla\Theta\left(X, Y, Z\right) + Po = \frac{\partial\Theta}{\Theta Fo}, \tag{4.103}$$

$$Ki = -\frac{\partial\Theta}{\partial N}, \tag{4.104}$$

$$\Theta\left(X, Y, Z\right) = \Theta_i. \tag{4.105}$$

4.3.3 Electricity and Magnetism

The following example concerns electrical engineering. The procedure based on integral analogues is used to determine the electrodynamic similarity criteria in a series electric *RCL* circuit. Another example concerns magneto-hydrodynamics. Using the same procedure, the physical similarity criteria which describe the plasma flow behavior while passing through an electromagnetic field are obtained.

Example 4.11 *Electricity. Electrodynamic circuit*

It is required to determine the number and form of the electrodynamic similarity criteria which describe transient phenomena in a series electric circuit consisting of resistance *R,* inductance *L* and capacity *C,* connecting the circuit to alternating voltage *U* with angular frequency ω. In addition, the quantities time τ and current *I* appear (Fig. 4.2).

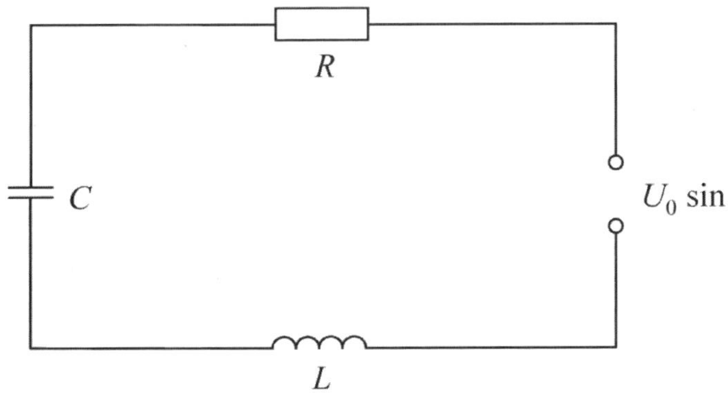

Fig. 4.2. Diagram of a series electric *RCL* circuit.

Determination of the number of criteria by the Pi theorem:	
Total number of dimensional physical quantities:	N=7
Number of dimensionally different quantities:	n =7
Number of basic and complementary dimensions:	r = 4 (L, M, T, I)
Total number of dimensionless criteria:	$k = N - r = 3$
Number of simple criteria:	$k_s = N - n = 0$
Number of composed criteria:	$k_k = n - r = 3$

The transient phenomenon in an electric circuit is described by a mathematical model with a differential equation

$$L\frac{\mathrm{d}I}{\mathrm{d}\tau} + RI + \frac{1}{C}\int I\mathrm{d}\tau = U_0 \sin \omega\tau. \tag{4.106}$$

$$\downarrow \quad \downarrow \quad \downarrow \quad \downarrow$$

$$\underbrace{L\frac{I}{\tau}}_{A} \quad \underbrace{RI}_{B} \quad \underbrace{\frac{I\tau}{C}}_{D} \quad \underbrace{U\omega\tau}_{E}$$

The corresponding similarity criteria are obtained using the given suggested procedure and forming the ratios of individual members. The first criterion follows directly from Eq. (4.106) and is

$$\pi_1 = \omega\tau,$$

this is the Hall criterion for an electrodynamic circuit. Further criteria are determined by creating the following ratios

$$\frac{E}{A} \rightarrow \frac{U\omega\tau^2}{LI} \Rightarrow \pi_2 = \frac{U\tau}{LI},$$

$$\frac{D}{A} \rightarrow \frac{\tau^2}{CL} \Rightarrow \pi_3 = \frac{R^2C}{L},$$

$$\frac{B}{A} \rightarrow \frac{R\tau}{L} \Rightarrow \pi_4 = \frac{\tau}{RC}.$$

These criteria characterize the electrodynamic circuit and have physical significance. For example, criterion π_4 represents the dimensionless time and is often called the Fourier electric number. The solution results in a generalized description of the process of unsteady electric current propagation in an *RCL* circuit in the form

$$\varphi\left(\frac{U\tau}{LI}, \frac{R^2C}{L}, \frac{\tau}{RC}, \omega\tau\right) = 0. \tag{4.107}$$

Using a suitable combination of the previous criteria one can obtain the composed criteria which express three fundamental parameters of an *RCL* circuit in the dimensionless form

$$\tau_R = \frac{IR}{U},\ \pi_C = \frac{CR}{\tau},\ \pi_L = \frac{L}{R\tau}. \tag{4.108}$$

Example 4.12 *Magneto-hydrodynamics. Plasma flow in an electromagnetic field*

The number and form of the physical similarity criteria which describe the plasma flow behavior in passing in an electromagnetic field are to be determined. The plasma is assumed to be incompressible and no temperature influence on the physical parameters is considered. The magnetic field intensity H, current density J, electric field intensity E, permeability μ, time τ, specific electric conductivity γ, velocity w, density ρ, pressure p, dynamic viscosity η, dimensional coordinates x, y, z, and reference quantities L, H_{ref}, J_{ref}, E_{ref}, w_{ref} and p_{ref} are the quantities to be considered.

Determination of the number of criteria by the Pi theorem:	
Total number of dimensional physical quantities:	$N = 19$
Number of dimensionally different quantities:	$n = 11$
Number of basic and complementary dimensions:	$r = 4$ (L, M, T, I)
Total number of dimensionless criteria:	$k = N - r = 15$
Number of simple criteria:	$k_s = N - n = 8$
Number of composed criteria:	$k_k = n - r = 7$

In the case considered, the magneto-hydrodynamic process is described by a mathematical model expressed by the following set of equations

$$\operatorname{rot} H = J,\ \operatorname{rot} E = -\mu\frac{\partial H}{\partial \tau},\ \operatorname{div} J = 0, \tag{4.109}$$

$$\operatorname{div} H = 0,\ J = \gamma\big(E + \mu(wH)\big),\ \operatorname{div} w = 0, \tag{4.110}$$

$$\rho\left(\frac{\partial w}{\partial \tau} + (w\nabla)w\right) = -\nabla p + \eta\nabla^2 w. \tag{4.111}$$

Individual members in the equations can be expressed by quantities

$$\underbrace{\frac{H_{ref}}{L}}_{A},\ \underbrace{J_{ref}}_{B},\ \underbrace{\frac{E_{ref}}{L}}_{C},\ \underbrace{\frac{\mu H_{ref}}{\tau}}_{D},\ \underbrace{\frac{J_{ref}}{L}}_{F},$$

$$\underbrace{\frac{H_{ref}}{L}}_{A},\ \underbrace{J_{ref}}_{B},\ \underbrace{\gamma E_{ref}}_{G},\ \underbrace{\gamma\mu w_{ref} H_{ref}}_{I},\ \underbrace{\frac{w_{ref}}{L}}_{K},$$

$$\underbrace{\frac{\rho w_{ref}}{\tau}}_{M},\ \underbrace{\frac{\rho(w_{ref})^2}{L}}_{N},\ \underbrace{\frac{p_{ref}}{L}}_{O},\ \underbrace{\frac{\eta w_{ref}}{L^2}}_{P}.$$

The composed magneto-hydrodynamic similarity criteria are obtained by the suggested procedure from the fundamental set of Eqs. (4.109)–(4.111) by forming suitable ratios of dimensional expressions

$$\frac{C}{D} \to Sh_e = \frac{E_{ref}\tau}{\mu H_{ref} L} = \frac{E_{ref}\tau}{B_{ref} L} = \frac{w_e \tau}{L}$$ – Strouhal electric number, where w_e is drift velocity,

$$\frac{I}{A} \to Re_m = \gamma\mu w_{ref} L = \frac{w_{ref} L}{\nu_m}$$ – Reynolds magnetic number, where ν_m is magnetic viscosity,

$$\frac{G}{A} \to \mathrm{Re}_e = \frac{\gamma E_{ref} L}{H_{ref}}$$ – Reynolds electric number,

$$\frac{N}{P} \to \mathrm{Re} = \frac{w_{ref} L}{\nu}$$ - Reynolds hydrodynamic number,

$$\frac{B}{A} \to \pi_m = \frac{J_{ref} L}{H_{ref}}$$ – magnetic field criterion,

$$\frac{O}{N} \rightarrow Eu = \frac{p_{ref}}{\rho\left(w_{ref}\right)^2} \text{ – Euler hydrodynamic number,}$$

$$\frac{P}{M} \rightarrow Fo_h \equiv Zh = \frac{\nu\tau}{L^2} \text{ – Fourier (Zhukhovski) hydrodynamic number.}$$

In addition to the seven composed criteria, eight simple criteria are used to describe the process

$$X = \frac{x}{L},\ Y = \frac{y}{L},\ Z = \frac{z}{L},\ \bar{H} = \frac{H}{H_{ref}},$$

$$\bar{J} = \frac{J}{J_{ref}},\ \bar{E} = \frac{E}{E_{ref}},\ W = \frac{w}{w_{ref}},\ P = \frac{p}{p_{ref}}.$$

Provided the plasma is incompressible and the physical parameters do not depend on the temperature, the resulting dimensionless equation describing the flowing plasma behavior in passing an electromagnetic field has the following form

$$\varphi\left(Sh_e, Re_m, Re_e, Re, \pi_m, Eu, Fo_h, X, Y, Z, \bar{H}, \bar{J}, \bar{E}, W, P\right) = 0. \qquad (4.112)$$

4.3.4 Physical Chemistry

In the following examples, the physically chemical similarity criteria are deduced by means of integral analogues. The first of them describes a gas adsorption process in passing through an adsorbent layer; the second one describes an unsteady chemical process of heat and mass transfer.

Example 4.13 *Physical chemistry. Gas adsorption*

The number and form of the criteria for physically chemical similarity describing the gas adsorption process in passing through an adsorbent layer (e.g. activated carbon) will be determined. The diffusion coefficient D, time τ, coordinates $x, y, z,$ flow velocity $w,$ gas concentration in pores c (kg m^{-3}), volume gas mass ρ_v (kg m^{-3}) adsorbed in the adsorbent, the relative volume of pores expressed as the ratio of the pore volume V_p and the adsorbent volume V_a are the quantities considered in the process. In addition, the reference quantities L, c_{ref}, $\rho_{v,ref}$ are added.

Determination of the number of criteria by the Pi theorem:	
Total number of dimensional physical quantities:	$N = 12$
Number of dimensionally different quantities:	$n = 6$
Number of basic and complementary dimensions:	$r = 3$ (L, M, T)
Total number of dimensionless criteria:	$k = N - r = 9$
Number of simple criteria:	$k_s = N - n = 6$
Number ofcomposed criteria:	$k_k = n - r = 3$

The mathematical model of the process is expressed by the equation

$$\frac{\partial \rho v}{\partial \tau}+\frac{V_p}{V_a}\frac{\partial c}{\partial \tau}=-w\nabla c+\frac{V_p}{V_a}D\nabla^2 c. \tag{4.113}$$

$$\begin{array}{cccc} \downarrow & \downarrow & \downarrow & \downarrow \\ \underbrace{\frac{\rho_{v,ref}}{\tau}}_{A} & \underbrace{\frac{c_{ref}}{\tau}}_{B} & \underbrace{\frac{wc_{ref}}{L}}_{E} & \underbrace{\frac{Dc_{ref}}{L^2}}_{F} \end{array}$$

Creating the ratios, one obtains the required composed criteria

$$\frac{F}{A}\rightarrow Fo_a=\frac{Dc_{ref}\,\tau}{Q_{v,ref}L^2}\ \text{– Fourier adsorption number,}$$

$$\frac{F}{B}\rightarrow Fo_d=\frac{D\tau}{L^2}\ \text{– Fourier diffusion number,}$$

$$\frac{E}{F}\rightarrow Pe_d=\frac{wL}{D}\ \text{– Péclet diffusion number.}$$

The dimensionless coordinates X, Y, Z, the relative volume of pores κ, concentration C, and adsorbed mass R are simple criteria

$$X=\frac{x}{L},\ Y=\frac{y}{L},\ Z=\frac{z}{L},$$

$$K=\frac{V_p}{V_a},\ C=\frac{c}{c_{ref}},\ R=\frac{\rho_v}{\rho_{v,ref}}.$$

The linear gas adsorption process in passing through a porous adsorbent layer is described by three composed criteria and six simple ones in the following form

$$\varphi\left(Fo_a, Fo_d, Pe_d, X, Y, Z, \kappa, C, R\right)=0. \tag{4.114}$$

Then, the original dimensional equation can be converted, by a procedure utilizing the reference quantities, into the following dimensionless form

$$\frac{\partial R}{\partial Fo_a} + K\frac{\partial C}{\partial Fo_d} = -Pe_d \nabla C + \kappa \nabla^2 C. \tag{4.115}$$

Example 4.14 *Physical chemistry. Heat and mass transfer in physical processes*

The number and form of the similarity criteria for an unsteady chemical process with inner mass and an energy source will be determined. The coordinates x, y, z, characteristic length L, mass concentration of the ith component ρ_i (kg m^{-3}), stechiometric number α_i, chemical reaction velocity u (kg m^{-3} s^{-1}), flow velocity w, diffusion coefficient of the ith component D_i, time τ, pressure p, kinematic viscosity ν, specific enthalpy h, specific heat capacity c, thermal conductivity λ, temperature T, reference temperature T_{ref} and density ρ are the acting quantities.

Determination of the number of criteria by the Pi theorem:	
Total number of dimensional physical quantities:	$N = 18$
Number of dimensionally different quantities:	$n = 13$
Number of basic and complementary dimensions:	$r = 4$ (L, M, T, Θ)
Total number of dimensionless criteria:	$k = N - r = 14$
Number of simple criteria:	$k_s = N - n = 5$
Number of composed criteria:	$k_k = n - r = 9$

The dimensionless coordinates X, Y, Z and the dimensionless temperature Θ are simple parametric criteria

$$X = \frac{x}{L},\ Y = \frac{y}{L},\ Z = \frac{z}{L},\ \Theta = \frac{T}{T_{ref}}. \tag{4.116}$$

Additionally, the stechiometric number α_i is assigned to the simple criteria.

The composed criteria can be determined from the analysis of the mathematical process model of the mass, momentum, and heat transfer, which is described by a set of three partial differential equations

$$\frac{\partial \rho_i}{\partial \tau} = -\alpha_i u - \operatorname{div}(w\rho_i) + \operatorname{div}(D_i \operatorname{grad} \rho_i), \tag{4.117}$$

$$\frac{\partial w}{\partial \tau} = -\frac{1}{\rho}\operatorname{grad} p - (w\,\nabla)w + \nu\nabla^2 w + \frac{\nu}{3}\operatorname{grad}\operatorname{div} w, \tag{4.118}$$

$$\frac{\partial(c\rho T)}{\partial \tau} = hu - \operatorname{div}(c\rho T\, w) + \operatorname{div}(\lambda \operatorname{grad} T). \tag{4.119}$$

Using the integral analogue method, the following expressions are obtained from individual members of the equation set

$$\underbrace{\frac{\rho_i}{\tau}}_{A},\ \underbrace{u,}_{B}\ \underbrace{\frac{w\rho_i}{L}}_{C},\ \underbrace{\frac{D_i\rho_i}{L^2}}_{D},$$

$$\underbrace{\frac{w}{\tau}}_{E},\ \underbrace{\frac{p}{qL}}_{F},\ \underbrace{\frac{w^2}{L}}_{G},\ \underbrace{\frac{\nu w}{L^2}}_{H},\ \underbrace{\frac{\nu w}{L^2}}_{H},$$

$$\underbrace{\frac{c\rho T}{\tau}}_{I},\ \underbrace{hu,}_{J}\ \underbrace{\frac{c\rho T w}{L}}_{K},\ \underbrace{\frac{\lambda T}{L^2}}_{M}.$$

Corresponding physically chemical similarity criteria are determined by forming ratios of the corresponding members, e.g. for the dimensionless time of diffusion, flow and heat transfer, one obtains

$$\frac{D}{A} \to Fo_m = \frac{D_i \tau}{L^2}$$ – Fourier mass number,

$$\frac{H}{E} \to Fo_h = Zh = \frac{\nu \tau}{L^2}$$ – Fourier hydrodynamic number (Zhukhovski number),

$$\frac{M}{I} \to Fo = \frac{\lambda \tau}{c\rho L^2} = \frac{a\tau}{L^2}$$ – Fourier number.

Similarly, other basic criteria of chemical diffusion, thermal, and hydrodynamic similarity can be deduced

$$\frac{B}{C} \to Da_1 = \frac{uL}{w\rho_i}$$ – Damköhler number (1.) hydrodynamical,

$$\frac{B}{D} \to Da_2 = \frac{uL^2}{D\rho_i}$$ – Damköhler number (2.) diffusion,

$$\frac{J}{K} \to Da_3 = \frac{quL}{c\rho w T_{ref}}$$ – Damköhler number (3.) (1. heat),

$$\frac{J}{M} \to Da_4 = \frac{quL^2}{\lambda T_{ref}}$$ – Damköhler number (4.) (2. heat),

$$\frac{G}{H} \to Da_5 \equiv Re = \frac{wL}{\nu} \quad \text{– Damköhler number (5.) (1. hydrodynamical)(Reynolds),}$$

$$\frac{F}{G} \to Eu = \frac{p}{\rho w^2} \quad \text{– Euler pressure number.}$$

The composed chemical reaction process can be described by functional relation consisting of nine composed criteria and five simple ones

$$\varphi\left(Fo_m, Fo_h, Fo, Da_1, Da_2, Da_3, Da_4, Da_5, Eu, X, Y, Z, \Theta, \alpha_i\right) = 0 . \quad (4.120)$$

In burning, activation molar energy E and fluid potential energy, given by the product of the molar gas constant R and temperature T, appear as additional physical quantities. The ratio of these energies leads to the further similarity criterion Ah_1. The criterion Le is added to this and characterizes the mutual relation between mass diffusion and heat transfer in burning

$$Ah_1 = \frac{E}{RT} \quad \text{– Arrhenius energetic number,}$$

$$Le = \frac{a}{D_i} = \frac{\lambda}{c\rho D_i} = Sc\,Pr^{-1} \quad \text{– Lewis number.}$$

4.3.5 Rheology and Tribology

In the following example, the mechanical–dynamic similarity criteria of wave propagation of mechanical tensions are obtained by making use of integral analogues. In the next example, the hydrodynamic similarity criteria describing pressure conditions in a fluid layer between moving surfaces are determined using the same procedure.

Example 4.15 *Rheology. Non-linear tension*

The number and form of mechanical–dynamic similarity criteria for wave propagation of mechanical tension in a viscous-elastic bar, joined to an elastic bar, have to be determined. The quantities considered are as follows: tension σ, dynamic viscosity η, modulus of elasticity E, density Q, wave velocity c, coordinate x, and times τ_1, τ_2. To these, reference length L and reference tension σ_{ref} are added. Both semi-limited bars are excited by a shock wave.

Determination of the number of criteria by the Pi theorem:	
Total number of dimensional physical quantities:	$N = 10$
Number of dimensionally different quantities:	$n = 6$
Number of basic and complementary dimensions:	$r = 3$ (L, M, T)
Total number of dimensionless criteria:	$k = N - r = 7$
Number of simple criteria:	$k_s = N - n = 4$
Number of composed criteria:	$k_k = n - r = 3$

The mathematical model of elastic bar vibration is represented by an ordinary differential 2nd-order equation. The stress propagation in a viscous–elastic bar corresponds to the Maxwell rheological model and is expressed by the wave equation

$$\frac{\partial^2\sigma}{\partial x^2} = \frac{1}{c^2}\frac{\partial^2\sigma}{\partial\tau_2^2} + \frac{E}{\eta c^2}\frac{\partial\sigma}{\partial\tau_1}, \text{ where } c = \sqrt{\frac{E}{\rho}}. \tag{4.121}$$

$$\downarrow \quad \downarrow \quad \downarrow \quad \downarrow \quad \downarrow$$

$$\underbrace{\frac{\sigma}{L^2}}_{A} \quad \underbrace{\frac{\sigma}{c^2\tau_2^2}}_{B} \quad \underbrace{\frac{E\sigma}{\eta c^2\tau_1}}_{C} \quad \underbrace{c}_{D} \quad \underbrace{\sqrt{\frac{E}{\rho}}}_{E}$$

The composed similarity criteria are obtained by making following ratios

$$\frac{A}{C} \to Fo_1 = \frac{\eta c^2\tau_1}{E L^2} \quad \text{– Fourier number of mechanical-viscous stress vibration.}$$

$$\frac{A}{B} \to Fo_2 = \frac{c^2\tau_2^2}{L^2} \quad \text{– Fourier number of wave propagation of mechanical tensions,}$$

$$\frac{D}{E} \to \pi_3 = c\sqrt{\frac{\rho}{E}} \quad \text{– dimensionless wave velocity}$$

To establish the composed criteria, dimensionless coordinate X, relative tension S and relative time T are described by the expressions

$$X = \frac{x}{L},\ S = \frac{\sigma}{\sigma_{ref}},\ T = \frac{\tau_1}{\tau_2}.$$

The process of mechanical stress wave propagation in the viscous-elastic bar is described by the equation

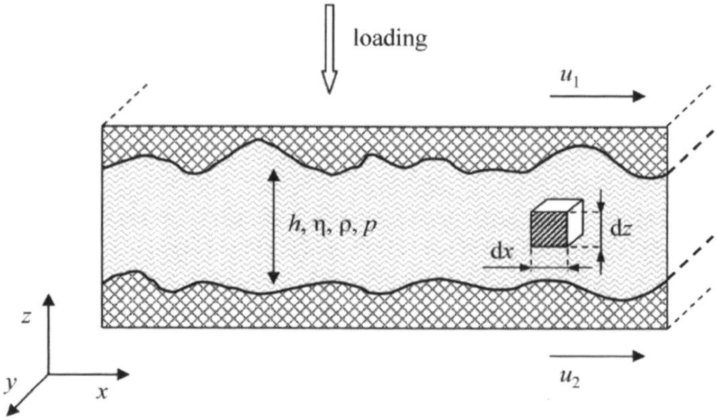

Fig. 4.3. Diagram for solving sliding friction.

$$\rho\left(Fo_1, Fo_2, \pi_3, X, S, T\right) = 0. \tag{4.122}$$

Example 4.16 *Tribology. Hydrodynamics of sliding friction*

It is required to determine the number and form of the hydrodynamic similarity criteria describing the pressure conditions in a fluid layer between two moving rough surfaces (Fig. 4.3). Because the fluid layer is thin enough and friction forces are substantially higher than inertia ones, the effect of inertia can be neglected. The considered quantities are as follows: rectangular coordinates x, y, density ρ and dynamic viscosity η of the fluid, local pressure p inside the layer, layer thickness h, sliding velocities u_1 and u_2 of the upper and bottom planar surface respectively, time τ, reference length L, and reference pressure p_{ref}.

Determination of the number of criteria by the Pi theorem:	
Total number of dimensional physical quantities:	$N = 11$
Number of dimensionally different quantities:	$n = 6$
Number of basic and complementary dimensions:	$r = 3$ (L, M, T)
Total number of dimensionless criteria:	$k = N - r = 8$
Number of simple criteria:	$k_s = N - n = 5$
Number of composed criteria:	$k_k = n - r = 3$

In the case of sliding friction dynamics ignoring the inertia forces, the general Navier–Stokes equation can be written in the general form of the Reynolds equation

$$\frac{\partial}{\partial x}\left(\frac{\rho h^3}{\eta}\frac{\partial p}{\partial x}\right)+\frac{\partial}{\partial y}\left(\frac{\rho h^3}{\eta}\frac{\partial p}{\partial y}\right)=$$

$$\downarrow \qquad\qquad \downarrow$$

$$\underbrace{\frac{\rho h^3 p_{ref}}{\eta L^2}}_{A} \qquad \underbrace{\frac{\rho h^3 p_{ref}}{\eta L^2}}_{B}$$

$$=6(u_1-u_2)\frac{\partial(\rho h)}{\partial x}+6\rho h\frac{\partial(u_1+u_2)}{\partial x}+12\frac{\partial(\rho h)}{\partial \tau}. \tag{4.123}$$

$$\downarrow \qquad\qquad \downarrow \qquad\qquad \downarrow$$

$$\underbrace{\frac{u\rho h}{L}}_{B} \qquad\qquad \underbrace{\frac{u\rho h}{L}}_{B} \qquad \underbrace{\frac{\rho h}{\tau}}_{C}$$

The composed criteria are obtained by establishing the ratios of individual terms

$$\frac{B}{A}\to So=\frac{\eta u L}{p_{ref}\,h^2} \text{ – Sommerfeld number,}$$

$$\frac{A}{C}\to Fo_H=\frac{h^2 p_{ref}\tau}{\eta L^2}=\frac{\nu\tau}{L^2} \text{ – Fourier hydrodynamic number (Zhukhovski number).}$$

In addition to the two composed criteria mentioned, hydrodynamic sliding friction is also expressed by simple criteria. Among them are the dimensionless coordinates X, Y, dimensionless pressure P, dimensionless layer thickness H and relative velocity U

$$X=\frac{x}{L},\ Y=\frac{y}{L},P=\frac{p}{p_{ref}},\ H=\frac{h}{L},\ U=\frac{u_1}{u_2}.$$

Therefore, the hydrodynamics of sliding friction can be described by an equation comprising two composed criteria and five simple ones

$$\varphi(So,Fo_H,X,Y,P,H,U)=0. \tag{4.124}$$

4.3.6 Technology

The following example shows the application of reference quantities to obtain similarity criteria for the magneto-hydro-thermo-mechanical process in electroslag remelting.

Example 4.17 *Technology. Magneto-hydrodynamics of electroslag remelting*

The number and form of similarity criteria for the magneto-hydro-thermo-mechanical process originating in electroslag remelting (Fig. 4.4) will be determined. Electric current sets the conductive liquid slag and molten metal in motion. The coordinates x, y, z, flow velocity w, electrode motion velocity w_s, time τ, density ϱ, pressure p, kinematic viscosity ν, gravitional acceleration g, volume thermal expansion coefficient β, temperature difference ΔT, magnetic permeability μ, current density J_A, and magnetic field intensity H are the quantities which act in the process. The reference length L, reference velocity w_{ref} and reference pressure p_{ref} can be introduced advantageously. Therefore, the composed process is described by the functional relation of eighteen variable physical quantities.

Determination of the number of criteria by the Pi theorem:	
Total number of dimensional physical quantities:	$N = 18$
Number of dimensionally different quantities:	$n = 12$
Number of basic and complementary dimensions:	$r = 5$ (L, M, Θ, T)
Total number of dimensionless criteria:	$k = N - r = 13$
Number of simple criteria:	$k_s = N - n = 6$
Number of composed criteria:	$k_k = n - r = 7$

Simple parametric criteria can be determined directly and are the following: dimensionless coordinates X, Y, Z, dimensionless velocities W, W_s, and dimensionless pressure P, which are expressed as the ratio of the dimensional quantity to the reference one

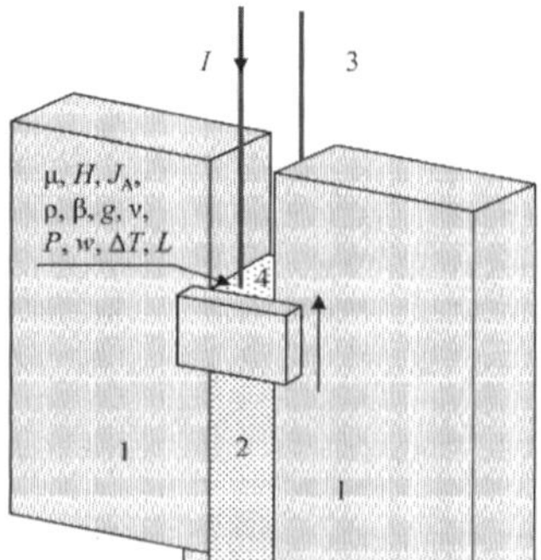

Fig. 4.4. Diagram of electroslag remelting (1 – welded material, 2 – weld, 3 – electrode, 4 – molten metal).

$$X=\frac{x}{L}, Y=\frac{y}{L}, Z=\frac{z}{L},$$

$$W=\frac{w}{w_{ref}},\; w_s=\frac{w_{s,}}{w_{ref}},\; P=\frac{p}{p_{ref}}.$$

The composed criteria can be determined from the mathematical model analysis of the process given by the expanded Navier-Stokes equation presented in the form

$$\underbrace{\frac{\partial w}{\partial \tau}}_{A}=\underbrace{-\frac{1}{\rho}\nabla p}_{B}-\underbrace{(w_s\nabla)w}_{C}+\underbrace{\nu\nabla^2 w}_{D}-\underbrace{g\beta\Delta T}_{E}+\underbrace{\frac{\mu}{\rho}(J_A H)}_{G}. \tag{4.125}$$

Equation (4.125) expresses the mathematical model of the magneto-hydro-thermomechanical process in welding. The first member A represents the time change of the slag and melted metal flow velocity, B is the influence of pressure, C is the influence of the mutual motion of molten metal and the electrode, D is the influence of the velocity vector of the viscous fluid flow, E is the influence of natural convection, and G represents the influence of electromagnetic convection.

After inserting the reference quantities into Eq. (4.125) and modifying it, one obtains

$$\underbrace{\frac{\partial W}{\partial\left(\frac{\nu\tau}{L^2}\right)}}_{Fo_h}-\underbrace{\frac{p_{ref}}{\rho w_{ref}^2}}_{Eu}\underbrace{\frac{w_{ref}L}{\nu}}_{Re}\nabla P-\underbrace{\left(W_s\frac{w_{ref}L}{\nu}\nabla\right)}_{Re}W+\nabla^2 W-$$

$$\underbrace{-\beta\Delta T}_{Gc}\underbrace{\frac{w_{ref}L}{\nu}}_{Re}\underbrace{\frac{gL}{w_{ref}^2}}_{Fr^{=1}}+\underbrace{\frac{w_{ref}L}{\nu}}_{Re}\underbrace{\frac{\mu H^2}{g\,w_{ref}^2}}_{Eu_m}\underbrace{\frac{J_A L}{H}}_{Am}. \tag{4.126}$$

There are seven composed criteria in Eq. (4.126)

$Fo_h = Zh = \dfrac{\nu\tau}{L^2}$ – Fourier hydrodynamic number (Zhukhovski number),

$Eu = \dfrac{p_{ref}}{\rho\, w_{ref}^2}$ – Euler number,

$Re = \dfrac{w_{ref}\, L}{\nu}$ – Reynolds number,

$$Fr = \frac{w_{ref}^2}{g\,L} \text{ – Froude number,}$$

$$Gc = \beta\,\Delta T \text{ – Gay-Lussac number,}$$

$$Eu_m = \frac{\mu\,H^2}{\rho\,w_{ref}^2} \text{ – Euler magnetic number,}$$

$$Am = \frac{J_A L}{H} \text{ – Ampére number.}$$

The original Eq. (4.125) can be rewritten into a dimensionless form

$$\frac{\partial W}{\partial Fo_h} = -Eu\,Re\nabla P - \left(W_s\,Re\nabla\right)W + \nabla^2 W - Gc\,Re\,Fr^{-1} + \\ + Re\,Eu_m\,Am. \tag{4.127}$$

The physical significance of individual criteria and terms in Eq. (4.127) implies the dependence of the non-stationary flow of molten slag and metal during welding on inertia force, pressure force, friction force, thermoelectric convection force, and electro-hydro-dynamic volume force. If the influence of the geometric dimensions of the moving electrode on the temperature field should also be looked at during the electroslag welding process, the material thermal diffusivity a must also be incorporated into the acting physical quantities. The dimensionless quantity

$$Pe = \frac{w\,L}{a} \text{ – Péclet number,}$$

becomes the characteristic criterion expressing the relation among the geometric dimensions, heat diffusion, and electrode movement velocity. Other criteria are included in accordance with the kind of boundary condition.

4.3.7 Ecology and Medicine

In the following example, the similarity criteria describing an unstable diffusion process of mixing and dispersion in tanks of waste water cleaning plants are determined using integral analogues. The next example is focused on obtaining the similarity criteria expressing the process of oxygen transfer in blood and its absorption in live tissue by utilizing the same procedure.

Example 4.18 *Ecology. Mass diffusion. Waste water cleaning*

The number and form of the physical similarity criteria describing an unsteady diffusion process of mixing and dispersion in tanks of waste water cleaning plants will be determined. In the process, the acting quantities are the impurities concentration c (kg m^{-3}), time τ, space coordinates x, y, z, dispersion diffusion coefficient D (m^2 s^{-1}), flow velocity w, the substance distribution coefficient k according the 1st-order reaction, density of volume water flow q_1 (m^2 s^{-1}) related to the depth unit, impurities concentration change Δc_1 (kg m^{-3}) in the investigated space, and surface area A of the investigated space. In addition, the reference length L and the reference concentration c_{ref} appear in the model.

Determination of the number of criteria by the Pi theorem:	
Total number of dimensional physical quantities:	$N = 12$
Number of dimensionally different quantities:	$n = 7$
Number of basic and complementary dimensions:	$r = 3$ (L, M, T)
Total number of dimensionless criteria:	$k = N - r = 9$
Number of simple criteria:	$k_s = N - n = 5$
Number of composed criteria:	$k_k = n - r = 4$

The mathematical model of the unstable waste water cleaning process is expressed by the following partial differential equation

$$\frac{\partial c}{\partial \tau} = D\nabla^2 c - w\nabla c - kc + \frac{q_1 \Delta c_1}{A}. \tag{4.128}$$

$$\downarrow \quad \downarrow \quad \downarrow \quad \downarrow \quad \downarrow$$

$$\underbrace{\frac{c_{ref}}{\tau}}_{B} \quad \underbrace{\frac{D c_{ref}}{L^2}}_{E} \quad \underbrace{\frac{w c_{ref}}{L}}_{F} \quad \underbrace{k\, c_{ref}}_{G} \quad \underbrace{\frac{q_1 \Delta c_1}{L^2}}_{H}$$

By making the ratios, the required composed similarity criteria are obtained

$$\frac{E}{B} \rightarrow Fo_d = \frac{D\tau}{L^2} \text{ – Fourier diffusion number,}$$

$$\frac{F}{E} \rightarrow Pe_d = \frac{wL}{D} \text{ – Péclet diffusion number,}$$

$$\frac{G}{E} \rightarrow Pr_d = \frac{k\,L^2}{D} \text{ – Prandtl diffusion number,}$$

$$\frac{H}{E} \rightarrow Ar_d = \frac{q_1\,\Delta c_1}{D\,c_{ref}} \text{ – Archimedes diffusion number.}$$

Simple parametric criteria such as dimensionless coordinates X, Y, Z, impurities concentration C, and impurities concentration change ΔC_1 are added to the composed criteria mentioned

$$X = \frac{x}{L},\ Y = \frac{y}{L},\ Z = \frac{z}{L},\ C = \frac{c}{c_{ref}},\ \Delta C_1 = \frac{\Delta c_1}{c_{ref}}.$$

The unstable waste water cleaning process in tanks of cleaning plants is described by four composed criteria and five simple ones

$$\varphi\left(Fo_d, Pe_d, Pr_d, Ar_d, X, Y, Z, C, \Delta C_1\right) = 0. \tag{4.129}$$

Further, the criteria corresponding to the boundary and initial conditions must be added. Then, the original dimensional equation can be converted into the following dimensionless form by utilizing reference quantities

$$\frac{\partial C}{\partial Fo_d} = \nabla^2 C - Pe_d \nabla C - Pr_d\, C + Ar_d. \tag{4.130}$$

Example 4.19 *Medicine. Physiology of breathing. Oxygen diffusion in tissue*

The number and form of similarity criteria expressing the oxygen transfer in blood capillaries and its constant absorption into live tissue will be determined. The quantities considered are: dimensionless oxygen concentration in tissue, usually defined as the oxygen volume V_k in the tissue volume V_t, non-stationary diffusion time r, coordinates x, y, z, diffusion coefficient D, volume oxygen consumption q_V in the tissue, defined as the oxygen volume for the unit tissue volume in unit time, and surface oxygen consumption q_S, defined as the oxygen volume for the surface unit in unit time. Reference length L is added to the quantities.

Determination of the number of criteria by the Pi theorem:	
Total number of dimensional physical quantities:	$N = 10$
Number of dimensionally different quantities:	$n = 6$
Number of basic and complementary dimensions:	$r = 3$ (L, M, T)
Total number of dimensionless criteria:	$k = N - r = 7$
Number of simple criteria:	$k_s = N - n = 4$
Number of composed criteria:	$k_k = n - r = 3$

The linear diffusion process is described by a partial differential 2nd-order equation

$$\frac{\partial \frac{V_k}{V_t}}{\partial \tau} = D\nabla^2 \frac{V_k}{V_t} - q_V. \tag{4.131}$$

$$\downarrow \qquad \downarrow \qquad \downarrow$$

$$\underbrace{\frac{1}{\tau}}_{A} \qquad \underbrace{\frac{D}{L^2}}_{B} \qquad \underbrace{q_V}_{E}$$

The similarity criteria are obtained by establishing the ratios of individual members

$$\frac{B}{A} \rightarrow Fo_d = \frac{D\tau}{L^2} \text{ – Fourier diffusion number,}$$

$$\frac{E}{B} \rightarrow Po_d = \frac{q_V L^2}{D} \text{ – Pomerantsev diffusion number.}$$

The boundary condition is usually expressed in the form of the Fick law

$$q_S = D\nabla \frac{V_k}{V_t}. \tag{4.132}$$

$$\downarrow \qquad \downarrow$$

$$\underbrace{q_S}_{F} \qquad \underbrace{\frac{D}{L}}_{G}$$

From the ratio of both members in Eq. (4.132), one obtains

$$Ki_d = \frac{q_S L}{D} \text{ – Kirpitchev diffusion number.}$$

The simple parametric criteria of dimensionless coordinates X, Y, Z and dimensionless oxygen concentration C in the tissue are added to the composed criteria mentioned

$$X = \frac{x}{L},\ Y = \frac{y}{L},\ Z = \frac{z}{L},\ C = \frac{V_k}{V_t}.$$

The non-stationary oxygen concentration in the live tissue is described by the following dimensionless equation

$$\varphi\left(Fo_d, Po_d, Ki_d, X, Y, Z, C\right) = 0, \tag{4.133}$$

in which the Fourier number Fo_d is the dimensionless diffusion time, the Pomerantsev number Po_d denotes the dimensionless internal oxygen consumption in the tissue, and the Kirpitchev number Ki_d signifies the dimensionless flow at the border of the investigated zone. The original dimensional mathematical model (4.131) and the boundary 2nd-kind condition (4.132) can be rewritten into a dimensionless form by utilizing the reference quantities

$$\frac{\partial C}{\partial Fo_d} = \nabla^2 C - Po_d, \tag{4.134}$$

$$\nabla C = Ki_d. \tag{4.135}$$

"Science does not try to explain: it hardly even interprets: science places models above all."

John von Neumann (*28.12.1903–8.2.1957*)

5

Mathematical Models

The power of mathematics is in its ability to express ideas and complex relations with symbols, while keeping simplicity and rationality in the expression at the same time. *Mathematical models* are examples which represent, from a certain point of view, the abstract expression of an object and the processes in progress therein. They result from the long-term development of abstraction in human perception and thinking in relation to Nature and the effort to describe phenomena objectively. In present-day computer modeling, the mathematical model represents a necessary precondition to the creation of a simulation model. In this chapter, the important aim is to point out that there are various kinds of mathematical models, not only *exact* or *approximate asymptotic ones,* but *approximate phenomenological ones* as well. These come into application where the development of asymptotic models has hot reached as yet or such models are too complex and less credible. This holds true about models in structurally complex systems, and about those in which the boundary conditions or physical parameters or other physical influences are difficult to determine. This concerns coupled processes especially. Here, a *phenomenological,* sometimes called *experimental mathematical model,* created on the basis of experimental results, is very important. It has high credibility but its accuracy depends on the credibility of its measurements. Moreover, it often requires higher experimental expenses and measurability of quantities.

The book [1.2] should be mentioned from the wide range of literature dealing with asymptotic mathematical models in science and engineering. For material engineering the book [8.23] and for

J. Kuneš, *Similarity and Modeling in Science and Engineering*,
DOI: 10.1007/978-1-907343-78-0_5,

composite materials the book [5.3] should be cited, too. Other numerous publications concern mathematical physics [5.19], thermo-mechanics [5.25], chemical engineering [1.4] and also other fields such as gyroscopic techniques [5.2]. In such fields as medicine [5.1], biology [5.7] and microbiology [5.5], epidemiology [5.8], immunology [5.18], and others in which modeling has become an increasingly significant factor, mathematical models have increasing importance and application. The above-mentioned citations represent a minute survey only of the widespread literature about mathematical modeling in diverse scientific and engineering fields.

Heretofore, the systematic processing of phenomenological mathematical models has been limited mostly to classic criterion equations for processes with few variable quantities and criteria equations in the form of the product of criteria. It appears that the systematic and generalized concept of the criterion equation, as an approximate phenomenological model, is lacking for complex systems or processes with many acting quantities.

5.1. Characterization of Mathematical Models

A mathematical model expresses a problem formally in physical and mathematical terms. It is represented by a set of mathematical relations describing unambiguously the examined phenomenon or process. It concerns a wide field in which various algebraic, ordinary or partial differential equations and sets thereof are found, as are various relations of the theory of sets, algebra, probability theory, mathematical logic and other areas of mathematics. However, these equations and relations become a mathematical model only if unambiguously assigned to a certain process or phenomenon. For example, the Poisson equation, expressed by a general variable, is not a mathematical model as it is not unambiguously assigned to a specific process or phenomenon. At most, it can be regarded as a mathematical model of a certain class of physical phenomena, e.g., in the case of steady-state physical fields with internal sources or sinks. However, from the point of view of the previous definition of a model, there is no unambiguous assignment to an original involving boundary and initial conditions, shape of the investigated region and physical parameters. Nevertheless, it becomes a mathematical model if we are speaking, e.g., about the Poisson equation of torsion or a temperature field with inner sources. In this, there is an analogy between the concepts of calculation and simulation. Distinguishing between these concepts consistently is fundamentally important in modeling.

A mathematical model can be:

- *phenomenological*, expressing the results of empirical research analytically or numerically,
- *asymptotic*, expressing the unambiguous mathematical law of a system or process behavior, with the law being understood as a partial case of a general scientific theory (e.g. the Laplace equation as the simplest case of the general physical field theory),
- *complex*, characterizing external, possibly even internal, behavior of complex dynamic systems.

Each of these mathematical models can be deterministic or stochastic. At the same time, the superiority of certain model types characterizes the degree of the development of a scientific field. Essentially, contemporary physics is a logically bound set of mathematical, asymptotic models. For cybernetics, complex models are characteristic.

Mathematical modeling represents only one of the periods in the overall strategy of examination by models. However, this period is often so exhilarating that the solver becomes carried away easily to artificial abstract forms, separated from reality, by making use of the mathematical language. Due to this, the modeling loses its basic properties – its credibility and purposefulness.

5.2 Asymptotic Mathematical Models

In solving various tasks in technical as well as non-technical fields, the partial differential equations are principally important, of which those of mathematical physics, represented by 2nd-order partial differential equations, are usually the most important. They describe physical processes and fields in distributed-parameter systems. Coupled physical processes and fields are examples of where the sets of these equations are used.

5.2.1 Equations of Mathematical Physics

A *general equation of mathematical physics* with real parameters can be written in the form of a 2nd-order partial differential equation

$$\sum_{i,j=1}^{n} a_{ij} \frac{\partial^2 u}{\partial x_i \, \partial x_j} + \sum_{i,=1}^{n} b_i \frac{\partial u}{\partial x_i} + cu + f = 0, \tag{5.1}$$

in which the parameters a_{ij}, b_i, c, f denote general continuous functions of the dependent variable u and of independent variables x_1, x_2, x_n, which are the geometric coordinates, usually x, y, z, and time τ.

Depending on the nature of the parameters, the following cases can occur:

• *linear parameters,* if a_{ij} (x, y, z, τ), $b_i(x, y, z, \tau)$, $c\,(x, y, z, \tau)$, and $f\,(x, y, z, \tau)$ are valid, which means that they are not functions of the dependent variable u,

• *non-linear parameters,* if a_{ij} (x, y, z, τ, u), $b_i(x, y, z,\tau, u)$, $c\,(x, y, z, \tau, u)$, and $f\,(x, y, z, \tau, u)$ are valid, which means that they are functions of the dependent variable u,

• *homogenous parameters,* if $a_{ij}(\tau, u)$, $b_i(\tau, u)$, $c(\tau, u)$, and $f(\tau, u)$ are valid, which means that they are not functions of the coordinates x, y, z,

• *non-homogenous parameters,* if a_{ij} (x, y, z, τ, u), $b_i(x, y, z, \tau, u)$, $c\,(x, y, z, \tau, u)$, and $f\,(x, y, z, \tau, u)$ are valid, which means that they are functions of the coordinates x, y, z,

• *isotropic parameters,* if the conditions $a_{ij}(x, y, z, \tau, u) = a_{ji}(x, y, z, \tau, u)$, and $b_i(x, y, z, \tau, u) = b_j(x, y, z, \tau, u)$ are valid for $i, j = 1,\ldots,n$,

• *anisotropic parameters,* if the conditions $a_{ij}(x, y, z, \tau, u) = a_{ji}(x, y, z, \tau, u)$, and $b_i(x, y, z, \tau, u) = b_j(x, y, z, \tau, u)$ are not valid for $i, j = 1,\ldots,n$.

These characteristics can be combined so that the parameters can be linear homogenous isotropic, etc.

To further simplify consideration, we consider Eq. (5.1) with two independent variable coordinates x, y only. Let us presume that the parameters a_{ij} are functions only of the coordinates. Let us denote them with capitals so that $A \equiv a_{11}$, $2B \equiv a_{12} + a_{21}$, $C \equiv a_{22,}$ with remaining members being denoted as $H\left(x, y, u, \frac{\partial u}{\partial x}, \frac{\partial u}{\partial y}\right)$. Then Eq. (5.1) becomes

$$A\frac{\partial^2 u}{\partial x^2} + 2B\frac{\partial^2 u}{\partial x \partial y} + C\frac{\partial^2 u}{\partial y^2} = H\left(x, y, z, \frac{\partial u}{\partial x}, \frac{\partial u}{\partial y}\right). \tag{5.2}$$

This equation describes the so-called Cauchy problem, which is important to explain the concept of the characteristics and to classify the 2nd-order partial differential equations. Depending on the kind of the characteristics, these equations can be divided into:

• *elliptic equations,* if the characteristics are complex in the given range, i.e. if the discriminant $\Delta = B^2 - AC < 0$,

• *hyperbolic equations,* if the characteristics are real and different, i.e. if the discriminant $\Delta = B^2 - AC > 0$,

• *parabolic equations,* if the characteristics are at the interface between both preceding types, i.e. if the discriminant $\Delta = B^2 - AC = 0$.

Next, some elliptic, parabolic, and hyperbolic 2nd-order equations are summarized and their physical significance is explained. Finally, some partial 4th-order differential equations are given as well.

Elliptic equations

The *Laplace equation*

$$\nabla^2 u = 0. \tag{5.3}$$

The *Poisson equation*

$$\nabla^2 u - F(x, y, z) = 0. \tag{5.4}$$

The *Tricomi equation* for $y > 0$

$$y\frac{\partial^2 u}{\partial x^2} + \frac{\partial^2 u}{\partial y^2} = 0, \tag{5.5}$$

where this equation becomes a hyperbolic one if $y < 0$.

The *elliptic equations describe a steady-state physical field with or without internal sources.* The *Laplace equation* (5.3) describes many steady-state physical fields with neither internal sources nor sinks. It is about electric, thermal, hydrodynamic, and filtration fields, and about a field of main strains and other physical fields in elasticity. A steady-state field with inner sources or sinks is described by the *Poisson equation* (5.4). This equation describes electric, electrostatic, and magnetic fields, flow of a perfect incompressible fluid with internal sources and sinks, vortex flow, viscous incompressible flow, filtration of fluids through porous materials, mechanical strains under torsion, mechanical deformations of a membrane loaded by internal overpressure, temperature fields with internal sources or sinks, neutron and other fields.

Parabolic equations

The *Fourier equation*

$$\nabla^2 u + B\frac{\partial u}{\partial \tau} = 0. \tag{5.6}$$

The *heat conduction equation*

$$\nabla^2 u + B\frac{\partial u}{\partial \tau} - F(x, y, z, \tau) = 0. \tag{5.7}$$

The *extended heat conduction equation*

$$\nabla^2 u + Au + B\frac{\partial u}{\partial \tau} - F(x, y, z, \tau) = 0. \tag{5.8}$$

The *Fourier-Kirchhoff equation*

$$\nabla^2 u + B\frac{\partial u}{\partial \tau} \pm H\nabla u - F(x, y, z, \tau) = 0. \tag{5.9}$$

The *equation of atmospheric diffusion of particles*

$$K\nabla^2 u + H\nabla u + B\frac{\partial u}{\partial \tau} = 0. \tag{5.10}$$

The *parabolic equations* describe the unsteady-state physical fields of mass diffusion or energy conduction. The Fourier equation (5.6) relates to the fields without inner sources. Electric, electromagnetic, thermal, hydromechanical, diffusive and filtration fields are amongst them. Compared to the Fourier equation, the heat conduction equation (5.7) involves an internal source. In the case of a temperature field, this can be, for example, internal heat originating due to electric current passage through a conductor or heat generated in a reactor fuel cell during nuclear transformations. The *extended heat conduction equation* (5.8) involves a member which represents the heat loss from a thin conductor surface into the surroundings.

The Fourier-Kirchhoff equation (5.9) is the most general linear parabolic equation which describes motion in temperature, diffusion and other physical fields. The motion member expresses movement of a region border, interface of phases, or source. Amongst the motion fields are the following: the temperature or diffusion field in flowing fluids, temperature or diffusion processes originated in the mixing fluids, melting, melting down, welding, cutting, burning, ablation and sublimation. This equation is the fundamental equation in metallurgy and machine engineering but its significance is also increasing in other important fields.

Equation (5.8) is a *special case of the Fourier-Kirchhoff equation.* This equation describes the diffusion of particles into the atmosphere. Using this equation, many ecological problems of air pollution can be solved

Hyperbolic equations

The *Helmholtz equation*

$$\nabla^2 u + Au = 0, \quad A(x, y, z) \neq 0. \tag{5.11}$$

The *Schrödinger equation*

$$\nabla^2 u + Au = 0,\ \ A = f(u). \tag{5.12}$$

The *incomplete wave equation*

$$\nabla^2 u + C\frac{\partial^2 u}{\partial \tau^2} = 0,\ \ C(x,y,z) < 0. \tag{5.13}$$

The *d'Alambert equation*

$$\frac{\partial^2 u}{\partial x^2} + C\frac{\partial u^2}{\partial \tau^2} = 0,\quad C(x) < 0. \tag{5.14}$$

The *complete wave equation*

$$\nabla^2 u + C\frac{\partial^2 u}{\partial \tau^2} - F(x,y,z) = 0. \tag{5.15}$$

The *extended complete wave equation*

$$\nabla^2 u + B\frac{\partial u}{\partial \tau} + C\frac{\partial^2 u}{\partial \tau^2} - F(x,y,z) = 0. \tag{5.16}$$

The *incomplete telegraph equation*

$$\nabla^2 u + Au + B\frac{\partial u}{\partial \tau} + C\frac{\partial^2 u}{\partial \tau^2} = 0. \tag{5.17}$$

The *abridged incomplete telegraph equation*

$$\nabla^2 u + Au + C\frac{\partial^2 u}{\partial \tau^2} = 0. \tag{5.18}$$

The *complete telegraph equation*

$$\nabla^2 u + Au + B\frac{\partial u}{\partial \tau} + C\frac{\partial^2 u}{\partial \tau^2} - F(x,y,z) = 0. \tag{5.19}$$

The *extended complete telegraph equation*

$$\nabla^2 u + D\nabla u + Au + B\frac{\partial u}{\partial \tau} + C\frac{\partial^2 u}{\partial \tau^2} - F(x,y,z) = 0. \tag{5.20}$$

The *higher order equation with respect to time*

$$C_1 + \frac{\partial^2}{\partial \tau^2}\nabla^2 u + B_1\frac{\partial}{\partial \tau}\nabla^2 u + \nabla^2 u. \tag{5.21}$$

The *hyperbolic equations* have basic significance in the theory of energy wave propagation in the solid, liquid and gaseous materials. This concerns, for example, problems of acoustics, supersonic aerodynamics, heat wave propagation in intensive thermal processes, seismic wave propagation during earthquakes and other geological investigations, mechanical vibration of ideally flexible and homogeneous strings and membranes, torsional vibration of rods, wave propagation in water levels, waves in connection with the vibration of elementary particles, electromagnetic wave propagation in space and in waveguides, and electrical signal transmission in power lines.

The *Helmholtz equation* (5.11) describes periodical characteristic vibration of strings and membranes or the heat transfer in thin wires and foils with consideration of external heat transfer into the surroundings. The *Schrődinger equation* (5.12) is the basic equation of quantum mechanics.

The *incomplete wave equation* (5.13) describes non-damped waves with no external forces influence. If the problem concerns periodic waves at a certain known frequency, the Helmholz equation (5.11) is used, with the solution of which the periodic wave amplitude can be obtained. In a one-dimensional case, the incomplete wave equation is usually called the *d'Alambert equation* (5.14). The *complete wave equation* (5.15) involves the influence of external forces or sources. When considering the influence of wave process damping in addition, one obtains the *extended wave equation* (5.16). The *incomplete telegraph equation* (5.17) is the general one for transmission of various kinds of energy in conductors, specifically by diffusion or wave propagation if the convective energy transfer into surroundings is being considered. It is the basic equation in communication engineering because it describes the transmission of messages with signal and information loss being taken into account. With the influence of diffusion member omitted, it changes to an *incomplete abridged form* (5.18). Compared to the incomplete form of the telegraph equation (5.17), the *complete telegraph equation* (5.19) involves internal energy sources or sinks. The *complete extended telegraph equation* (5.20) has in addition a motion member expressing, e.g., the motion of energy source, zone borders, etc. A set of six telegraph equations forms a *general system of Maxwell equations* describing the non-stationary energy transmission in space. From it, essentially all other cases of physical fields, represented by previous partial differential equations, are deduced. The *higher order equation with respect to time* (5.21) describes diverse physical processes, the substance of which is often not yet fully known.

4th-order partial differential equations

The *homogenous biharmonic equation*

$$\nabla^4 u = \frac{\partial^4 u}{\partial x^4} + 2\frac{\partial^4}{\partial x^2 \partial y^2} + \frac{\partial^4 u}{\partial y^4} = 0. \tag{5.22}$$

The *non-homogenous biharmonic equation*

$$\nabla^4 u = F(x, y), \tag{5.23}$$

$$\nabla^4 u + Au = F(x, y), \tag{5.24}$$

$$\nabla^4 u + B\frac{\partial u}{\partial \tau} = F(x, y, \tau), \tag{5.25}$$

$$\nabla^4 u + C\frac{\partial^2 u}{\partial \tau^2} = F(x, y, \tau), \tag{5.26}$$

$$\nabla^4 u + B\frac{\partial u}{\partial \tau} + C\frac{\partial^2 u}{\partial \tau^2} = F(x, y, \tau), \tag{5.27}$$

The partial 4th*-degree differential equations* represent basic equations in elasticity and dynamics, but they have many other applications as well. The *homogenous bi-harmonic equation* (5.22) is very important in the field of elasticity to solve static deflection of beams and plates which are not loaded by external force. With the solution, the biharmonic function is obtained, which is, for example, the static, elastic or thermoelastic potential. In the case of the homogenous equation (5.22), the biharmonic function $u(x, y)$ can be decomposed by a simple relation to two harmonic functions v (x, y) and $w(x, y)$, which can be defined in the same region as

$$u(x, y) = xv(x, y) + w(x, y), \tag{5.28}$$

if the function u is even with respect to x. Solving of the function u leads to finding the two harmonic functions v and w.

The analogous cases, such as the homogenous biharmonic equation (5.22), are described by Eq. (5.23), but with the external static loading being considered. The undamped vibration of beams and other bodies at eigenfrequency is expressed by Eq. (5.24) and the damped vibration is expressed by Eq. (5.25) for loading by shock force F. With inertia forces being taken into account, the undamped vibration is described by Eq. (5.26) and the damped vibration by Eq. (5.27).

5.2.2 Conditions of Unambiguity

The *complete mathematical model* is given by the basic equation or possibly by a set of basic equations and, in addition, by conditions of unambiguity. The unambiguity conditions consist of initial and boundary ones, of restrictive and optimization ones, and also of physical parameters and the geometric shape of the zone. The knowledge of these is necessary for solving, but correct determination of them is often a difficult problem.

Initial conditions

The *initial condition* defines the value of the dependent variable function u in arbitrary place of the zone x, y, z in the time $\tau = 0$

$$u\ (x,\ y,\ z,\ 0) = u^*\ (x,\ y,\ z), \tag{5.29}$$

where $u^*(x,\ y,\ z)$ is the function prescribed for the whole zone of the model.

Boundary conditions

The *boundary conditions* express the relation of the examined zone to the external surroundings. They can be divided into fundamental and composed conditions (variable in time, non-linear, and combined ones, etc.). Each of them can be arbitrarily variable.

The basic boundary conditions are:

the 1st *-kind, Dirichlet boundary condition,* expressing the known value of the function u on the zone surface at the point $x,\ y,\ z$

$$u(x,y,z,\tau) = u^*(x,y,z,\tau), \tag{5.30}$$

the 2nd*-kind, Neumann boundary condition,* expressing the gradient of the function u on the zone surface at the point $x,\ y,\ z$

$$-a(x,y,z,\tau)\frac{\partial u(x,y,z,\tau)}{\partial n} = q^*(x,y,z,\tau), \tag{5.31}$$

the 3rd-kind, Fourier, boundary condition, expressing the linear relation between the function u and its derivation in a normal direction on the zone surface at the point $x,\ y,\ z$

$$-a(x,y,z,\tau)\frac{\partial u(x,y,z,\tau)}{\partial n} = b^*(x,y,z,\tau)\left(u^*(x,y,z,\tau) - u(x,y,z,\tau)\right), \tag{5.32}$$

the 4th-kind boundary condition, expressing the condition for surface energy fluxes at the point $x,\ y,\ z$ at the interface between both zones, having different physical properties, in perfect contact

$$a_1(x,y,z,\tau)\frac{\partial u(x,y,z,\tau)}{\partial n_1} = a_2(x,y,z,\tau)\frac{\partial u(x,y,z,\tau)}{\partial n_2}. \quad (5.33)$$

In the equations of basic boundary conditions, $a(x, y, z, \tau)$ denotes the zone physical parameter, $b(x, y, z, \tau)$ denotes the physical parameter of external surroundings, $q\ (x, y, z, \tau)$ denotes the surface energy flux, and the index* denotes the prescribed function.

Restrictive and optimization conditions

The *restrictive conditions* occur in optimalization tasks. They define the extent of physical and geometric parameters of the zone, or possibly of external surroundings, which must not be exceeded in optimalization.

The *optimization conditions* express maximum, minimum or other values of a controlled quantity in order to reach an optimal process Thus, the optimization conditions can be, for example, minimum energy consumption (the energetic point of view), maximum utilization of material mechanical properties (the rigidity point of view) or minimum time of process duration (the time point of view), etc.

5.2.3 Mathematical Model Transformations

In modeling, mathematical model transformation is often very important Its main purpose is to simplify and facilitate the task solution The following are the most important transformations:

- coordinate system transformations,
- conformal transformations of the zone,
- integral transformations of the continuous mathematical model by the time variable,
- integral transformations of the continuous mathematical model by the space coordinates,
- differential transformations of a discrete mathematical model,
- linear functional transformations of a discrete mathematical model.

An outline of individual transformation principles and their important physical significance for modeling follow, including fundamental expressions for the most important transformations used. A more detailed survey of mathematical model transformations is presented in many publications

Coordinate system transformations

Relatively simple transforming relations between rectangular, cylindrical, spherical, and possibly curvilinear coordinate systems exist, which comprise only basic mathematical functions. These relations are well known.

In a general case, the *coordinate transformation* can be expressed as

$$x_i' = f_i(x_1, x_2, \cdots, x_n), \tag{5.34}$$

for $i = 1,...,m$, where x_i are the coordinates in the original coordinate system, x_i' are the coordinates in the new system, and f_i are the transforming functions where $m \neq n$ can be valid in a general case.

The inverse function $f_i^{-1}(x_1', x_2',...,x_m')$, and converting the coordinates x_i' to coordinates x_i is called the *inverse coordinates transformation.*

Conformal transformations of the zone

Sometimes in solving boundary tasks, the conformal transformation can be used advantageously instead of the coordinate transformation. Essentially, it means to transform a more complex planar zone to a simpler one. Here, the conformal transformation method is used most frequently, and is among the most effective instruments of the complex variable theory wherever the task can be transformed to a harmonic or biharmonic boundary one. Usually, the solution is simplified substantially if a function is known which maps, for example, the considered zone to a circle, an annular ring or a plate. An example from aerodynamics is the conformal transformation of the by-passing of a planar blade cascade with complex-shape blade profiles to a cascade composed of a set of circular cylinders or plates.

Essentially, the principle of the conformal transformation method is simple. Let us consider a planar physical field expressed in rectangular coordinates. With the relations between individual points of the original complex plane being mapped $z(x,y) = x + iy$ and the corresponding points of the transformed plane $w(u,v) = u + iu$ being defined unambigously, the following is valid

$$w(u,v) = f[z(x,y)], \tag{5.35}$$

and the functional relation $f(z)$ represents the *conformal transformation* of the original zone $z(x, y)$ to the zone $w(u, v)$. The corresponding complex potentials $z = x + iy$ and $w = u + iv$ represent orthogonal networks of lines in both assigned zones. If one succeeds in determining these orthogonal networks by experiment or calculation, one obtains the conformal mapping, because two arbitrary mutually assigned points (homologues) $M(x,y)$ and $M(u,v)$ are determined by equal values of the complex potential. The conformal transformation is selected so that it may be calculated at each zone point and the simplest boundary conditions, e.g. constant, may exist at the border of the examined zone.

Integral transformations of a continuous mathematical model by a time variable

Continuous dynamic systems behavior is described by differential equations or by sets of them. To solve them, several methods of integral transformations by time are used, which include the Laplace transformation and Carson transformation.

The *Laplace transformation* is a calculating procedure which enables the transformation of the function $f(\tau)$ of a real variable τ to the function $F(p)$ of a complex variable p. The function $f(\tau)$ is called the *original,* or model or object, whereas the function $F(p)$ is called the *image.* The Laplace transformation expresses the assignment of the image

$$F(p)=\int_0^{\infty} f(\tau)e^{-p\tau}\mathrm{d}\tau \tag{5.36}$$

to the original $f(\tau)$, where p is a complex number. The expression $e^{-p\tau}$ is called the kernel. It is presumed that the function $f(\tau)$ can be integrated into each finite interval $0 \leq a \leq b < \infty$. The transformation is called unilateral in the case of integral limits $(0, \infty)$, bilateral in the case $(-\infty, \infty)$, or finite in the case $(0, T)$.

The *inverse Laplace transformation* expresses the assignment of the original

$$f(\tau)=\lim_{y\to\infty}\frac{1}{2\pi i}\int_{x-iy}^{x+iy} F(p)e^{p\tau}\mathrm{d}p \tag{5.37}$$

to the image $F(p)$. In practice, the simple relation between the Laplace image of the function $f(\tau)$ and its derivation $f'(\tau)$ is used

$$\int_0^{\infty} f'(\tau)e^{-p\tau}\mathrm{d}\tau = pF(p)-f(0). \tag{5.38}$$

Provided $f(0) = 0$, the algebraic operation of $F(p)$ multiplying by a complex number p corresponds to the original derivation operation. The operator calculus has been based on this principle. The operation of multiplication by p corresponds to the original derivation, and multiplication by p^{-1} corresponds to the original integration.

Integral transformations of a continuous mathematical model by space coordinates

The Fourier, Hankel and Legendre transformations are among the integral transformations by space coordinates. Above all, the choice of the transformation depends on the geometrical arrangement of the zone and the kind of boundary conditions.

The *Fourier transformation,* one-dimensional bilateral infinite, expresses the assignment of the image

$$F(y)=\frac{1}{\sqrt{2\pi}}\int_{-\infty}^{\infty} f(x)e^{-ixy}\mathrm{d}x \tag{5.39}$$

to the original $f(x)$ in the interval $(-\infty, \infty)$, where y is a real number.

The *inverse Fourier transformation* expresses the assignment of the original

$$f(x)=\frac{1}{\sqrt{2pi}}\int_{-\infty}^{\infty} F(y)e^{ixy}dy \tag{5.40}$$

to the image $F(y)$.

The Fourier transformation is executed also for the functions $f(x)$in the semi-infinite interval $(0, \infty)$ or in the finite one (a, b).

Differential transformations of a discrete mathematical model

These can be used to advantage in examining discrete dynamic systems. In comparison to integral transformations, the main difference is that the original is transformed to the image by derivation instead of integration. The simpler differential transformations can be done by making use of the Taylor series. The *transformation T* expresses the assignment of the image

$$F(k)=M(k)\left[\frac{\partial^k q(x)f(x)}{\partial x^k}\right]_{x=x_i} \tag{5.41}$$

to the original $f(x)$, which is a continuous function of the variable x for $k = 0, 1, 2,...$, which are non-negative integers $M(k) \neq 0$, $q(x) = 0$. The function $M(k)$ is a weight factor and the function $q(x)$ is called the kernel. The image $F(k)$ is called the spectrum of the function $f(x)$ at the point $x = x_i$.

The *inverse transformation T* consists of assignment of the original

$$f(x)=\frac{1}{q(x)}\sum_{k=0}^{\infty}\frac{(x-x_i)^k}{k!}\frac{F(k)}{M(k)} \tag{5.42}$$

to the image $F(k)$.

In special cases, the functions $M(k)$ and $q(x)$ are chosen either equal to one

$$M(k) = 1; \quad q(x) = 1 \tag{5.43}$$

for all corresponding k and x, or are used in the form

$$M(k) = \frac{H^k}{k!}; \quad q(x) = 1 \tag{5.44}$$

for all relevant k and x, where H is a constant.

Linear functional transformations of a discrete mathematical model

These are utilized in examining discrete dynamic systems as well. In comparison to integral transformations, the original is transformed to an image by a linear function.

The *transformation Z* transforms the sequence of finite numbers, either real or generally complex, $\{f_n\}$ to their image conforming to the relation

$$F(z) = \sum_{n=0}^{\infty} f_n z^{-n}. \tag{5.45}$$

The assignment is unambiguous and the following relation holds for the *inverse transformation Z*

$$f_n = \frac{1}{2\pi i} \oint_C F(z) z^{n-1} \mathrm{d}z, \tag{5.46}$$

for $n = 0, 1, 2,\ldots$, where C is a circle $z = \rho e^{-it}$, $t \in (0, 2\,\pi)$.

The Laplace transformation of continuous models is an analogue of the transformation Z of discrete models.

Physical significance of mathematical model transformations

In transforming the mathematical model, not only the actual model changes in relation to the space or time, but important similarity and information aspects appear as well. In the case of physical processes and fields described by partial 2nd-order differential equations, e.g., the task to solve a non-stationary field is transformed by the *Laplace transformation* to a task, described by a ordinary 2nd-order differential equation, to solve a dynamic circuit. In this way, the similarity of an internal process passes to the similarity in external behavior. Similarly, the basic equations of dynamic circuits, expressed by ordinary differential equations, are transferred to algebraic equations.

The finite integral transformations by geometric coordinates, such as, e.g. the *Fourier transformation,* are characterized by containing the information about the geometric shape and character of the physical process. Considering various integral transformations of a mathematical model leads to averaging of the sought physical quantity in space. These transformations are used to solve tasks in non-limited, semi-limited and even limited zones. Forexample, in heating at hick-walled body, the average integral temperature in the volume of the body, which is the characteristic quantity for the thermal stress determination, is obtained by the Fourier transformation. From a physical point of view, e.g., the task of heating a thick-walled body is transferred, by the Fourier transformation, to the task of heating a thin wall which simplifies the solution substantially. Compared to the Laplace transformation, simpler resulting expressions are the main advantage of the Fourier transformation, among others. Above all, the Fourier transformation is used in cases of bodies in rectangular and cylindrical coordinates. The Hankel transformation is used for cylindrical coordinates and the Legendre transformation for spherical ones.

The differential *T transformation* is used to solve complicated tasks, especially those which are described by sets of equations with the solution being completed suitably with the Laplace and Fourier transformations. With difference mathematical models used, e.g. in numerical filters and models theory, the *Z transformation* is applied.

5.2.4 Methods of Tasks Solution

The above-mentioned equations of mathematical physics, or the asymptotic mathematical models by another name, can be solved by various methods after being completed with unambiguity conditions. These methods can be divided into *analytical* and *numerical* ones.

Analytical methods

The *analytical methods* enable one to obtain the task solution in the form of a mathematical expression for the searched quantity as a function of space coordinates and time. The solution must correspond to certain equation and unambiguity conditions. Usually in most tasks, the mathematical process model must be simplified so that the task may be solved. In using the analytical methods especially, the principal problem is to define the degree of mathematical model simplification while retaining its credibility.

The *method of variable separation* (also called the Fourier method), the *Green functions method*, and the *method of potentials* are among the classic analytical methods. Another group of analytical methods consists of *integral transformation methods,* based on the principle of mathematical transformation of variables. The *Laplace* and *Fourier transformations,* which are described in more detail in sub-section

5.2.3, are among most standard ones. The application of all the above-mentioned accurate analytical methods are limited to linear tasks with linear boundary conditions and simpler zones. Another group of analytical methods involves *variation methods,* which are suitable even for the approximate solution of non-linear tasks, in contrast to the previously mentioned methods. They are based on the principle that a variation task is solved by finding an extreme of a functional in an integral shape characterizing the given process, instead of solving a differential mathematical model of a physical field. Usually, it is the *minimum of the functional of energy. The Ritz, Kantorovich, and Biot methods* are among this group. Other analytical methods convert the tasks with boundary conditions to other types of equations and tasks.

Numerical methods

The *numerical methods,* the principle of which consists of discretization of variables, can be widely applied in developing computer modeling. The numerical methods enable one to obtain the task solution in a finite number of discrete points, called nodes, of a chosen computational grid. These methods consist of the *finite difference method* (FDM), the *finite element method* (FEM), the *finite volume method* (FVM), and the *boundary element method* (BEM). The fundamental principles of the numerical methods are described in section 8.2. These methods are suitable even for non-linear tasks with complex boundary conditions in zones of complicated geometry.

5.3 Phenomenological Mathematical Models

These are approximate mathematical models expressing the results of experimental examination in analytical or numerical forms. They can be called *experimental mathematical models.* A criterion equation, originated according to the similarity theory as a product of individual similarity criteria with corresponding parameters as exponents, is the simplest example of an experimental mathematical model.

Each of these exponents expresses the weighted part of the corresponding criterion or physical quantity influence on the examined process and is determined by experiment.

The phenomenological models are used wherever it is not possible or is extremely difficult to use asymptotic models. This involves highly structured systems or processes or dynamic systems with strong influence of memory, inertia, and damping, etc. Of course, new mathematical physics fields, for which the asymptotic mathematical models are not known, are among this group.

5.3.1 Phenomenological Model Formation and Evaluation

Dimensional analysis enables one to determine the number and the form of similarity criteria and an approximate mathematical model of the process in the criterion equation form. This procedure can be widely applied, especially where the general basic equation of the examined process is not known. The criterion equation is an algebraic relation in which dimensionless criteria and numerical parameters appear. Classic criterion equations are products of criteria and parameters, but more complicated cases require a general criterion equation form in which there are additive members and possibly even other functions as well.

In the general form of a criterion equation, it is far more difficult to determine the relation only by classic dimensional analysis. The procedure by so-called polynomial development of functions can be more useful in some ways. To reach the necessary accuracy of the function approximation, the satisfactory-degree polynomial must be determined, the coefficients of which are determined by experiment. It is better to consider a higher polynomial degree because a faulty relation can be obtained with the degree less than necessary. In case the polynomial degree is higher than necessary, such coefficients, at redundant polynomial members, will result in their value being less by several orders than that of other coefficients. With a general form of a criterion equation, usually many more experiments are necessary to determine the coefficients sought.

Determination of the number and the form of dimensionless similarity criteria

Let us use N dimensional physical quantities $x_1, x_2, \ldots, x_N$ characterizing the examined system or process. By means of the Pi theorem, the number of simple and composed similarity criteria is determined, by which the original physical quantities can be replaced.

By means of a dimensional matrix, the forms of corresponding similarity criteria Π_1, Π_2, Π_{N-r} are obtained by a matrix solution. Attention must be paid to finding a suitable similarity criteria set which will contain most of the dimensionless quantities already introduced.

Determination of the form and parameters of criterion equation

From the similarity criteria found, those which are important to describe the examined process in a system are selected. Statistical analysis wil show which irrelevant criteria have been involved in the dimensionless relation sought. On the contrary, the non-inclusion of an important similarity criterion will appear in evaluating the agreement of the criterion equation with the experiment.

In a fundamental case, the classic criterion equation form can be considered in the form of a similarity criteria product with certain powers. The similarity criteria powers are the parameters of the criterion equation and are determined on the basis of experimentally established data of the examined process.

The classic product form of a criterion equation is

$$\Pi_{N-r} = k_{N-r} \prod_{i=1}^{N-r-1} \Pi_i^{k_i}. \tag{5.47}$$

However, if the classic criterion equation form is not satisfactory, a more complicated form of the functional relation must be used. Unfortunately, the dimensional analysis theory does not answer the question as to what this form should be.

By extending the classic criterion equation form with additive members, the diverse dynamic systems with distributed parameters and complex structure can be described, in which the strong influence of inertia, damping, etc. appear.

The general criterion equation form is

$$\varphi(\Pi_1, \Pi_2, \cdots, \Pi_{N-r}) = 0. \tag{5.48}$$

Next, we are going to discuss the evaluation of the general criterion equation form. The exception is a statistical t-test of individual similarity criteria in the criterion equation that is limited to the classic-product-form criterion equation only.

Evaluation of agreement of the criterion equation with experiment

The agreement of the criterion equation with the experiment is evaluated by using the *determination index* defined in regression analysis by the following relation

$$R^2 = 1 - \frac{\sum_{i=1}^{n} (y_i - \hat{y}_i)^2}{\sum_{i=1}^{n} (y_i - \overline{y})^2}. \tag{5.49}$$

The greater the value of the determination index R^2 is, the better the agreement of quantity $\hat{y}_i$ calculated from the criterion equation with the measured quantity y_i is. The number of measurements, i.e. the number of all measured time levels in all experiments applied to obtain the criterion equation, is given by the quantity n.

Evaluation of the suitability of the criteria choice and the criteria equation form

Incorrect choice of the functional relation form or the inclusion of irrelevant similarity criteria in the dimensionless relation sought is detected by means of two tests of statistical analysis.

The correctness of the proposed form of criterion equation, i.e. how well the equation can express the examined relation, is determined by the *F-test,* defined by the following relation in the regression analysis

$$F = \frac{R^2}{p_{kr}} \frac{n - p_{ko}}{1 - R^2}, \tag{5.50}$$

where p_{kr} is the number of the similarity criteria in the criterion equation and p_{ko} is the number of searched criterion equation coefficients. With the equation in the classic product form, $p_{ko} = p_{kr} + 1$ is valid.

It is determined how many times the obtained value F is greater than the quantile $f_{1\text{-}a}$ of the distribution F with p_{kr} and $n–p_{ko}$ degrees of freedom. The value of F should be at least 4 to 5 times greater than the corresponding quantile, so that the criterion equation may be denoted as a representative of the examined relation at the significance level of $1–\alpha$, i.e. with the probability of $1–\alpha$. In order to denote the criterion equation as a suitable representative and to use it to describe the process, the condition of a satisfactorily large *determination index* R^2 must be fulfilled.

The ability to apply individual similarity criteria in a classic-product–form criterion equation, i.e. the orientation significance of the similarity criteria in the examined relation, is determined by means of the *t-test* for the jth similarity criterion, defined in the regression analysis by the following relation

$$t_j = \frac{k_j}{\sqrt{a_{jj}}\sqrt{\dfrac{\sum_{i=1}^{n}(y_i - \hat{y}_i)^2}{n - p_{ko}}}}, \qquad A = \left(X^T X\right)^{-1}, \tag{5.51}$$

where k_j is the exponent of the corresponding j^{th} similarity criterion and a_{jj} are the diagonal elements of the matrix $\boldsymbol{A}$. The matrix $\boldsymbol{X}$ is the measurement one, i.e. the matrix row is one measured time level in every experiment from which the criterion equation is evaluated. The individual row elements of the matrix $\boldsymbol{X}$ are the numerical values of the similarity criteria in the examined criterion equation for the relevant time measurement level.

The absolute value $|t_j|$ is compared with the quantile t_1–α/2 of the Student distribution t with n–p_{ko} degrees of freedom. With the $|t_j|$ greater than the relevant quantile, the similarity criterion can be denoted as useful in the corresponding criterion equation at the significance level of 1–α.

5.4 Applications

The following examples involve mathematical models for relatively complicated tasks the exact solution of which is extraordinarily complex or even impossible. It is important to demonstrate how some of these tasks can be solved approximately even by applying an analytical mathematical model, but first of all it will be shown how experimental (phenomenological) mathematical models can be utilized. Other commonly used deterministic and stochastic models are presented in Chapters 8 and 9.

The complexity of tasks is given by the number of variable quantities and their physical character, and often also by the complexity of the internal system structure. Usually in these tasks, it is very difficult to express, e.g., the boundary conditions and physical parameters which complicate the solution substantially. In this case, the approximate phenomenological models can be used, of which the main advantage is that they are based on real conditions and their accuracy depends on that of measurement only. Therefore, the credibility of these models is relatively high in relation to the examined object.

5.4.1 Electrical Engineering

The following example illustrates the fact that asymptotic mathematical models can be utilized even in the solution of complicated tasks of thermal processes during the short-circuit in electric machines.

Example 5.1 *Electric machines. Short-circuit in a turbo-alternator* shaft

A short-circuit in electric machines and other equipment is accompanied by a heat process which is among the most complicated thermal tasks due primarily to the lack of unambiguity conditions. The electric arc originating in an electric short-circuit on a turbo-alternator shaft can cause a turbo-alternator breakdown and affect the shaft material considerably. To be specific, a very intensive momentary thermal source and generated hot plasma up to 3500°C are acting. In addition to shaft surface damage, even the material structure can be changed, which influences its mechanical properties. The structural change is closely related to the creation of hardened structures which can cause cracking during machine operation. The aim of the solution is to establish the nature and depth of the thermal influences on the shaft material after the short-circuit.

Problem structure

Thermal system:	plasma → rotor → surroundings
Thermal process:	three-dimensional unsteady state heat conduction with ablation of the material on the rotor surface
Thermal task:	direct approximate
Mathematical model:	analytical with numerical solution procedure

Physical principle of the problem

During a short-circuit between the alternator collector rings, the current is going over the rotor shaft. The thermal process originated in the short-circuit is accompanied by ablation, with which the welded-on metal is melted off from the shaft surface by the centrifugal force. The determination of the amount of melted metal and a final profile of the damaged surface with various sized craters are unambiguous. The deepest crater is the most significant, under which the material is most substantially damaged. The determination of ablation velocity and time is not unambiguous. The incompleteness of the mathematical model, added to this, causes uncertainty in solving the task. The procedure to determine the credible physical process and the mathematical model needs a detailed physical process analysis. It follows from this that the solution is influenced substantially by the size of the thermal source and by its changing place during the gradual melting off. The influence of non-equilibrium heat propagation is limited to the shaft surface and cannot affect substantially internal diffusion heat conduction. The solution in the general form cannot be precisely found and therefore an approximate but sufficiently credible model must be used.

Mathematical model

With a moving heat source and material ablation, the thermal process can be described by the fundamental equation

$$\underbrace{\operatorname{div}(\lambda \operatorname{grad} T)}_{A}+\underbrace{\mathbf{w} \operatorname{grad}(c\rho T)}_{B}\pm\underbrace{q_V}_{C}=\underbrace{c\rho\frac{\partial T}{\partial \tau}}_{D}, \tag{5.52}$$

in which the individual members are: A – heat increment by conduction, B – heat increment by movement in the system, C – specific volume heat source (+) or sink (–), D – time change of inner energy. In solving short-time melting and ablation, the thermal process in the material, relatively near the solid phase–molten phase boundary, can be considered to be linear and the previous equation becomes

$$\nabla^2 T(x,y,z,\tau)+\frac{\mathbf{w}}{a}\nabla T(x,y,z,\tau)\pm\frac{q_V}{\lambda}=\frac{1}{a}\frac{\partial T}{\partial \tau}, \tag{5.53}$$

and in dimensionless form

$$\nabla^2\Theta(M,Fo)+Pe\nabla\Theta(M,Fo)\pm Po(M,Fo)=\frac{\partial\Theta}{\partial Fo}, \tag{5.54}$$

where $\mathbf{w}$ is the ablation velocity and $M(X, Y, Z)$ is the point in space.

With ablation without any inner source or sink, the Pomerantsev number $Po = 0$ is valid. In some cases, ablation velocity can be neglected and then the Péclet number will be $Pe = 0$.

In a general case in which heat removal by convection, radiation and metal ablation is considered, the boundary condition has the following form

$$Ki=-\frac{\partial\Theta}{\partial N}+Po+Bi(1-\Theta_S)+Sf\left(1-\Theta_S^4\right). \tag{5.55}$$

This equation expresses, by the dimensionless Kirpitchev number, the heat flux on the shaft surface which is difficult to determine. Sometimes, the influence of electric arc radiation (Stefan number $Sf = 0$) or convection (Biot number $Bi = 0$) or possibly even heat removal with melted off metal (Pomerantsev number $Po = 0$) can be neglected due to the momentary nature of the process.

The momentary spot nature of the process enables one to consider the rotor as a semi-space into which energy E is delivered by pulsed heating. Heat conduction can be considered linear out of the zone in close proximity to the surface source. Under the mentioned preconditions and with homogeneous initial rotor temperature, the analytical solution of dimensionless heating Θ at distance x of the source spot can be determined, as follows

$$\Theta=\frac{\lambda}{\alpha}x^3\left(T-T_0\right)E^{-1}=\frac{1}{8\pi Fo_x\sqrt{\pi Fo_x}}\exp\left(-\frac{1}{4Fo_x}\right). \tag{5.56}$$

For the local Fourier number $Fo_x = a\tau x^{-2}$. The expression (5.56) is given in a table and expressed graphically.

Analytical model

Despite the fact that heat conduction only is considered in Eq. (5.56), the task of melting and ablation can be solved approximately by distributing the total energy into individual doses corresponding to the energy of the melting of elementary volumes. The process of gradual melting-off of the metal in doses can be expressed by repeated use of the gradual momentary action of a thermal source. The progressive crater formation, its shape and size (Fig. 5.1), can be comprehended by space distribution of thermal energy centres and the time sequence thereof. The simplicity of solving even a multiple task, which can scarcely be solved otherwise, is remarkable.

The proposed model is approximate because the action of the accumulated energy of molten metal on the heat conduction in the shaft is considered. The results of the solution, in compliance with the proposed model, deliver some higher temperatures. On the other hand, the bare shaft, after the metal has been melted off, enables the heat transfer into the shaft surface by electric arc radiation. However, this energy can hardly be determined quantitatively. Rather, this energy can be estimated to be approximately equal to that contained in a molten drop.

In the selected analytical model, the number of discharges is very important. The discharges deepen a crater and follow one after another

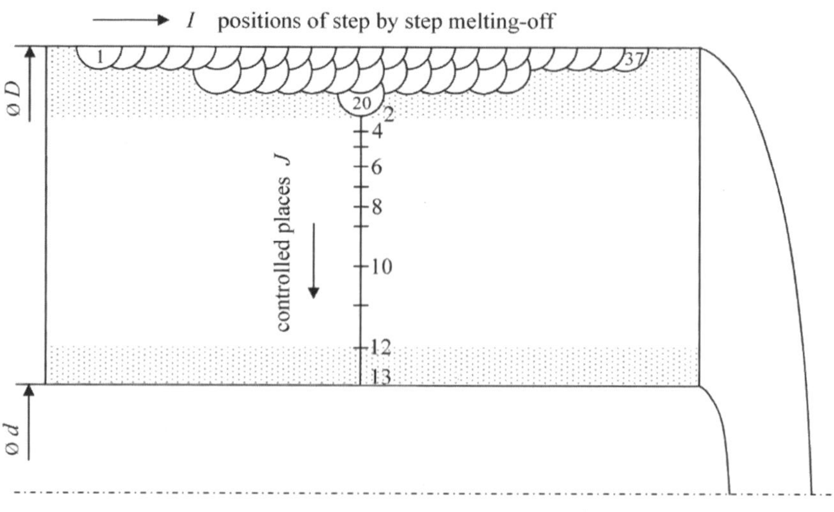

Fig. 5.1. Procedure diagram of expected gradual melting-off of metal from the shaft.

as fast as corresponds to fictive time τ. The discharges always melt off the material part to a constant depth. The shifting of the electric arc from one place to another could cause a certain temperature field uniformity in an exposed place, and therefore, a reduction of the thermal gradient in the crater bottom. The main principle of the procedure of thermal process simulation in a shaft is the calculation of the temperature increments from partial energy doses in individual places and at selected time intervals.

Simulation results

The results of temperature simulation in the most exposed place in the bottom and adjoining zone of the deepest crater are presented in Fig. 5.2. The time dependence of temperature shows, that the depth of temperature penetration into the shaft is relatively small. In surveyed points, the influence of more distant electric arcs appears only after approximately 20 seconds. Not only longitudinal but also peripheral crater deepening were considered in this case. A metallographic test was carried out to verify the simplified analytical mathematical and simulation model. According to this, the critical temperature for structural changes was not reached at the distances greater than 2.1 mm from the bottom surface of the deepest crater.

The solution results show that the application of the analytical mathematical model can be advantageous even when solving some complicated tasks, especially if relevant information should be obtained.

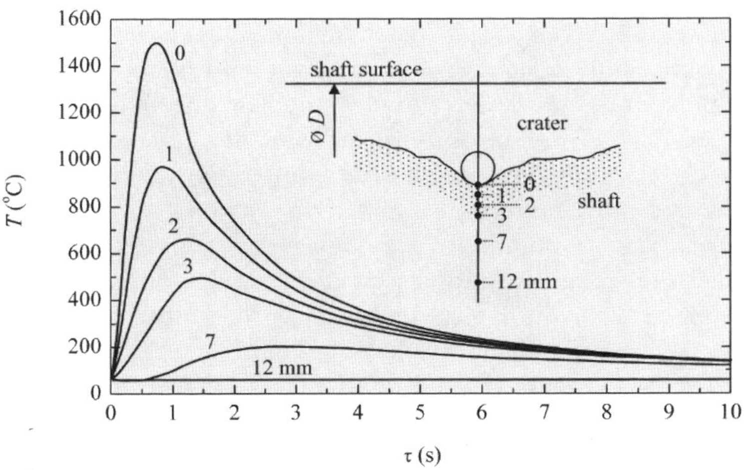

Fig. 5.2. Time dependence of temperature in the most exposed points at different distances from the bottom of the deepest crater.

Of course, this is possible provided the problem's physical principle has been analyzed in detail previously.

5.4.2 Physical Technology

The following examples take advantage of phenomenological mathematical models. The dependence of the sample surface temperature on spraying time during plasma spraying is determined in the first example. The maximum rise of surface temperature and the time duration of molten material during the laser interaction with a material are derived.

Example 5.2 *Physical technology. Plasma spraying of a layer on a substrate surface*

The plasma spraying of layers will be analyzed, the number and form of similarity criteria determined and criterion equations describing the examined process derived. In contrast to the following examples, this concerns a simple system on which a physically complicated process is acting.

Detailed physical analysis of acting quantities in plasma spraying determined originally 57 quantities. The complexity of the physical process and the large number of acting quantities connected with it led to a narrowed selection of acting quantities, i.e. to dimensional matrix simplification. In the selection, the intensity of the influence of individual quantities in the process and the possibility of measuring them were decisive. The selection includes the following quantities: specimen surface temperature T_S, initial specimen surface temperature T_{ini}, reference temperature T_{ref}, powder melting point T_T, burner travelling velocity in individual axes w_x, w_y, w_z, burner reference travelling velocity w_{xref}, source output power P, powder thermal conductivity λ_P, substrate thermal conductivity λ_S, powder specific thermal capacity c_P, substrate specific thermal capacity c_S, powder density ρ_p, substrate density ρ_S, specific latent heat of powder melting l_P, characteristic powder dimension L_P, characteristic sample dimension L_{ref}, transverse spacing of burner tracks *d,* number of spraying passes *j,* overhang of the burner track over the sample *s,* burner axis distance from the sample *z,* burner axis distance from the measurement point δ, time *t*, mass flow of Ar, H_2 m_{Ar}, m_{H2}, carrier gas mass flow m_{np}, reference mass flow m_{ref} of the gas, carrier gas pressure p_{np} and relative powder feeding angle β.

Determination of the number of criteria by the Pi theorem:	
Total number of dimensional physical quantities:	$N = 28$
Number of dimensionless input physical quantities:	$N_b = 2$
Number of dimensionally different quantities:	$n = 11$
Number of basic and complementary dimensions:	$r = 4\ (L, M, T, \Theta)$
Total number of dimensionless criteria:	$k = N - r + N_b = 26$
Number of simple criteria:	$k_s = N - n + N_b = 19$
Number of composed criteria:	$k_k = n - r = 7$

Determination of the form of similarity criteria

Determination of the physical quantities occurring in the examined process of plasma spraying and setting of the number of similarity criteria is followed by the deduction of the similarity criteria form. This is determined by dimensional analysis, by solving the dimensional matrix composed of physical quantities, in which the solver's experience and intuition play an important role. In this, the highest possible number of the already introduced and generally applied similarity criteria must be obtained. The general procedure for constructing and solving the dimensional matrix is described in Chapter 3.

The dimensional matrix **A** will be assembled thus

$$
\begin{array}{c|cccccccccccccc}
 & T_s & T_{ini} & T_{ref} & T_T & w_x & w_y & w_z & w_{x,ref} & P & \lambda_P & \lambda_S & c_P & c_S & \rho_P \\
\hline
L & 0, & 0, & 0, & 0, & 1, & 1, & 1, & 1, & 2, & 1, & 1, & 2, & 2, & -3, \\
M & 0, & 0, & 0, & 0, & 0, & 0, & 0, & 0, & 1, & 1, & 1, & 0, & 0, & 1, \\
T & 0, & 0, & 0, & 0, & -1, & -1, & -1, & -1, & -3, & -3, & -3, & -2, & -2, & 0, \\
\Theta & 1, & 1, & 1, & 1, & 0, & 0, & 0, & 0, & 0, & -1, & -1, & -1, & -1, & 0,
\end{array}
$$

$$
\begin{array}{cccccccccc|cccc}
\rho_S & l_P & L_P & L_{ref} & d & s & z & \delta & t & m_{Ar} & m_{H_2} & m_{np} & m_{ref} & p_{np} \\
-3, & 2, & 1, & 1, & 1, & 1, & 1, & 1, & 1, & 0, & 0, & 0, & 0, & -1 \\
1, & 0, & 0, & 0, & 0, & 0, & 0, & 0, & 0, & 1, & 1, & 1, & 1, & 1 \\
0, & -2, & 0, & 0, & 0, & 0, & 0, & 0, & 0, & -1, & -1, & -1, & -1, & -2 \\
0, & 0, & 0, & 0, & 0, & 0, & 0, & 0, & 0, & 0, & 0, & 0, & 0, & 0
\end{array}.
$$

Using the matrix solution procedure, the solution matrix **B** is obtained. The dimensionless similarity criteria can be determined directly from the solution matrix **B**, where P_1 to P_{19} are the simple criteria and π_1 to π_7 the composed ones

$$P_1 = \Theta_S = \frac{T_s}{T_{ref}} \text{ – sample surface temperature,}$$

$$P_2 = \Theta_{ini} = \frac{T_{ini}}{T_{ref}} \text{ – initial sample surface temperature,}$$

$$P_3 = \Theta_T = \frac{T_T}{T_{ref}} \text{ – powder melting temperature,}$$

$$P_4 = W_x = \frac{w_x}{w_{x,\,ref}} \text{ – burner velocity in the } x\text{-axis,}$$

$$P_5 = W_y = \frac{w_y}{w_{x,\,ref}} \text{ – burner velocity in the } y\text{-axis,}$$

$$P_6 = W_z = \frac{w_z}{w_{x,\,ref}} \text{ – burner velocity in the } z\text{-axis,}$$

$$P_7 = \Lambda_P = \frac{\lambda_P}{\lambda_S} \text{ – powder thermal conductivity,}$$

$$P_8 = C_P = \frac{c_P}{c_S} \text{ – powder specific thermal capacity,}$$

$$P_9 = Ro_P = \frac{\rho_P}{\rho_S} \text{ – powder density,}$$

$$P_{10} = Zr = \frac{L_P}{L_{ref}} \text{ – characteristic powder particle size,}$$

$$P_{11} = \Delta = \frac{\delta}{L_{ref}} \text{ – burner axis distance from the measurement point,}$$

$$P_{12} = Z = \frac{z}{L_{ref}} \text{ – burner axis distance from the sample,}$$

$$P_{13} = D = \frac{d}{L_{ref}} \text{ – transversal distance of the burner tracks,}$$

$$P_{14} = S = \frac{s}{L_{ref}} \text{ – burner track overhang,}$$

$$P_{15} = M_{\mathrm{Ar}} = \frac{m_{\mathrm{Ar}}}{m_{ref}} \text{ – argon mass flow,}$$

$$P_{16} = M_{\mathrm{H2}} = \frac{m_{\mathrm{H2}}}{m_{ref}} \text{ – hydrogen mass flow,}$$

$$P_{17} = M_{np} = \frac{m_{np}}{m_{ref}}$$ – carrier gas mass flow,

$$P_{18} = j$$ – number of spraying,

$$P_{19} = \beta$$ – relative powder feeding angle,

$$\pi_1 = Fo = \frac{\lambda_S t}{c_S \rho_S L_{ref}^2}$$ – Fourier number,

$$\pi_2 = Ki = \frac{P}{\lambda_S T_{ref} L_{ref}}$$ – Kirpitchev (Pomerantsev) number,

$$\pi_3 = Pe = \frac{c_S \rho_S}{\lambda_S} L_{ref} w_x = \frac{w_x L_{ref}}{a_S}$$ – Péclet number,

$$\pi_4 = N_4 = \frac{c_S T_{ref}}{w_x^2}$$ – dimensionless number of the sample thermal capacity,

$$\pi_5 = N_5 = \frac{p_P L_{ref}^2}{w_x m_{ref}}$$ – dimensionless number of the powder feeding velocity,

$$\pi_6 = N_6 = \frac{p_P L_{ref}^3 c_S \rho_S}{m_{ref} \lambda_S}$$ – dimensionless number of the powder thermal properties,

$$\pi_7 = N_7 = \frac{l_P m_{ref}}{p_P w_x L_{ref}^2}$$ – dimensionless number of the heat necessary to melt the powder.

Formation of the experimental mathematical model

Because a complete mathematical model of plasma spraying and heat transfer into the sample is difficult to build up, an experimental mathematical model in the form of a criterion equation is assembled and used to determined the sample surface temperature by means of plasma spraying parameters such as the travelling velocity, discharge gas flow, and spraying time.

Criterion equations have been derived from many plasma spraying experiments involving longitudinal burner tracks over the sample without a powder and with parameters expressing spraying of layers of a thermal barrier coating and diverse burner travelling velocities. The *basic criterion equation* describes the dependence of surface temperature on spraying time and burner travelling velocity

$$\Theta_S = 10.4755 Fo^{0.4655} Pe^{-0.1775}. \qquad (5.57)$$

The *extended criterion equation* describes the dependence of surface temperature on spraying time, burner travelling velocity, and the deposition of various layers of a thermal barrier coating

$$\Theta_S = 613140 Fo^{0.4720} Ki^{-2.0546} Pe^{-0.1750} M_{\mathrm{Ar}}^{1.3101} M_{\mathrm{H_2}}^{1.4920}. \qquad (5.58)$$

In comparison with the extended criterion equation, the *modified criterion equation* involves the influence of different distances of the burner tracks overhang from the sample border. This equation also enables one to compare various arrangements of the criteria because $P = \frac{M_{\mathrm{H_2}}}{M_{\mathrm{Ar}}} Pe$,

$$\Theta_S = 3.2958 Fo^{0.4667} Pe^{-0.1269} S^{-0.1877} P^{0.2404}. \qquad (5.59)$$

The evaluation of agreement of the criterion equations with the experiment, suitability of selection of the criteria and the criteria equation forms is summarized in Table 5.1. The criterion equations are compared by means of the determination index R^2, the whole-equation F-test value, and the t-test values of individual similarity criteria at the significance level of 95 %.

According to the derived F-test values and the determination index, all criterion equations can be considered as suitable representatives of the dependence of the examined sample surface temperature on the layer spraying parameters. Their usefulness is confirmed by statistical t-tests of individual criteria in the basic criterion equation. It can be seen that the composed criterion P can be used in the modified equation, containing the combination of criteria Ki, M_{Ar}, and M_{H2}, even though the use of these individual criteria in the extended criterion equation is not recommended. The comparison of the t-test values for various criteria shows that the criterion Fo delivers the largest amount of information into the model. This is also confirmed by the exponent of the criterion Fo, which changes only slightly in the criterion equations. The criterion Pe, expressing the burner motion, appears to be another important variable.

Comparison of the basic criterion equation with the experiment is presented in Fig. 5.3 for two burner traveling velocities in testing of the spraying parameters. The temperature obtained from the basic criterion equation is a satisfactory approximation of the sample surface temperature. The difference of the measured sample surface temperature from the surface temperature from the criterion equation can be considered to be small, especially over shorter times. For times

Tab. 5.1. Statistical evaluation of criterion equations

Criterion equation	Basic	Extended	Modified
R^2	0.955	0.958	0.951
F	8.073×10^5	4.170×10^5	8.009×10^5
$\lvert t_{Fo} \rvert$	1.619×10^3	1.987×10^3	3.247×10^3
$\lvert t_{Ki} \rvert$	—	1.217×10^{-7}	—
$\lvert t_{Pe} \rvert$	1.53×10^2	1.53×10^2	1.62×10^2
$\lvert t_{M_{Ar}} \rvert$	—	1.642×10^{-7}	—
$\lvert t_{M_{H_2}} \rvert$	—	2.692×10^{-7}	—
$\lvert t_S \rvert$	—	—	4.09×10^2
$\lvert t_P \rvert$	—	—	8.8×10^1
$F_{0.95}$	3.09	2.31	2.46
$t_{0.975}$	1.984	1.984	1.984

over four minutes long, the influence of the thermal inertia and thermal memory of the system, appearing at low burner travelling velocity of 1 m·min^{-1}, is obvious. It appears that the phenomena of the inertia and thermal memory of dynamic systems can play an important role not only in large mass systems, as further mentioned in Example

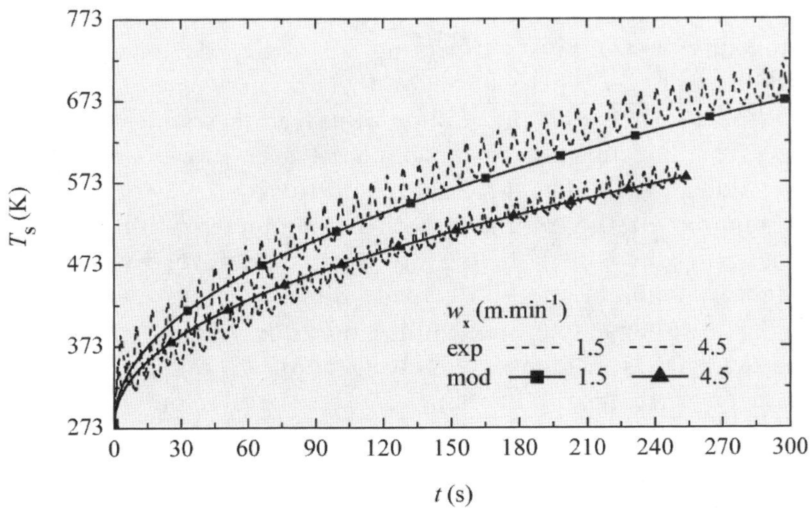

Fig. 5.3. Substrate surface temperature during the experiment and that obtained from the basic criterion equation for different burner travelling velocities.

5.5, but also in relatively small thermal systems with fast proceeding processes. Therefore, the application of the classic criterion equation in the product form is limited in such a case.

The derived criterion equations build up an approximate experimental mathematical model of the thermal process with longitudinal burner tracks over the sample without transversal shifts and without powder feeding. A similar procedure can be used to derive the criterion equation for the plasma spraying process. The surface temperature obtained from the criterion equations describes the average surface temperature value with neither minima nor maxima of the surface temperature being considered. The experimental mathematical model is characterized by good accuracy even when considering selected quantities acting in plasma spraying.

Example 5.3 *Intensive heat sources. Laser beam interaction with the material surface*

The processes taking place during laser beam interaction with the material will be analyzed physically. This involves the processes of laser radiation absorption, heating, melting, vaporizing and heat propagation in the material, thermal stress induction and deformation in the material solid phase, molten material movement in the molten phase, and parameters of pulsed laser modification of the surface. The number and form of the similarity criteria describing these processes will be determined. The criterion equations wil lbe derived, describing the maximum rise of surface temperature and the time durign which the molten material remains on the material surface under the effect of the laser.

Detailed physical analysis of acting quantities defines a total of 75 physical quantities which describe the processes in progress. By means of the Pi theorem and dimensional analysis, 71 similarity criteria are reduced in full, from which 54 are simple ones and 17 composed ones [4.4]. The physical complexity of processes and the large number of the acting quantities connected with them lead to a more stringent selection thereof and, therefore, to a reduced but satisfactory number of the derived dimensionless criteria. This involves the following quantities: pulse duration time t_p, time during which the material is molten t_{melt}, melting point T_m, maximum surface temperature T_{max}, reference temperature T_{ref}, laser power output density ϕ impacting the material surface, and specimen thickness L_s. The selected thermophysical properties are: specific thermal capacity c_s of the solid phase, specific thermal capacity c_{sl} of the solid phase at the melting point, solid phase heat conductivity λ_s, solid phase heat conductivity

λ_{sl} at the melting point, solid phase thermal diffusivity a_s, solid phase thermal diffusivity a_{sl} at the melting point, latent heat of melting l_m and solid phase surface reflectivity R_s.

Determination of the number of criteria by the Pi theorem:	
Total number of dimensional physical quantities:	$N = 14$
Number of dimensionless input physical quantities:	$N_b = 1$
Number of dimensionally different quantities:	$n = 8$
Number of basic and complementary dimensions:	$r = 4$ (L, M, T, Θ)
Total number of dimensionless criteria:	$k = N - r + N_b = 11$
Number of simple criteria:	$k_s = N - n + N_b = 7$
Number of composed criteria:	$k_k = n - r = 4$

Determination of the form of similarity criteria

The form of the similarity criteria can be determined after determining the physical quantities occurring in the examined laser beam interaction with the material surface and the number of similarity criteria. This form will be determined by dimensional analysis procedure by solving the dimensional matrix composed of physical quantities, in which the solver's experience and intuition play an important role. The main requirement for this is to obtain the maximum possible number of introduced, i.e. commonly used, similarity criteria. The general procedure for setting up and solving the dimensional matrix is described in Chapter 3.

The following dimensional matrix **A** is set up

$$\begin{array}{c|ccccccccccc|cccc} & t_p & t_{melt} & T_m & T_{\max} & T_{ref} & \phi & L_s & c_s & c_{sl} & \lambda_s & \lambda_{sl} & a_s & a_{sl} & l_m & R_s \\ \hline L & 0, & 0, & 0, & 0, & 0, & 0, & 1, & 2, & 2, & 1, & 1, & 2, & 2, & 2, & 0 \\ M & 0, & 0, & 0, & 0, & 0, & 1, & 0, & 0, & 0, & 1, & 1, & 0, & 0, & 0, & 0 \\ T & 1, & 1, & 0, & 0, & 0, & -3, & 0, & -2, & -2, & -3, & -3, & -1, & -1, & -2, & 0 \\ \Theta & 0, & 0, & 1, & 1, & 1, & 0, & 0, & 0, & 0, & -1, & -1, & 0, & 0, & 0, & 0 \end{array}$$

The solution matrix **B** is obtained using the matrix solving procedure. The dimensionless similarity criteria can be determined directly from the solution matrix **B,** where P_1 to P_7 are simple criteria and π_1 to π_4 are composed ones.

$$P_1 = R_s \text{ – surface reflectivity of the material solid phase,}$$

$$P_2 = \hat{t}_{melt} = \frac{t_{melt}}{t_p} \text{ – molten material duration time,}$$

$$P_3 = \Theta_m = \frac{T_m}{T_{ref}} \text{ – melting temperature,}$$

$$P_4 = \Theta_{max} = \frac{T_{max}}{T_{ref}} \text{ – maximum surface temperature,}$$

$$P_5 = C_{sl} = \frac{c_{sl}}{c_s} \text{ – specific thermal capacity of the material solid phase,}$$

$$P_6 = \Lambda_{sl} = \frac{\lambda_{sl}}{\lambda_s} \text{ – thermal conductivity of the material solid phase,}$$

$$P_7 = A_{sl} = \frac{a_{sl}}{a_s} \text{ – thermal diffusivity of the material solid phase,}$$

$$\pi_1 = Fo = \frac{a_s t_p}{L_s^2} \text{ – Fourier number,}$$

$$\pi_2 = K_i = \frac{\phi L_s}{\lambda_S T_{ref}} \text{ – Kirpitchev (Pomerantsev) number,}$$

$$\pi_3 = K_1 = \frac{c_s T_{ref} L_s^2}{a_s^2} \text{ – thermal properties of the material solid phase,}$$

$$\pi_4 = K_2 = \frac{c_s T_{ref}}{l_m} \text{ – melting enthalpy.}$$

Construction of the experimental mathematical model

Because of the difficulties in building up a complete mathematical model for the effect of laser radiation on the material surface, experimental mathematical models have been created in the form of a criterion equation. These models can be used to describe the maximum rise of the material surface temperature and the melting time of the material in the laser beam spot. The experimental arrangement and further details of the models and results are presented in [4.4].

The criterion equation for the maximum rise of the material surface temperature has been derived from many experiments with the effect of laser radiation on materials using Nd:YAG (pulse time of 5 ns at FWHM, λ = 266 nm) and KrF (pulse time of 27 ns at FWHM, λ = 248 nm) lasers. The *maximum rise of the material surface temperature under the action of a nanosecond pulse* can be written as follows

$$\Theta_{max} = 1.7425\, Fo^{0.5400} \left[\left(1 - R_s\right) Ki \right]^{0.9963}. \tag{5.60}$$

The criterion equation for the melting time of the material on the surface has been derived on the basis of many experiments with the effect of laser radiation on materials using a KrF laser (pulse time of 27 ns at FWHM, λ = 248 nm). The *time during which the material on the surface is molten under the action of a nanosecond-pulse* can be written

$$\hat{t}_{melt} = 8.5535 \times 10^{-4} \left[\left(1 - R_s\right) Ki \right]^{2.2800} K_1^{-0.5067} \Lambda_{sl}^{-0.7721} C_{sl}^{-2.5560}. \tag{5.61}$$

The evaluation of the agreement of the criterion equation with the experiment and of criteria selection and the suitability of the form of the criterion equation are summarized in Tab. 5.2. The criterion equations are compared by means of the determination index R^2, the F-test values of the whole equation, and the t-test values of individual similarity criteria at the significance level of 95 %.

Comparison of the values of the maximum temperature rise on the material surface, determined by the experiments, with those taken from the criterion equation (5.60), is shown in Fig. 5.4. Here, the experiments are aranged according to the measured maximum temperature rise. The maximum surface temperature rise values measured and predicted by the criterion equation are identical. The determination index expresses a very good correlation between the measured values and those calculated from the criterion equation. According to the F-test, the whole criterion equation can be denoted as significant and useful to describe the experimental data. The high values of the t-tests of individual criteria in the criteria equation also confirm their significance.

The time during which the material on the surface is molten, established in the experiments and calculated from the criterion equation (5.61), is the same, as can be seen in Fig. 5.5. The experiments are arranged in a sequence according to the measured time during which the material is molten. The determination index and the F-test of the criterion equation show that the criterion equation can be regarded as a suitable representative of the examined dependence of the time during which the material is molten on the process parameters. According to statistical t-tests of individual criteria, all of them can be considered significant and useful.

The obtained criteria equations form an approximate experimental mathematical model for the laser beam action on the material surface. This task is interesting due to the fact that the criteria equation in the

Tab. 5.2. Statistical evaluation of criterion equations

Criterion equation	Maximum temperature rise	Time during which material is molten
R^2	0.951	0.935
F	9.849×10^3	1.621×10^3
$\lvert t_{Fo} \rvert$	3.041×10^3	–
$\lvert (1-R_s)K_i \rvert$	4.514×10^3	5.110×10^2
$\lvert t_{K_1} \rvert$	–	8.580×10^2
$\lvert t_{\Lambda_{sl}} \rvert$	–	2.820×10^2
$\lvert t_{C_{sl}} \rvert$	–	6.800×10^1
$F_{0.95}$	2.996	2.372
$t_{0.975}$	2.700	2.700

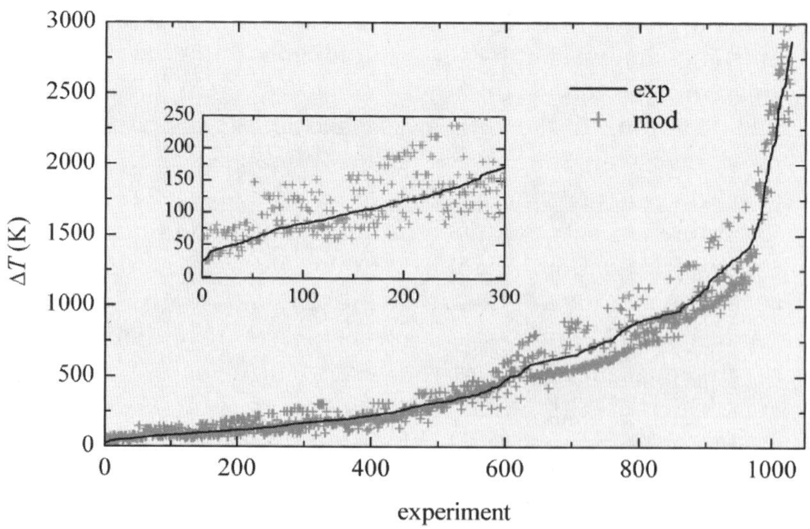

Fig. 5.4 Maximum surface temperature rise caused by a nano-second pulse established by experiment and by criterion equation.

classic product form can be applied with sufficient accuracy even to describe very fast phenomena occurring under the laser action.

5.4.3 Mechanical Engineering

The following examples utilize phenomenological mathematical models. In the first example that is completely different from others, oil warming in a gear box working under extreme arctic conditions is determined.

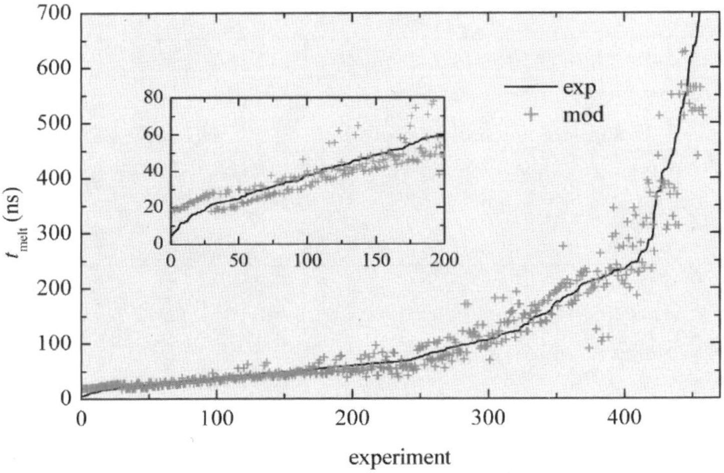

Fig. 5.5. Time during which material on the surface is molten as a result of the effect of a nanosecond pulse determined by experiments and by the criterion equation.

The second example derives the spindle elongation of a machine tool that lowers machining accuracy. The important influence of dynamic memory and system inertia requires the criterion equation in the classic product form to be extendend using additive members.

Example 5.4 *Machine systems. Gear box working under extreme arctic conditions*

The operating modes of a gear box in extreme conditions will be analyzed. The number and form of similarity criteria will be stated and an approximate experimental mathematical model, in the shape of a criterion equation for gear box thermal behavior, will be constructed. A large gear box is a part of the long distance belt conveyer in arctic regions. The whole system operates in outdoor temperatures which vary from +40 to –60°C in the local arctic region. Permanent operating loads occur at the outdoor temperature of – 40°C. Below this temperature, the maintaining operation takes place at reduced velocity. The gear box oil plays the key role in solving the task. Its viscosity changes rapidly due to decreasing external temperature by several orders and can stop operation of the belt conveyer. The aim is to determine an approximate mathematical model limiting the conditions for gear box operation.

Determination of the number of criteria by the Pi theorem:	
Total number of dimensional physical quantities:	$N = 27$
Number of dimensionally different quantities:	$n = 14$
Number of basic and complementary dimensions:	$r = 4$ (L, M, T, Θ)
Total number of dimensionless criteria:	$k = N - r = 23$
Number of simple criteria:	$k_s = N - n = 13$
Number of composed criteria:	$k_k = n - r = 10$

The following physical quantities should be considered in the processes: temperature change of external surroundings, of the wall, oil and bearing ΔT_P, ΔT_S, ΔT_{ol} and ΔT_{lo}, reference temperature $T_{ref} = T_{max} - T_{min}$, power losses P, reference power losses P_{ref}, revolutions of the input shaft n, reference revolutions n_{ref}, moment of forces M on the input shaft, oil velocity w in the gear box, relative velocity w_{ref}, oil level height h, characteristic length (maximum oil level) L, time τ, specific volume heat source q_V, oil viscosity η, oil and material thermal diffusivity α_{ol}, α_{mat}, oil and material thermal conductivity λ_{oL}, λ_{mat}, oil and material specific thermal capacity c_{oL}, c_{mat}, oil and material density ρ_{ol}, ρ_{mat}, heat transfer coefficients α_{ol}, a_P from the oil into the material and from the material into the surroundings.

Determination of the form of similarity criteria

After determining the physical quantities occurring in the examinined gear box system, and the number of similarity criteria, it is neceaary to derive their form. This is determined by the dimensional analysis procedure, by solving the dimensional matrix composed of physical quantities, in which the solver's experience and intuition are very important. The main requirement in this case is to obtain the largest possible number of introduced, i.e. generally used, similarity criteria. The general procedure of assembling and solving the dimensional matrix is described in Chapter 3.

Dimensional matrix **A** is set up as follows

$$\begin{array}{c|ccccccccccccc} & \Delta T_P & \Delta T_S & \Delta T_{ol} & \Delta T_{lo} & T_{ref} & P & P_{ref} & n & n_{ref} & M & w & w_{ref} & h \\ \hline L & 0, & 0, & 0, & 0, & 0, & 2, & 2, & 0, & 0, & 2, & 1, & 1, & 1, \\ M & 0, & 0, & 0, & 0, & 0, & 1, & 1, & 0, & 0, & 1, & 0, & 0, & 0, \\ T & 0, & 0, & 0, & 0, & 0, & -3, & -3, & -1, & -1, & -2, & -1, & -1, & 0, \\ \Theta & 1, & 1, & 1, & 1, & 1, & 0, & 0, & 0, & 0, & 0, & 0, & 0, & 0, \end{array}$$

$$\begin{array}{cccccccccc|cccc} L & \tau & q_V & \eta & a_{ol} & a_{mat} & \lambda_{ol} & \lambda_{mat} & c_{ol} & c_{mat} & \rho_{ol} & \rho_{mat} & \alpha_{ol} & \alpha_P \end{array}$$

$$\left[\begin{array}{cccccccccc|cccc} 1, & 0, & -1, & -1, & 2, & 2, & 1, & 1, & 2, & 2, & -3, & -3, & 0, & 0 \\ 0, & 0, & 0, & 1, & 0, & 0, & 1, & 1, & 0, & 0, & 1, & 1, & 1, & 1 \\ 0, & 1, & -3, & -1, & -1, & -1, & -3, & -3, & -2, & -2, & 0, & 0, & -3, & -3 \\ 0, & 0, & 0, & 0, & 0, & 0, & -1, & -1, & 0, & 0, & 0, & 0, & -1, & -1 \end{array}\right].$$

The solution matrix **B** is obtained by the matrix solution procedure. The dimensionless similarity criteria can be determined directly from solution matrix **B**, where P_1 to P_{13} are simple criteria and π_1 to π_{10} composed ones.

$$P_1 = \Delta\Theta_P = \frac{\Delta T_P}{T_{ref}} \text{ – temperature change of the external surroundings,}$$

$$P_2 = \Delta\Theta_S = \frac{\Delta T_S}{T_{ref}} \text{ – temperature change of the wall,}$$

$$P_3 = \Delta\Theta_{ol} = \frac{\Delta T_{ol}}{T_{ref}} \text{ – temperature change of the oil,}$$

$$P_4 = \Delta\Theta_{lo} = \frac{\Delta T_{lo}}{T_{ref}} \text{ – temperature change of the bearing,}$$

$$P_5 = P_p = \frac{P}{P_{ref}} \text{ – power losses,}$$

$$P_6 = P_n = \frac{n}{n_{ref}} \text{ – revolutions of the input shaft,}$$

$$P_7 = P_w = \frac{w}{w_{ref}} \text{ – oil velocity in the gear box,}$$

$$P_8 = P_h = \frac{h}{L} \text{ – oil level height,}$$

$$P_9 = P_a = \frac{a_{mat}}{a_{ol}} \text{ – relative thermal diffusivity,}$$

$$P_{10} = P_\lambda = \frac{\lambda_{mat}}{\lambda_{ol}} \text{ – relative thermal conductivity,}$$

$$P_{11} = P_c = \frac{c_{mat}}{c_{ol}} \text{ – relative thermal capacity,}$$

$$P_{12} = P_e = \frac{\rho_{mat}}{\rho_{ol}} \text{ – relative density,}$$

$$P_{13} = P_\alpha = \frac{\alpha_{mat}}{\alpha_{ol}} \text{ – relative heat transfer coefficient,}$$

$$\pi_1 = Fo = \frac{a_{ol}\tau}{L^2} \text{ – Fourier number,}$$

$$\pi_2 = Po_B = \frac{P}{\lambda_{ol} T_{ref} L} \text{ – Pomerantsev number,}$$

$$\pi_3 = Po_V = \frac{q_V L^2}{\lambda_{ol} T_{ref}} \text{ – Pomerantsev number,}$$

$$\pi_4 = Br = \frac{\eta a_{ol}^2}{\lambda_{ol} T_{ref} L^2} \text{ – Brinkman number,}$$

$$\pi_5 = Cl = \frac{\rho_{ol} a_{ol}^3}{\lambda_{ol} T_{ref} L^2} \text{ – Clausius number,}$$

$$\pi_6 = Ec = \frac{w^2}{c_{ol} T_{ref}} \text{ – Eckert number,}$$

$$\pi_7 = Nu = \frac{\alpha_{ol} L}{\lambda_{ol}} \text{ – Nusselt number,}$$

$$\pi_8 = Bi = \frac{\alpha_p L}{\lambda_{mat}} \text{ – Biot number,}$$

$$\pi_9 = N_n = \frac{nL^2}{a_{ol}} \text{ – thermal power losses,}$$

$$\pi_{10} = N_M = \frac{M\, a_{ol}}{\lambda_{ol} T_{ref} L^3} \text{ – relative moment of forces.}$$

Construction of the experimental mathematical model

Modeling scales

Due to large dimensions, the experiments had to be executed on a reduced physical model, located in a special cooling box. For the modeling scales expressing the relation of the quantity on the model to that on the original have been chosen equal to one for temperatures, thermal diffusivity, and thermal conductivity. Other scales depend mostly on the selection of the length scale $\mu_L = L_M L_D^{-1}$, for which $\mu_L = 0.2$ has been chosen with regard to the gear box several meters long.

Because in the length scale $\mu_L \neq 1$ there are usually some problems keeping similarity conditions for all other quantities, one must proceed to make sure that the strength of the influence of individual quantities on the examined process is considered.

Experimental model arrangement

The model gear box is located in a specially arranged cooling chamber. It is driven by an electric motor with controlled speed. Compressed nitrogen from a pressure bottle is used to ensure sufficient cooling power and low temperature control. The nitrogen amount and, therefore, the temperature inside the cooling chamber is controlled with a regulating valve. To take measurements on the dynamic model on the gear box shaft, an electric torsion-meter was inserted between the input gear box shaft and the driving motor shaft. The signal from the rotating torsion-meter shaft is led through a signal converter to a program-controlled measuring system. In measurements, the influence of different external temperatures in the range of –60 to +40°C and that of different revolutions on the oil temperature and other places of the exposed gear box was studied. The measurements were taken for two different arctic oils. The oil has a decisive influence on the gear box function reliability in the extreme arctic conditions. One of the examined oils did not freeze in any are of the gear box, even at the limiting temperature of –60°C and revolutions of 1000 min^{-1}.

Experimental mathematical model

The gear box is an articulated system with complex heat transfer and internal sources. In it, however, there are no alternative dynamic changes, inertia, or memory. Therefore, the classic criterion equation in the product form can be applied. However, most of the similarity criteria found do not influence substantially the independent variables. So, fur basic criterion equations were derived from the measurement results. These equations describe the gear box thermal behavior under various conditions. The first equation

$$P_P = k_1 \Delta\Theta_P^{k_2} P_n^{k_3}, \tag{5.62}$$

expresses the dependence of the gear box power losses on the environment temperature and the input shaft revolutions. The second equation

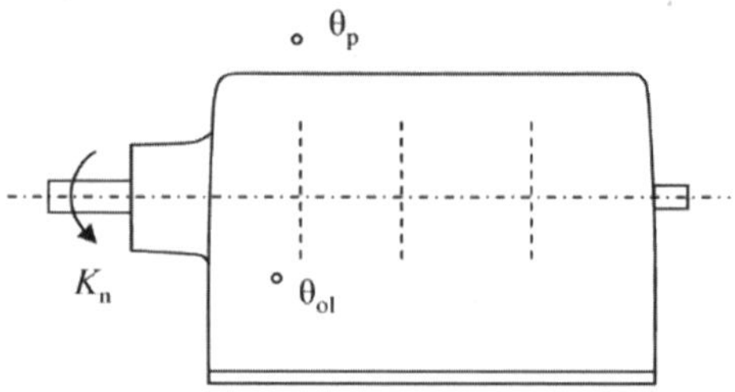

(a)

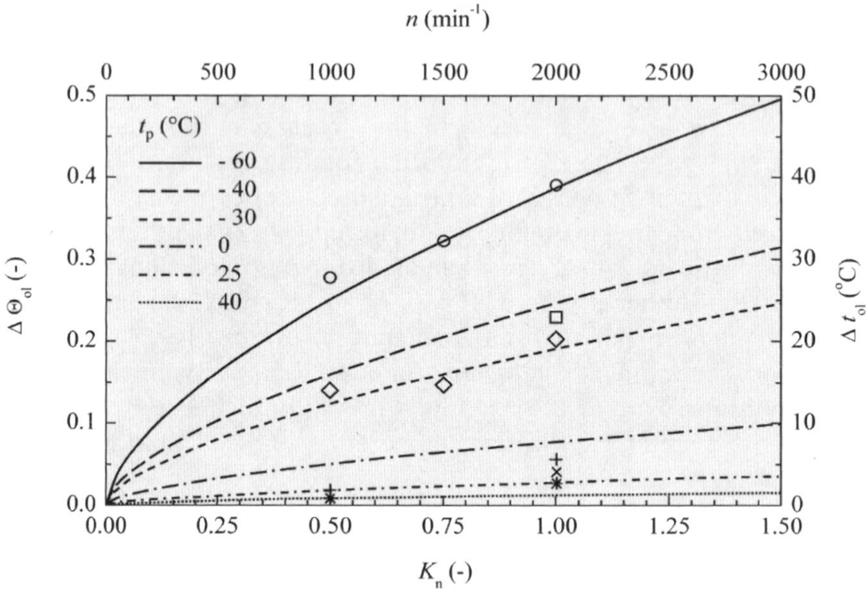

(b)

Fig. 5.6. Oil warming in a thermally exposed place of the gear box (a), and its dependence on the revolutions and external environment temperature changing from +40 to –60°C (b).

$$\Delta\Theta_{ol} = k_1 P_p^{k_2}, \tag{5.63}$$

expresses the dependence of gear box oil warming on the power losses. The third equation

$$N_M = k_1 \Delta\Theta_P^{k_2} P_n^{k_3}, \tag{5.64}$$

expresses the dependence of the input shaft moment of forces on environment temperature and shaft revolutions. The last equation

$$\Delta\Theta_{ol} = k_1 \Delta\Theta_P^{k_2} P_n^{k_3}, \tag{5.65}$$

expresses the dependence of gear box oil warming on environment temperature and input shaft revolutions.

The following criterion equation can be used to show how partial solutions of oil warming the in gear box depend on the revolutions and different external environment temperature

$$\Delta\Theta_{ol} = 0.78 \Delta\Theta_P^{3.4} P_n^{0.63}. \tag{5.66}$$

According to experimentally established criterion equation (5.66), the oil warming dependence on the revolutions and environmental temperature are presented in Fig. 5.6. Based on these results, the thermal condition and functionality of the gear box can be determined in advance under diversely changing operating conditions. A more detailed survey of the results of solving several alternatives and an evaluation of them are presented in [5.16].

Example 5.5 *Manufacturing machines. Thermal instability of a large machine tool*

The operating modes of a large machine tool – a horizontal boring machine – will be analyzed. The number and form of the similarity criteria will be determined and an approximate experimental mathematical model will be constructed in the form of a criterion equation describing the thermal stability of the machine tool. The machine structure is strongly articulated and its dimensions and weight are considerable. With the patterns of spindle and tool shift dynamic changes clarified in various machining modes, a corrective system can be entered into the machine control and the machining accuracy can be enhanced. However, because of the extraordinarily large weight and structure of the machine and rapid machining mode changes, the criterion equations in the classic product form can be used on a limited scale only as a mathematical model for the dynamic thermal behavior of the machine.

In the process, the following physical quantities appear: deformation z in the z-axis, temperature T in a characteristic place of the machine, external environment temperature T_p, oil temperature T_{ol}, initial machine temperature T_{ini}, reference temperature T_{ref}, time τ, main motor input power P, spindle run-out z_v, tailstock sleeve run-out z_p, characteristic length L, spindle revolutions n, heat in cutting process q_r, oil flow velocity w_{ol}, oil density ρ_{ol}, oil specific heat capacity c_{ol}, oil cooler power capacity P_{ch}, machine material thermal conductivity λ, machine material thermal diffusivity a, coefficient of heat transfer from the machine into the external environment α.

Determination of the number of criteria by the Pi theorem:	
Total number of dimensional physical quantities:	$N = 20$
Number of dimensionally different quantities:	$n = 13$
Number of basic and complementary dimensions:	$r = 4$ (L, M, T, Θ)
Total number of dimensionless criteria:	$k = N - r = 16$
Number of simple criteria:	$k_s = N - n = 7$
Number of composed criteria:	$k_k = n - r = 9$

Determination of the form of similarity criteria

After determining the physical quantities appearing in the examined machine tool system and the number of similarity criteria, it is necessary to deduce the similarity criteria form. This is carried by the dimensional analysis procedure with the solution of the dimensional matrix composed of physical quantities, where the solver's experience and intuition play a very important role. With this, the main requirement is to obtain the largest possible number of the already introduced, i.e. generally used, similarity criteria. The general procedure for assembling and solving the dimensional matrix is described in Chapter 3.

Dimensional matrix **A** is assembled as follows

$$
\begin{array}{c|cccccccccccccc}
 & z & T & T_p & T_{ol} & T_{ini} & T_{ref} & \tau & P & z_v & z_p & L & n & q_r & w_{ol} \\
L & 1, & 0, & 0, & 0, & 0, & 0, & 1, & 2, & 1, & 1, & 1, & 0, & 2, & 1, \\
M & 0, & 0, & 0, & 0, & 0, & 0, & 0, & 1, & 0, & 0, & 0, & 0 & 1, & 0, \\
T & 0, & 0, & 0, & 0, & 0, & 0, & 1, & -3, & 0, & 0, & 0, & -1, & -2, & -1, \\
\Theta & 0, & 1, & 1, & 1, & 1, & 1, & 0, & 0, & 0, & 0, & 0, & 0 & 0, & 0,
\end{array}
$$

$$
\begin{array}{cc|cccc}
\rho_{ol} & c_{ol} & P_{ch} & \lambda & a & \alpha \\
-3, & 2, & 2, & 1, & 2, & 0 \\
1, & 0, & 1, & 1, & 0, & 1 \\
0, & -2, & -3, & -3, & -1, & -3 \\
0, & 0, & 0, & -1, & 0, & -1
\end{array}.
$$

The solution matrix **B** can be obtained using the matrix solution procedure. The dimensionless similarity criteria can be determined

directly from solution matrix **B**, where P_1 to P_7 are simple criteria and π_1 to π_9 are composed ones.
Characteristic quantities:

$$P_1 = Z = \frac{z}{L} \text{ – thermal deformation in the } z\text{-axis,}$$

$$P_2 = \Theta = \frac{T}{T_{ref}} \text{ – characteristic temperature,}$$

$$P_3 = \Theta_P = \frac{T_P}{T_{ref}} \text{ – external environment temperature,}$$

$$P_4 = \Theta_{ini} = \frac{T_{ini}}{T_{ref}} \text{ – initial temperature,}$$

Working mode:

$$P_5 = Z_v = \frac{z_v}{L} \text{ – spindle run-out,}$$

$$P_6 = Z_p = \frac{z_p}{L} \text{ – tailstock sleeve run-out,}$$

$$\pi_1 = Kd_1 = \frac{nL^2}{a} \text{ – spindle revolutions,}$$

$$\pi_2 = Pk = \frac{P}{\lambda T_{ref} L} \text{ – main motor input power,}$$

Cutting process:

$$\pi_3 = K_1 = \frac{q_r a}{\lambda T_{ref} L^3} \text{ – heat from cutting process,}$$

Cooling:

$$P_7 = \Theta_{ol} = \frac{T_{ol}}{T_{ref}} \text{ – cooling oil temperature,}$$

$$\pi_4 = Pe = \frac{w_{ol} L}{a} \text{ – Péclet number (the heat led away by oil),}$$

$$\pi_5 = K_2 = \frac{\rho_{ol} a^3}{\lambda T_{ref} L^2} \text{ – oil density,}$$

$$\pi_6 = K_3 = \frac{c_{ol} L^2 T_{ref}}{a^2} \text{ – oil heat,}$$

$$\pi_7 = K_4 = \frac{P_{ch}}{\lambda T_{ref} L} \text{ – oil cooler output power,}$$

Other quantities:

$$\pi_8 = Fo = \frac{a\tau}{L^2} \text{ – Fourier number (the time),}$$

$$\pi_9 = Bi = \frac{\alpha L}{\lambda} \text{ – Biot number (heat transfer from the machine into the surroundings).}$$

Thermal deformation of the spindle Z in the z-axis is a dependent variable which reduces the machine accuracy. Other dimensional quantities include the influence of machine warming, the external environment temperature, the machine operating mode expressed by the input power, the revolutions, the shifting of the position of machine parts, and the heat generated in the cutting process. The effect of oil cooling plays an important role as expressed by oil temperature, the flow quantity, the oil's physical properties, and the cooler output. Other quantities are time and the physical properties of the materials. The spindle length is held as a characteristic longitudinal dimension for the z-axis deformation. The maximum or possibly minimum reachable oil temperature or warming thereof can be chosen as the reference or initial temperature.

Construction of the experimental mathematical model

The general classic criterion equation form, describing the relation between the dependent variable Z and other independent variables, can be written in the product form. However, the use of a model requiring the determination of more than 16 parameters is not feasible. Therefore, the model must be simplified on the basis of excluding inessential parameters, so that the model's credibility would be maintained.

As an example, let us consider a horizontal boring machine in which the thermal deformation, reducing the machining accuracy, is given by the sum of total axial elongation of the spindle, including the active part of the sliding screw Z_2 and the drive box rear wall shift Z_1, in which the sliding screw is fastened. The following equation is sufficient for the drive box deformation Z_1,

$$Z_1 = k_1 + k_2 \Theta_1^{k_3} \Theta_2^{k_4}, \tag{5.67}$$

in which k_1 to k_4 are the parameters determined from the experiment by non-linear optimization. Θ_1 is the drive box rear wall temperature and Θ_2 is the oil temperature in the spindle headstock. The influence of other acting quantities is negligible. However, the equations similar to the relation (5.67) are not valid generally for the spindle, but hold for its simple warming only with initial equilibrium conditions.

With the *thermal memory* and the *thermomechanical velocity* being introduced, a general mathematical model can be assembled in the form of an *extended criterion equation* which has the following form for, for instance, the total thermal spindle deformation

$$W_{Z2,i} = k_1\left(k_2 W_{\Theta 4,i} + k_3 W_{\Theta 2,i} + k_4 W_{\Theta 3,i}\right)^{k_5} + \left(1 - k_1\right) W_{Z2,i-1}. \quad (5.68)$$

This equation contains thermal deformation rate $W_{Z'}$ warming rate W_{Θ}, the external environment temperature Θ_3, and the rear radial bearing temperature Θ_4. The thermal memory is represented by parameter k_1. The thermomechanical velocities are expressed as the ratio of the change in thermal deformation, or warming, to the time step ΔFo, with the index i denoting the ith time step. The resulting relative spindle deformation rate relative to the workpiece can be determined as the sum of W_{Z1} and W_{Z2}.

Figure 5.7 shows the results for a *horizontal boring machine* of another type with a different thermal structure. The measured (points) and simulated (line) total elongation of the spindle Z_2 and its velocity W_{z2} are depicted. Individual experiments are separated by the vertical line and denoted by the letters A to G. In experiments A to C, the tailstock sleeve was shifted out by 1200 mm and in experiments D to G it was 200 mm, the spindle was shifted by 400 mm only with different revolutions. The total time, corresponding to dimensionless Fo, is 80 hours. The extended criterion equation has the following form

$$W_{Z2,i} = k_1\left(-0.407 W_{\Theta 4,i} - 0.280 W_{\Theta 2,i} - 0.091 W_{\Theta 3,i}\right)^{k_5} + \left(1 - 1.07\right) W_{Z2,i-1}, \quad (5.69)$$

where $k_1 = 0.82$ for $n \neq 0$ or $k_1 = 0.53$ for $n = 0$. The temperatures corresponding to the front and rear part of the tailstock sleeve Θ_5, Θ_7 and the temperature of the final transmission Θ_6 in its central part are independent quantities.

Comparison of the measured and simulated results shows the standard deviation of 3.6 % and the maximum deviation of 11.3 % of the maximum deformation. Nevertheless, it must be considered that the model also covers such faulty machine operations as cooling drop-outs (vibrations W_{Z2} in experiment C), and markedly changing environment temperature, etc. Due to this, the deviation mentioned can be held as acceptable.

In the considered case of a strongly structured massive dynamic system, the classic similarity theory enables one to determine a credible mathematical model only for the steady-state operating mode of the machine. The possibility of this theory being applied for dynamic

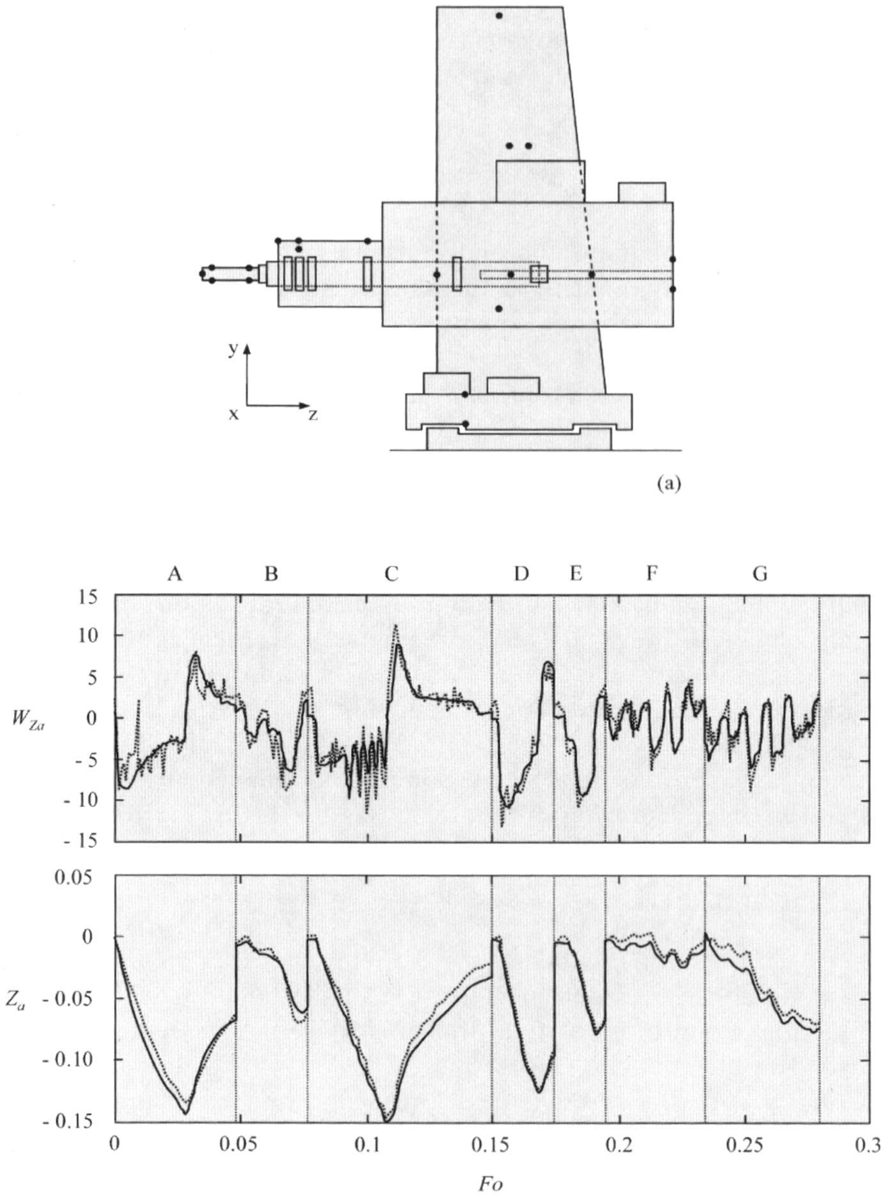

(b)

Fig. 5.7. The measured (points) and simulated (lines) total axial dimensionless thermal elongation of the spindle Z_2 and its velocity W_{z2} for experiments A to G.

machine operating conditions is strongly limited. Only with the system thermal memory and thermomechanical velocity included in the criterion equation can these limitations be eliminated, and a more general phenomenological mathematical model can be assembled. The credibility of this model is ensured even in the dynamic conditions. This important piece of knowledge can be also applied to other dynamic systems. More detailed information is presented in [5.20].

"...all knowledge starts from experience and ends in it. Propositions arrived at by purely logical means are completely empty as regards reality."

Albert Einstein *(14.3.1879–18.4.1955)*

6

Physical Models

These models have a special position in modeling because they have lasted so long, from the historical point of view, without losing anything any of their significance. They emerged in ancient times in a very primitive form, with some hints even in primeval ages. Early models followed from people's everyday encounters with nature's phenomena and their imitation of it. First of all, it was a simple geometrical similarity and models were used by the Babylonians (4th to 3rd century B.C.), by Aristotle (4th century B.C.) and others. However, there is no reference to modeling theory.

The singularity of physical modeling is related to the fact that it is nearest to an examined object in its informational content and similarity assignment. It serves as a certain standard to determine model credibility with abstract modeling increasing steadily. In determining model credibility, abstract mathematical or computer simuliation models need a feed-back to the examined object, with the said feedback being the direct object of investigation (a natural physical model) or a physical model. The experimenter's intuition is very important in experimental model design and realization.

However, in spite of the significance of physical modeling and its long duration, there are not many published books and other works treating it independently and systematically. Physical modeling has been used in many scientific fields, including aerodynamics, hydraulics, hydro-ballistics, mechanics and technology.

J. Kuneš, *Similarity and Modeling in Science and Engineering*,
DOI: 10.1007/978-1-907343-78-0 _6, © Cambridge International Science Publishing 2012

6.1 Characterization of Physical Models

Physical modeling is based on the principle of physical similarity between a model and an object. The *physical model* is always an artificial object. The *natural model* representing an object (an original) examined under various conditions, is a special case. According to the degree of model abstraction relative to the object, natural and physical models are nearest to the object from the informational point of view. When more abstract physically mathematical, mathematical, and cybernetic models are used, natural and physical models serve to verify their credibility. Despite the increasing significance of computer models, one can see the exceptional and often even irreplaceable position of natural and physical models wherever high solution credibility must be provided. This is valid even for system identification, as well as for creating a credible mathematical model.

However, natural models and physical models are usually more demanding in terms of cost and time than models based on the mathematical similarity principle. Hence, to *find a suitable relation between natural or physical models and computer models is one of the most important tasks in experimentation. The experimental efficiency depends to a significant extent on this.*

6.2 Physical Modeling Procedure

The experimenter's intuition and ability to distinguish essential phenomena from non-essential ones characterize physical modeling and make an 'experimental art' of it. Even modeling procedure is based, first of all, on good knowledge of the physical substance of the examined object and the processes under way. Further, it is based on fundamental similarity theorems in modeling and knowledge of the properties of similarity criteria. This includes similarity criteria transformation and modeling scale determination. Especially, modeling scale determination needs a significant degree of experimenter intuition and experience.

6.2.1 Similarity Theorems

In physical modeling, one proceeds from several *physical similarity theorems.*

The *first similarity theorem* expresses a condition for the model (M) assignment to the object (O) in the form of the equality of corresponding similarity criteria for a physically similar system or process

$$\pi_{i,M} = \pi_{i,O}, \qquad P_{j,M} = P_{j,O}, \tag{6.1}$$

$i, j = 1, 2, 3,...,$

where π_i are composed similarity criteria and P_j are simple ones which describe this system or process.

The *second similarity theorem* was formulated by M.V. Kirpitchev and A.A. Guchman. It expresses the condition for the equality requirement of mutually corresponding criteria for unambiguity conditions (initial and boundary conditions, restrictive and optimization conditions)

$$\pi_{i,M} = \pi_{i,O}, \qquad P_{j,M} = P_{j,O}, \tag{6.2}$$

i,j =.1, 2, 3... .

where π_i are composed similarity criteria and P_j are simple ones for the unambiguity conditions of this system or process.

The *third similarity theorem* expresses the condition for the functional relations of similarity criteria to be identical for a model and an object

$$f_M\left(\pi_{i,M}, P_{j,M}\right) = f_O\left(\pi_{i,O}, P_{j,O}\right), \tag{6.3}$$

$i, j = 1, 2\ 3,...,$

where π_i are composed similarity criteria and P_j are simple ones describing this system or process, and f_M, f_O are identical functional relations.

In classic similarity theory, the functional relation is a product of individual criteria. However, in many cases of dynamic processes, e.g., the system inertial influence must be taken into account, due to which the criterion equation is extended by additive members. Essentially, this functional relation of the similarity criteria, called a criterion equation, represents an approximate mathematical model determined from experimental results. Therefore, it is more suitable to speak about an experimental mathematical model.

6.2.2 Similarity Criteria Transformations

The simple and composed similarity criteria are distinguished as *determining* and *determined* ones, according to whether they contain independent or dependent variables.

The criteria are characterized by different practical properties, one of them is the ability to be transformed to a criterion of another form.

At the same time, the criteria can be multiplied or divided, raised to power or multiplied by an arbitrary dimensionless constant.

Therefore, the similarity criteria transformation is a kind of intermediate stage between independent similarity criteria and the criterion equation.

Let us use the arbitrary criteria π_1, π_2 and the constant k. Then, the following is valid

$$\pi_1 \pi_2 = \pi_3, \frac{\pi_1}{\pi_2} = \pi_4, \frac{1}{\pi_1} = \pi_5, k\pi_1 = \pi_6, \tag{6.4}$$

or, more generally

$$k \pi_1^{\frac{r}{t}} \pi_2^{\frac{s}{u}} = \pi, \tag{6.5}$$

where r, s are integers and t, u are non-zero integers.

Because behind a simple number its physical significance is hidden in each criterion, the physical transformation corresponds to the mathematical one. An example is given in Fig. 6.1.

Limitless similarity expresses a limiting physical process condition, in which the physical similarity criteria lose their original significance, either all criteria (total limitless similarity) or some of them only (partial limitless similarity). From the theoretical point of view, this means that in a mathematical model some members in the fundamental equation or some equations in an equation system, describing the investigated process, are not used. The limitless similarity shows itself by automatic similarity preservation, independently of system

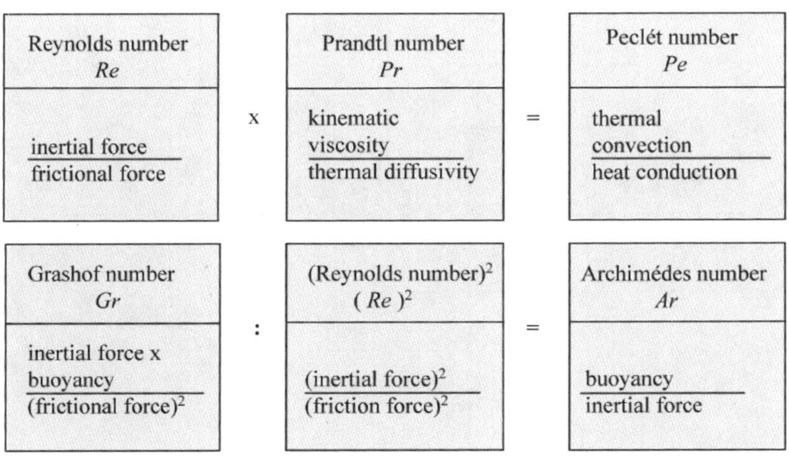

Fig. 6.1. Example of the physical transformation of the similarity criteria.

parameter changes, but with preservation of the same boundary and initial conditions. In similarity and modeling theory, the significance of limitless similarity is based on experimental simplification by limiting the number of criteria and their influence. This can be determined by experiment.

6.2.3 Scales Determination in Physical Modeling

In physical modeling, the similarity criteria are used to their full extent, from the beginning of the model design. By means of the similarity criteria and three similarity theorems, scale similarity indicators and, from them, a mutual coupling between scales of individual quantities characterizing an investigated process or system, are determined. This procedure is shown in Fig. 6.2.

In compliance with the first and second physical similarity theorems, the similarity model–object assignment can be obtained and the modeling scales can be expressed. The relation between the quantity on a model and an original is expressed by the modeling scale. With the quantity X_M on the model corresponding to the quantity X_O on the original (the object), the following relation is valid for the scale

$$\mu_X = \frac{X_M}{X_O}. \tag{6.6}$$

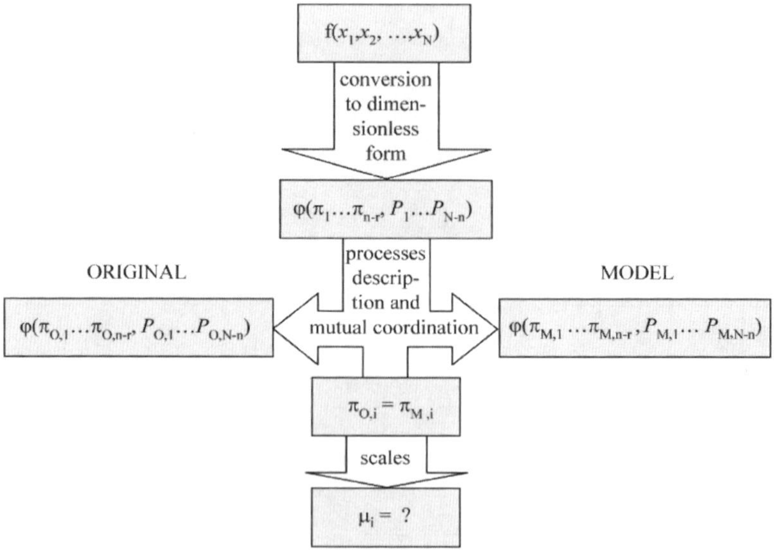

Fig. 6.2. Diagram of the procedure in model scales determination.

In this way, the modeling scale expresses the similarity criteria transformation or assignment of the mutually corresponding quantities of model and original which are given by the first and second similarity theorems.

6.2.4 Example

The following example illustrates the procedure of model scales determination in physical modeling.

Example 6.1 *Heat conduction in a solid body*

The modeling scales for the heat conduction thermal process in a solid body and, in addition, for the 2nd-kind and 3rd-kind boundary conditions have to be determined.

Determination of modeling scales for thermal process

For the thermal process time similarity, the equality of the Fourier numbers for a model and an object is valid

$$Fo_M = Fo_O \Rightarrow \frac{a_M \tau_M}{L_M^2} = \frac{a_O \tau_O}{L_O^2}. \tag{6.7}$$

After modification, the scale similarity indicator can be obtained in the following form

$$\frac{\mu_a \mu_\tau}{\mu_L^2} = 1, \tag{6.8}$$

and the *time scale* expressing the time on a model to that on an object can be expressed

$$\mu_\tau = \frac{\tau_M}{\tau_O} = \frac{a_O}{a_M}\left(\frac{L_M}{L_O}\right)^2 = \frac{\mu_L^2}{\mu_a}. \tag{6.9}$$

Similarly, the Pomerantsev numbers equality is valid

$$Po_M = Po_O \Rightarrow \frac{q_{V,M}\, L_M^2}{\lambda_M T_{ref,M}} = \frac{q_{V,O}\, L_O^2}{\lambda_O T_{ref,O}}. \tag{6.10}$$

After modification, the scale similarity indicator can be obtained

$$\frac{\mu_{qV}\, \mu_L^2}{\mu_\lambda \mu_T} = 1, \tag{6.11}$$

and from this one gets an *internal heat sources scale (scale of volume density of heat flux)*

$$\mu_{q_V} = \frac{q_{V,M}}{q_{V,O}} = \frac{\lambda_M}{\lambda_O}\left(\frac{L_O}{L_M}\right)^2 \frac{T_{ref,M}}{T_{ref,O}} = \frac{\mu_\lambda \mu_T}{\mu_L^2}. \tag{6.12}$$

Determination of modeling scales for thermal process boundary conditions

Similarly, as for the scales of the process in an investigated system, the modeling scales can be determined for boundary, initial, restrictive and optimization conditions. For example, for thermal tasks with 2nd-kind and 3rd-kind boundary conditions, the equality of Kirpitchev or Biot numbers is valid for a model and an object

$$K_{iM} = K_{iO} \Rightarrow \frac{q_{A,M} L_M}{\lambda_M T_{ref,M}} = \frac{q_{A,O} L_O}{\lambda_O T_{ref,O}}, \tag{6.13}$$

$$B_{iM} = B_{iO} \Rightarrow \frac{\alpha_M L_M}{\lambda_M} = \frac{\alpha_O L_O}{\lambda_O}. \tag{6.14}$$

After modification, the following scale similarity indicators are obtained

$$\frac{\mu_{qA}\mu_L}{\mu_\lambda \mu_T} = 1, \qquad \frac{\mu_\alpha \mu_L}{\mu_\lambda} = 1, \tag{6.15}$$

from which the *scale of surface density of heat flux* can be determined

$$\mu_{q_A} = \frac{q_{A,M}}{q_{A,O}} = 1, \qquad \frac{T_{ref,M}\lambda_M L_O}{T_{ref,O}\lambda_O L_M} = \frac{\mu_T \mu_\lambda}{\mu_L}, \tag{6.16}$$

and also the *scale of heat transfer coefficient*

$$\mu_\alpha = \frac{\alpha_M}{\alpha_O} = \frac{\lambda_M L_O}{\lambda_O L_M} = \frac{\mu_\lambda}{\mu_L}. \tag{6.17}$$

The practical procedure for using modeling scales in physical model design will be obvious from the following examples.

6.3 Applications

The following examples show the use of physical models in different areas of mechanics of solid bodies, fluid thermomechanics and mechanical engineering. In the examples, the physical modeling procedure and its results are presented. The examples are chosen not only so that they differ in terms of the kinds of tasks solved, but also that they illustrate the predominance of various forms of physical model implementation.

6.3.1 Mechanics of Solid Bodies

The two examples demonstrate physical models of dynamic stresses in a rotating circular ring and in a power hammer.

Example 6.2 *Experimental elasticity. Dynamic stress in a ring*

In the physical model of a rotating circular ring, the stress σ in the ring needs to be determined. The ring rotates at the frequency n, its cross section is A, the medium diameter is D, the material density is ρ, and gravitational acceleration is g. By using scale similarity indicators, the scale relations which enable one to recalculate the quantities from the model to the original will be determined.

According to the Pi theorem, the total number of dimensionless criteria $k = 3$ is determined because there are six dimensional quantities and three basic dimensions here. When applying the mathematical model analysis describing the stress in the ring, the following criteria can be specified

$$\pi_1 = \frac{A}{D^2}, \quad \pi_2 = \frac{g}{Dn^2}, \quad \pi_3 = \frac{\sigma}{g\rho D},$$

and then the scale similarity indicators

$$\frac{\mu_A}{\mu_L^2} = 1, \quad \frac{\mu_g}{\mu_L \mu_n^2} = 1, \quad \frac{\mu_\sigma}{\mu_g \mu_\rho \mu_L} = 1. \tag{6.18}$$

For $\mu_g = 1$ one obtains $\mu_L = \mu_n^{-2}$. Further, the following relation holds for the stress on the object,

$$\sigma_O = \frac{\sigma_M}{\mu_\sigma} = \frac{\sigma_M}{\mu_\rho \mu_L} = \frac{\sigma_M \mu_n^2}{\mu_\rho}. \tag{6.19}$$

Example 6.3 *Forming machines. Dynamic stress*

Dynamic stress σ, originating in a leading bar at a distance x, will be investigated for the physical model of a power hammer (Fig. 6.3). The F_G-heavy power hammer moves with acceleration a, the leading bar

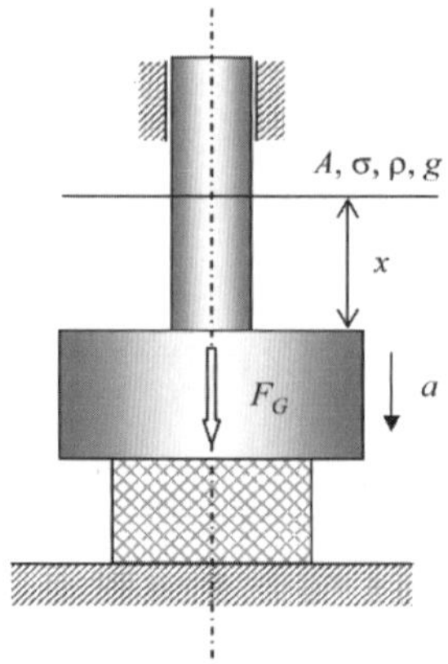

Fig. 6.3. Diagram of a physical power hammer model.

has density ρ and cross section A. The problem is described by seven variables in the following form

$$\varphi\left(F_G, A, \rho, g, a, x, \sigma\right) = 0. \tag{6.20}$$

For the total number $N = 7$ of dimensional quantities and the number of fundamental dimensions $r = 3$, one obtains the number of dimensionless criteria $k = N - r = 4$. These criteria are

$$\pi_1 = \frac{A}{x^2}, \quad \pi_2 = \frac{\rho g x^3}{F_G}, \quad \pi_3 = \frac{\sigma x^2}{F_G}, \quad P_1 = \frac{a}{g},$$

with corresponding scale similarity indicators

$$\frac{\mu_A}{\mu_L^2} = 1, \qquad \frac{\mu_\rho \mu_g \mu_L^3}{\mu_F} = 1, \qquad \frac{\mu_\sigma \mu_L^2}{\mu_F} = 1, \qquad \frac{\mu_a}{\mu_g} = 1. \tag{6.21}$$

Two scales can be chosen for the physical model, with two others being determined by the relations (6.21). The gravity-acceleration scale will be $\mu_g = 1$. The scale of forces μ_F and model material can be chosen, i.e. the density scale μ_ρ. With this, the *length scale* is given

$$\mu_L = \sqrt[3]{\frac{\mu_F}{\mu_\rho}}, \tag{6.22}$$

and so is the *stress scale*

$$\mu_\sigma = \frac{\mu_F}{\mu_L^2} = \sqrt[3]{\frac{\mu_F}{\mu_\rho^2}}. \tag{6.23}$$

To recalculate the stress from the model to the original, the following relation will be used

$$\sigma_O = \frac{\sigma_M}{\mu_\sigma}. \quad (6.24)$$

6.3.2 Thermomechanics of Fluids

The modeling scales of a physical model of heat exchanger are determined in the following example.

Example 6.4 *Thermomechanics. Heat transfer in an exchanger*

Heat transfer with a turbulent flow will be investigated on a physical exchanger model. In the original, the water flows inside pipes with diameter d. The modeling scales will be determined.

For the heat transfer in the original and model, thermal and hydrodynamic similarity is valid, which is represented by three dimensionless criteria, among which the following relation holds

$$Nu = f\,(Re,\ Pr). \quad (6.25)$$

The basic modeling conditions are expressed by the following relations

$$Re_O = Re_M,\ Pr_O = Pr_M,\ Nu_O = Nu_M. \quad (6.26)$$

From the condition $Re_O = Re_M$ one gets

$$\frac{w_O\, d_O\, \rho_O}{\eta_O} = \frac{w_M\, d_M\, \rho_M}{\eta_M}, \quad (6.27)$$

and from it the *dynamic viscosity scale* is obtained

$$\mu_\eta = \frac{\eta_M}{\eta_O} = \mu_w\, \mu_d\, \mu_\rho. \quad (6.28)$$

Similarly, from the condition $Pr_O = Pr_M$ one obtains

$$\mu_\eta = \frac{\mu_\lambda}{\mu_c}, \quad (6.29)$$

where μ_λ is the *thermal conductivity scale,* and μ_c is the *specific thermal capacity scale.* After comparing these last relations, one gets

$$\mu_w \mu_d = \frac{\mu_\lambda}{\mu_c \mu_\rho}. \quad (6.30)$$

The *heat transfer coefficient scale* is implied from the condition $Nu_O = Nu_M$, so that

$$\mu_\alpha = \frac{\mu_\lambda}{\mu_d}. \tag{6.31}$$

With the same fluid being used for the model and the original (object), $\mu_\lambda = \mu_c = \mu_\rho = 1$, from the conditions (6.30) and (6.31) one obtains

$$\mu_w \mu_d = 1, \qquad \mu_\alpha = \frac{1}{\mu_d}. \tag{6.32}$$

Therefore, for modeling, the pipe diameter and the flow velocity can be changed so that the following would be valid

$$w_M d_M = w_O d_O. \tag{6.33}$$

With a five-fold smaller model ($\mu_d = 0.2$), the velocity scale will be $\mu_w = 5$ and, therefore, the flow velocity will be $w_M = 5w_O$ on the model. For the heat transfer coefficient, $\mu_a = \alpha_M \alpha_O^{-1} = 5$ and, therefore, $\alpha_M = 5\alpha_O$.

6.3.3 Mechanical Engineering

In the following example, the power characteristics are examined on a physical model of a piston engine. The design of a physical model of a turbo-set rotor is illustrated in the next example, where the dynamic properties are analyzed. In the other example, the reaction of a steam pipeline during the operating conditions is determined on the physical model. The last example deals with utilization of a heat pipe as an active dynamic chill in steel casting.

Example 6.5 *Piston engines. Determination of output power characteristics by models*

Using a physical model of a piston machine, the output power characteristics will be studied and the relation between output power and longitudinal dimensions will be determined first of all.

The indicated output power of the piston engine is described by the following relation

$$P_i = k\, p_{is}\, S\, L\, n, \tag{6.34}$$

where k is the dimensionless constant, p_{is} is the average indicated pressure in the cylinder, S is the piston area, L is the piston stroke, and n is the number of revolutions. Equation (6.34) can be transformed into the following scale form

$$\mu_P = \frac{P_M}{P_O} = \mu_p \mu_L^3 \mu_n. \tag{6.35}$$

For a model with equal compression degree $\mu_p = P_M P_O^{-1} = 1$, the following holds

$$\frac{\mu_L^3}{\mu_P} = \frac{1}{\mu_n}, \tag{6.36}$$

i.e., the ratio of the cylinder volume to the machine output power is inversely proportional to the number of revolutions.

Also, the piston-machine *scale of output power* can be written in the following form

$$\mu_P = \mu_p \mu_L^3 \mu_n = \frac{\mu_\rho \mu_L^5}{\mu_\tau^3} = \mu_\rho \mu_w^3 \mu_L^2, \tag{6.37}$$

because it holds for the *revolution scale* $\mu_n = \mu_\tau^{-1}$, the *pressure scale* $\mu_p = \mu_\rho \mu_w^2$, and the *velocity scale* $\mu_w = \mu_L \mu_\tau^{-1}$. With the *density scale* $\mu_\rho = \rho_M \rho_O^{-1}$, the *pressure scale* is $\mu_p = \mu_w^2$. With the pressure relation $\mu_p = 1$ preserved, one obtains

$$\mu_w = \frac{\mu_L}{\mu_\tau} = 1, \quad \text{and then } \mu_\tau = \mu_L. \tag{6.38}$$

With equal pressures on the original and the model, equal velocities are maintained and time is proportional to longitudinal dimensions. With the conditions mentioned, the *revolution scale*

$$\mu_n = \frac{1}{\mu_\tau} = \frac{1}{\mu_L}, \tag{6.39}$$

is obtained, and with $\mu_w = \mu_\rho = 1$ the *output power scale* is

$$\mu_P = \mu_\rho \mu_w^3 \mu_L^2 = \mu_L^2. \tag{6.40}$$

With equal compression and velocities, the piston-engine output powers are related by second powers of longitudinal dimensions (cylinder diameters).

If equal revolutions and density on the model and the object are maintained, one obtains $\mu_\tau = \mu_\rho = 1$ and then

$$\mu_w = \frac{\mu_L}{\mu_\tau} = \mu_L, \qquad \mu_p = \mu_\rho \, \mu_w^2 = \mu_L^2. \tag{6.41}$$

For the *output power scale* of the engine, the following is valid

$$\mu_P = \mu_\rho \, \mu_w^3 \, \mu_L^2 = \mu_L^5. \tag{6.42}$$

Therefore, if $\mu_L = 10^{-1}$, the output power of the original is 10^5-times higher than the physical model output power.

Example 6.6 *Dynamics of machines. Critical revolutions of a turbo-set rotor*

A physical model of a turbo-set rotor will be designed to study its dynamic properties.

Characteristic quantities

In this task, there are eight characteristic quantities: distance L between bearings, shaft diameter d, rotor material density ρ, density m of the solitary disk, rotor moment of inertia I, product of the modulus of elasticity and cross-section modulus $E\,J$, support stiffness k and rotation frequency f.

Similarity criteria and scales

In compliance with the Pi theorem, five dimensionless similarity criteria ($N = 8$, $n = 7$, $r = 3$, $k_k = 4$, $k_s = 1$) can be created from the characteristic quantities

$$\pi_1 = \frac{\rho d^2 L^4 d^2}{E J}, \qquad \pi_2 = \frac{I}{\rho d^2 L^3},$$

$$\pi_3 = \frac{m}{\rho d^2 L}, \qquad \pi_4 = \frac{kL^3}{E J}, \qquad P_1 = \frac{d}{L}.$$

The *length scale* μ_L, *diameter scale* μ_d, *modulus-of-elasticity scale* μ_E and *density scale* μ_ρ are chosen

$$\mu_L = \frac{L_M}{L_O}, \qquad \mu_d = \frac{d_M}{d_O}, \tag{6.43}$$

$$\mu_E = \frac{E_M}{E_O}, \qquad \mu_\rho = \frac{\rho_M}{\rho_O}. \tag{6.44}$$

From the cross-section moment definition, the *cross-section moment scale* is

$$\mu_J = \mu_d^4. \tag{6.45}$$

Other scales will be determined from the equality of mutually corresponding criteria for the original and model. The *scale of frequency* is

$$\mu_f = \frac{\mu_d \sqrt{\mu_E}}{\mu_L^2 \sqrt{\mu_\rho}}, \tag{6.46}$$

the *inertia moment scale* is

$$\mu_I = \mu_\rho \mu_d^2 \mu_L^3, \tag{6.47}$$

the *scale of solitary mass*

$$\mu_m = \mu_\rho \mu_d^2 \mu_L, \tag{6.48}$$

the *scale of support stiffness*

$$\mu_k = \frac{\mu_E \mu_d^4}{\mu_L^3}, \tag{6.49}$$

and for the *diameter scale*, one obtains

$$\mu_d = \mu_L. \tag{6.50}$$

Model design (A)

The scale (6.50) provides complete geometric similarity. Its fulfillment causes some unfavourable results in model experimentation. So, for example, with the same material used both for the model and the original ($\mu_\rho = \mu_E = 1$), the following expression results from the relation for the scale of frequencies (6.46)

$$\mu_f = \frac{1}{\mu_L}, \tag{6.51}$$

and, therefore, in the reduced model the critical revolutions increase with an equal ratio. Measurements should be taken on the model at high revolutions, especially if the high-order critical revolutions are investigated. Under the conditions mentioned, other scales will be

$$\mu_I = \mu_L^5, \qquad \mu_m = \mu_L^3, \qquad \mu_k = \mu_L. \tag{6.52}$$

By maintaining $\mu_D = D_M D_O^{-1} = \mu_b = b_M b_O^{-1} = \mu L$, the solitary masses, shaped as wheels with diameter D and width b, have the mass–moment-of-inertia ratio

$$\mu_m = \frac{m_M}{m_O} = \frac{D_M^2 b_M}{D_O^2 b_O} = \mu_L^3, \tag{6.53}$$

$$\mu_I = \frac{I_M}{I_O} = \frac{D_M^4 b_M}{D_O^4 b_O} = \mu_L^5, \tag{6.54}$$

which agrees with Eq. (6.52). Therefore, with complete geometrical similarity maintained, the gyro effect of the wheels will be respected as well. Similarly, the influence of shear forces and rotational inertia of solitary masses will be preserved.

Model design (B)

If the critical revolutions should be reduced due to the reason mentioned, e.g. to the same value as that of the original ($\mu_f = 1$), different scales must be chosen for lengths and diameters ($\mu_d \neq \mu_L$). With the same material chosen for the object and the model ($\mu_\rho = \mu_E$), the following expression will be obtained from Eq. (6.46)

$$\mu_d = \mu_L^2, \tag{6.55}$$

and other scales will be

$$\mu_I = \mu_L^7, \qquad \mu_m = \mu_L^5, \qquad \mu_k = \mu_L^5. \tag{6.56}$$

From the mass ratio of solitary wheels with the diameter D and width b one obtains

$$\mu_m = \frac{m_M}{m_O} = \frac{D_M^2 b_M}{D_O^2 b_O} = \mu_L^5, \tag{6.57}$$

which can be fulfilled if choosing

$$\frac{D_M}{D_O} = \mu_d = \mu_L^2, \qquad \frac{b_M}{b_O} = \mu_b = \mu_L. \tag{6.58}$$

The diameters of solitary disks should be chosen in the diameter scale and their width in the length scale. Then, however, the similarity condition for moment of inertia ($\mu_I = \mu_L^7$) cannot be maintained because the moment-of-inertia ratio will be

$$\mu_I = \frac{I_M}{I_O} = \frac{D_M^4 b_M}{D_O^4 b_O} = \mu_d^4 \, \mu_L = \mu_L^9. \tag{6.59}$$

Model design (C)

If it is required to maintain the moment-of-inertia similarity (e.g., to provide the gyro effect of wheels), the following expressions should be chosen

$$\frac{D_M}{D_O} = \mu_d = \mu_L, \; \frac{b_M}{b_O} = \mu_b = \mu_L^3. \tag{6.60}$$

However, this cannot be done in many cases because of the small thickness of model wheels.

Which of the proposed possibilities will be chosen by an experimenter depends on the specific task, on the object and model dimensions, on the aim of the experiment and on which is possible in the laboratory.

Example 6.7 *Power engineering. Experimental elasticity. Reaction of a steam pipeline*

The forces and moments by which a steam pipeline heated to the operating temperature affects the anchoring point, will be determined. The pipeline has a complicated spatial shape and its ends are firmly anchored.

Model choice

The physical model of the pipeline approximately twenty times reduced is made of a steel tube. The displacements corresponding to heating on the model are made with a mechanical micrometric screw. To measure the reactions and moments in supports in the model, resistance strain gauges are used, which are arranged in a special volume measuring head. The task is solved as a curved bar, where the displacements in supports are known, meanwhile the responses and moments in these supports are being determined.

Characteristic quantities

The characteristic values of the original involve the outer diameter d, wall thickness s, medium diameter r of the pipeline, curvature radius R

of the elbow, cross-section moments S_x, S_y to the x and y axes, deviation cross-section moment S_{xy}, section moment J, polar section moment J_p, length L, distance l of supports, Young's modulus E, shear modulus G, thermal dilatation Δx, pipeline heating ΔT, force F and moment M in the support, support coordinates x_0 and y_0, Kármán dimensionless coefficient K expressing the influence of deformation of the cross-section in curved pipeline parts and the influence of inner over pressure, Kármán dimensionless coefficient K_k expressing the distortion of the curved pipeline part, inner overpressure p and thermal expansion coefficient α.

Fundamental relations

Thermal dilatation

$$\Delta x = l\,\alpha\,\Delta T. \tag{6.61}$$

The cross-section moments

$$S_x = \int K\,y^2 \mathrm{d}L, \quad S_y = \int K\,x^2 \mathrm{d}L, \quad S_{xy} = \int K\,x\,y\,\mathrm{d}L. \tag{6.62}$$

The Kármán coefficient

$$K = \frac{10+12\lambda^2}{1+12\lambda^2}\left[1-4\left(\frac{r}{s}\right)^3\frac{p}{E}\right], \quad \lambda = \frac{R_s}{r^2}. \tag{6.63}$$

The force in the support

$$F = E\,J\,\Delta x\frac{\sqrt{S_y^2+S_{xy}^2}}{S_x S_y - S_{xy}^2}. \tag{6.64}$$

The moment in the support

$$M = E\,J\,\Delta x\frac{y_0 S_y - x_0 S_{xy}}{S_x S_y - S_{xy}^2}. \tag{6.65}$$

Similarity criteria and scales

A geometric similarity requirement needs not be fulfilled. To make the model, an ordinary tube can be used and the similarity between inlets and outlets on the original and on the model must be maintained. That is, the similarity of thermal dilatations, forces and moments must be maintained.

The *force scale* is determined from the force criterion

$$\mu_F = \frac{F_M}{F_O} = \frac{\mu_E \mu_J \mu_{\Delta x}}{\mu_L^3}. \tag{6.66}$$

Similarly, the *moment scale* is obtained from the moment criterion

$$\mu_M = \frac{M_M}{M_O} = \frac{\mu_E \mu_J \mu_{\Delta x}}{\mu_L^2}. \tag{6.67}$$

The *length scale* will be fulfilled for the pipeline lengths, curvature radii and coordinate system

$$\mu_L = \frac{L_M}{L_O} = \frac{R_M}{R_O} = \frac{x_M}{x_O} = \frac{y_M}{y_O}. \tag{6.68}$$

With internal overpressure neglected, from the equality of the Kármán coefficients for both the original and the model it follows that

$$1 = \frac{\lambda_M}{\lambda_O} = \frac{R_M s_M r_O^2}{R_O s_O r_M^2} = \frac{\mu_L \mu_s}{\mu_r^2}. \tag{6.69}$$

From this, for the tube used for the model (given s_M, r_M, and even μ_x and μ_r in consequence), the length scale μ_L is determined, by which the equality of the Kármán coefficients is maintained for both systems. With the pipeline arranged in space, the equality of other model-original similarity criteria must be maintained

$$K_{k,M} = K_{k,O}, \tag{6.70}$$

$$\pi_{G,M} = \frac{E_M}{G_M} = \pi_{G,O} = \frac{E_O}{G_O}, \tag{6.71}$$

$$\pi_{J,M} = \frac{J_M}{J_{p,M}} = \pi_{J,O} = \frac{J_O}{J_{p,O}}. \tag{6.72}$$

Model design

A model tube with parameters s_M, r_M, E_M, G_M, J_M, and $J_{P,M}$ is selected. With this, the scales μ_s, μ_r, μ_E, and μ_J are determined. The length scale μ_L is chosen from Eq. (6.69) and with it the model dimensions. Thermal dilatation Δx_O in the original is known from the relation (6.61). Thermal dilatation Δx_M in the model is chosen so that the forces may be easily measured. With it, the scale $\mu_{\Delta x}$ is chosen and the relations (6.66) and (6.67) are used to recalculate the forces and moments.

Results

The spatial pipeline shape and the arithmetical average of several measurements for one solved task are presented in Fig. 6.4 and Table 6.1. The measurements are compared with the numerical solution results. In particular, the differences found are due to the necessary significant model reduction (w_L = 0.05) and the related inaccuracy of strain gauge measurements, and to not maintaining the geometrical similarity of the model tubes, but they are also partially due to the inaccuracy of the mathematical model and the numerical solution. However, from the practical need point of view, the conformity can be considered to be very good and t

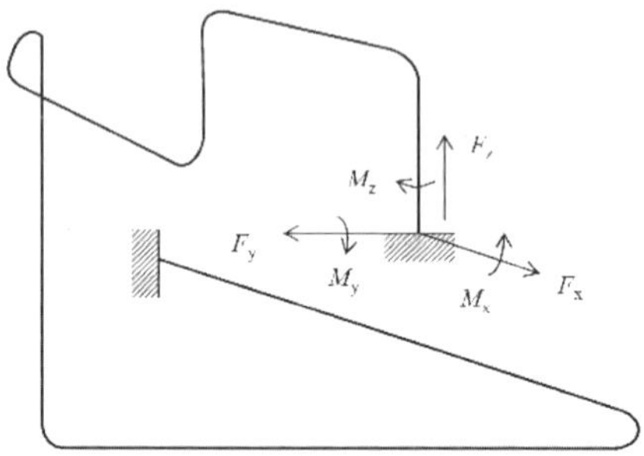

Fig. 6.4. Diagram showing pipeline reaction in a physical model.

Tab. 6.1. Sample of measurement results on a physical pipeline model

Quantity	F_x	F_y	F_z	M_x	M_y	M_z
Units	(N)	(N)	(N)	(N·m)	(N·m)	(N·m)
Model measurement	10.3	9.6	11.1	0.185	2.52	–2.29
Results for the object	1500	1400	1620	530	7270	–6580
Numerical calculation	1310	1320	1690	–125	7740	–6660
Difference (%)	–12.5	–5.7	+4.3	–	+7.2	+1.2

Example 6.8 *Foundry thermomechanics. Active dynamic chill for control of thermal process in the steel casting* [6.5]

The model experiment is concerned with the comparison of three different ways of intensified cooling of a plate casting in a sand mold. It deals with the use of a classical chill, a chill with a built-in heat pipe, and a heat pipe placed in the sand mold near the surface of the casting.

Model choice

The model of a casting composed of three T-sections displaced by

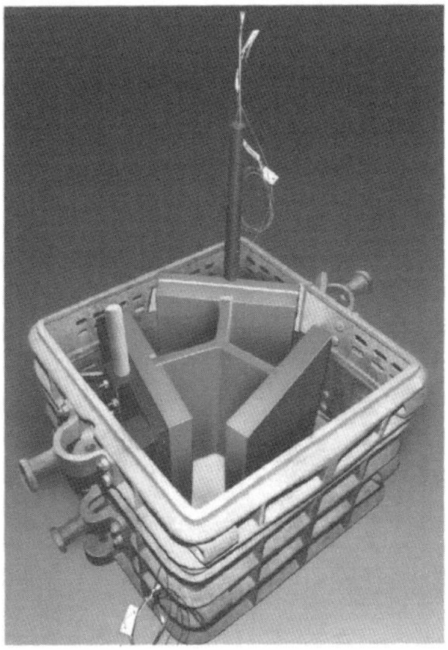

Fig. 6.5. Position of the physical model of three plates in a mould box.

120° with a common gate at the place of contact is shown in Fig. 6.5. The advantage of such an arrangement is that the same technological conditions are maintained, but mutual affecting of the thermal process is a disadvantage. The condensation part of the pipe was cooled by a water or air cooler.

Results

The solution of the indirect problem starts from the measured temperatures in three different sections: A, B and K (Fig. 6.6a). The position of the measuring points in these sections correspond with the requirements for solving indirect problems. The results of its solution for the heat flux density, expressed by means of the Kirpitchev number *Ki,* dependent upon time, represented by the Fourier number *Fo*, are shown in the Fig. 6.6b. From the plots K and B, the differing effect of an active chill comprising a heat pipe (K), compared with a passive chill without a pipe (B), can be seen. The plot A proves that a mere placing of a heat pipe in the sand mold, near the surface of the casting, does not even start up the probe.

Figuew 6.7 shows the microstructure after solidification and cooling for an unaffected material (a), for a material with a passive chill (b), and for a material cooled with a chill with a heat pipe (c). Comparison of the macrostructures reveals a significant difference in the content ratio of shrinkage porosities for the three cases mentioned above. With an increasing intensity of cooling, this ratio is reduced. Thus,

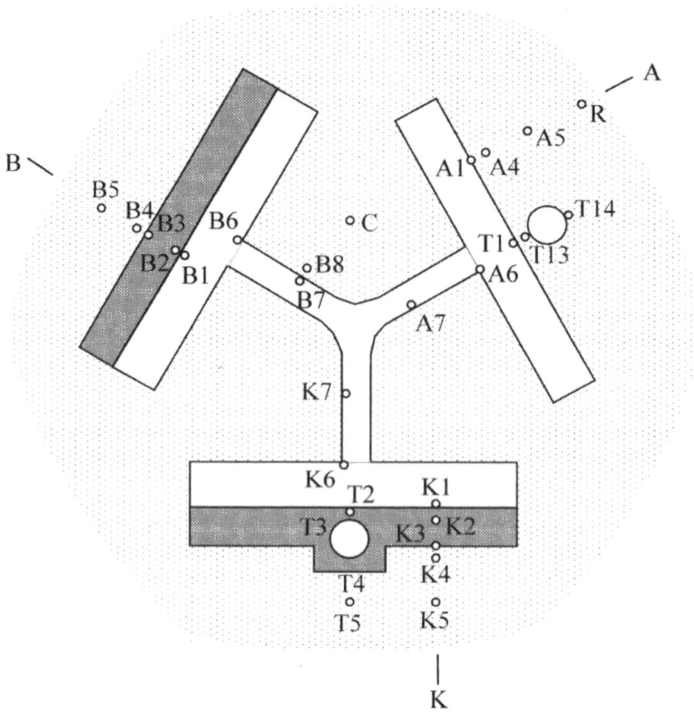

(a)

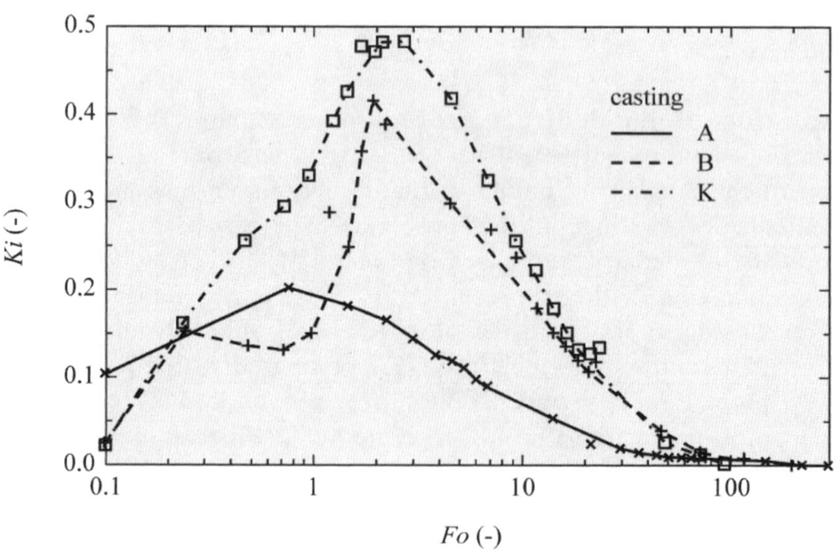

(b)

Fig. 6.6. Cross section of the physical model (a) and time-dependent plot of heat flux density for three options of cooling plate castings (b).

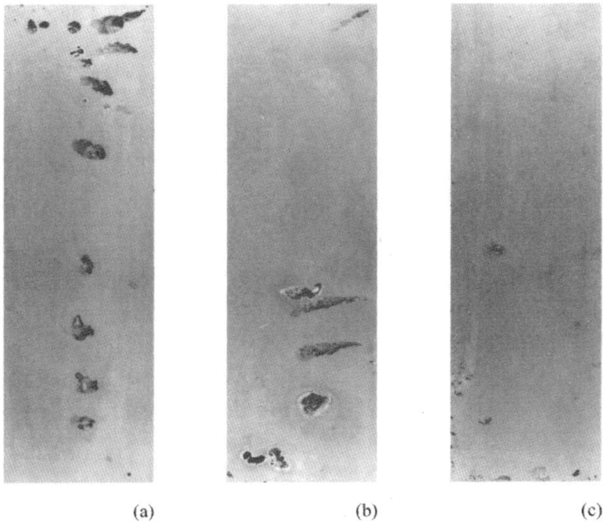

Fig. 6.7. Microstructure of a material thermally unaffected during cooling (a), cooled with a passive chill (b), and cooled with a chill with a heat pipe (c).

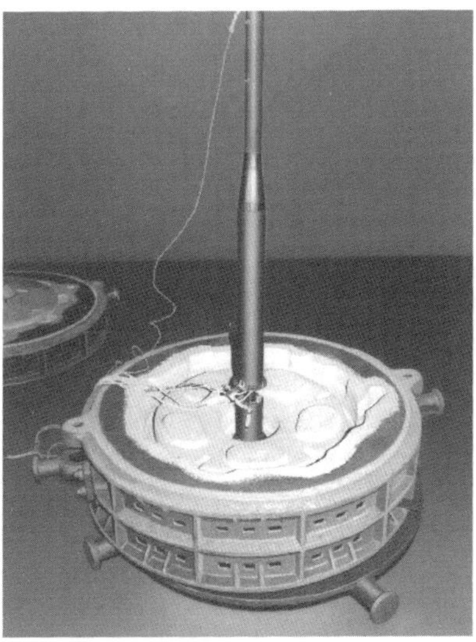

Fig. 6.8. The heat pipe inserted into the hub of a physical model of geared wheel casting.

the intensity of cooling acts especially on the inner compactness of the casting.

The experiments were performed also on five times smaller physical model (diameter of original was 4.5 m) a geared wheel casting with the high-temperature sodium heat pipe located in a coper chill, insulated on its periphery, which was inserted into the hub of the wheel (Fig. 6.8). In the first casting, the heat pipe was cooled by a water cooler. In the second one, a tube with air-cooled fins was used.

The aim of these experiments were to determine the effect of passive and active cooling of the wheel hub upon the reduction of the temperature gradient between the rim and the hub. In addition, they have revealed the necessity of correct dimensioning of the mass of the chill relative to the mass of the adjoining part of the casting. In this way it is possible to ensure that the start-up of the working of the heat pipe will include not only the region of cooling of the casting, but also that of solidification, so that the heat pipe can fulfill the function of an active element affecting the material structure.

The heat pipe can, in combination with a classical passive chill, form an active chill with a substantially higher and controllable cooling effect. However, it is always necessary to ensure suitable positioning of the heat pipe in the chill such that the maximum permissible temperature of the heat pipe is not exceeded. When designing the heat pipe and the chill, it is also necessary to take into account the working characteristic of the heat pipe, which, in turn, must meet the requirement of thermal process control in the region of solidification or cooling of the casting.

"I am especially fond of similarity, principal preceptor of Nature's secrets."

Johannes Kepler *(27.12.1571–15.11.1630)*

7

Physical Analogues

The *physical analogy* is based on the principle of the mathematical and physical similarity between a model and an object. Essentially, it extends the concept of a physical model to physically different phenomena which have analogous mathematical descriptions.

In the pre-computer era, physical analogues and physical models had the dominant position in modeling experiments. Due to the advent of computers in modeling the situation changed and computer models became more general and the most widespread. However, physical analogues retain their significant role in cases of single-purpose models and in other cases as well.

Among the many publications on analogue models and modeling, published mostly before the entry of computers into modeling, are the book [7.5], discussing passive electric models, and the books [7.4] and [7.6] on analogue modeling methods and analogue passive and active models. Analogue and hybrid models and computers are described in [7.1]. Many books are dedicated to applications in engineering, thermomechanics [1.7], heat and mass transfer [7.8], fluid mechanics [3.32] and many other fields.

7.1 Characterization of Physical Analogues

In the past, there were many different physical analogies applied especially in hydromechanics, thermomechanics, elasticity, electrical engineering, and other fields. In the pre-computer era, these analogies were, together with physical models, the main instruments in model

J. Kuneš, *Similarity and Modeling in Science and Engineering*,
DOI: 10.1007/978-1-907343-78-0_7,

development, but their evolution stagnated and was overcome by the entry of new and more progressive computer models. Most of them, e.g. hydraulic, hydrodynamic, pneumatic, mechanical, and thermal analogies, lost their significance as progressive modeling methods. Only a few of these physical analogies remained as the most suitable modeling means to solve some special tasks, primarily because of their possible improvement due to the entry of new technologies. This concerns, e.g., electrical, optical, chemical, and friction analogies. These analogies can be advantageous because of their simplicity, approach objectivity, immediateness, and unpretentiousness. In addition to the above-mentioned properties, the choice of an analogue is influenced by accuracy, availability, the nature of the task, and the possibility to measure the quantities on a model.

Physical analogues are advantageous because of their clarity, as they keep a considerable degree of physical similarity and relation to the real object despite being based on mathematical similarity and on a higher degree of abstraction in comparison with physical models. Therefore, they are applied where simple model preparation, measurement and an immediate idea about the model's behavior come to the fore. This occurs in education and in single-purpose model design, especially hybrid ones, in controlling systems and in many other areas. On the contrary, physical analogues are disadvantageous because they are primarily single-purpose.

However, new technologies of surface layer deposition, either as coatings or thin layers, enables one to make one-layer or multi-layer models to solve different equation types and their sets, and in this they open new possibilities for physical analogues. Another important sphere is the implementation of physical analogues in specialized computational systems.

In modeling utilizing physical analogues, first of all an appropriate physical analogue type is chosen. From the mathematical similarity of process equations on the object and on its analogue, relations for mutual recalculation of quantities among them are obtained. After converting the task and solving it on a physical analogue, the results obtained are then transferred back to the object.

7.2 Various Physical Analogies

There are many physical analogies proceeding from an analogous mathematical description. This holds both for the domain of equations of mathematical physics, involving elliptic, parabolic, and hyperbolic equations, and others. For example, in addition to a previously known and fairly widely-used membrane analogy expressing membrane deformation, there is a shear-stress analogy in a torsion-loaded shaft, a flowing analogy for viscous incompressible fluid in a pipeline, a vortex flow analogy of ideal incompressible fluid, a potential flow analogy for ideal incompressible fluid with sources and sinks, a fluid filtration analogy influenced by sources and sinks, a steady temperature field analogy with heat sources and sinks, a neutron field analogy, an electrostatic field analogy and analogies of other physical phenomena. All of them are described by the Poisson elliptic equation. Today, these analogies are demonstrations only of the perfect abstract interconnection of different phenomena and processes in Nature and serve mostly to clearly illustrate this perfection. Three analogies retaining as yet their practical applications can be mentioned.

Electrothermal analogy

The analogy between heat transfer and electric current passage through a conductor has been known since the year 1872, when Kohlrausch expressed analogous relations between heat transfer and electrical current conduction in a bar. Before this date, the Fourier law of the heat transfer in a homogenous thermal conductor was deduced and according to this, the Ohm law for electric current conduction had been derived. The electrothermal analogy began to be practically used only at the beginning of the 20th century.

Steady temperature and electric fields without internal sources are described by Laplace equations

$$\nabla^2 T = 0, \tag{7.1}$$

$$\nabla^2 U = 0, \tag{7.2}$$

where T denotes temperature and U electric voltage.

Steady temperature and electric fields with internal sources (+) *or sinks* (–) are described by Poisson equations

$$\nabla^2 T \pm \frac{q_V}{\lambda} = 0, \tag{7.3}$$

$$\nabla^2 U \pm \frac{J_V}{\gamma} = 0, \tag{7.4}$$

where q_V is the volume heat flux density from an internal heat source or a sink, J_V is volume current density, λ is thermal conductivity, and γ is specific electrical conductivity.

Unsteady temperature and electric fields are described by Fourier equations

$$\nabla^2 T = \frac{1}{a}\frac{\partial T}{\partial \tau}, \tag{7.5}$$

$$\nabla^2 U = \rho_e c_e \frac{\partial U}{\partial \tau_e}, \tag{7.6}$$

where τ is the thermal process time and τ_e is the electric process time, a is thermal diffusivity, ρ_e is specific electric resistance and c_e is specific volume electric capacitance.

Unsteady temperature and electric fields with internal sources or sinks are expressed by the following basic heat conduction equations

$$\nabla^2 T \pm \frac{q_V}{\lambda} = \frac{1}{a}\frac{\partial T}{\partial \tau}, \tag{7.7}$$

$$\nabla^2 U \pm \frac{J_V}{\gamma} = \rho_e c_e \frac{\partial U}{\partial \tau_e}. \tag{7.8}$$

Motion unsteady fields with internal sources are described by the Fourier-Kirchhoff equation, which has the following form for temperature and electric fields

$$\nabla^2 T \pm \frac{\mathbf{w}}{a}\nabla T \pm \frac{q_V}{\lambda} = \frac{1}{a}\frac{\partial T}{\partial \tau}, \tag{7.9}$$

$$\nabla^2 U \pm \mathbf{w_e}\, \rho_e c_e \nabla U \pm \frac{J_V}{\gamma} = \rho_e c_e \frac{\partial U}{\partial \tau_e}, \tag{7.10}$$

where $\mathbf{w}$, $\mathbf{w_e}$ denote the velocity in a thermal or electric system, border motion velocity, etc.

Similarly, mathematical models for analogous wave fields can be expressed with or without heat diffusion. All linear equations mentioned can also be written in a non-linear form. The analogous forms of equations for boundary and initial conditions can be written similarly.

Chemical–thermal analogy

The chemical–thermal analogy offers relatively significant possibilities, not fully utilized as yet, to establish approximately temperature and other fields. The method is based of a formal analogy between the heat flux and diffusion flux of a diluted substance in a solution due to the chemical activity gradient. In general, this analogy holds for the elementary heat amount and for the elementary flux of the diluted substance *A*

$$\mathrm{d}Q = -\lambda S \frac{\mathrm{d}T}{\mathrm{d}x}\mathrm{d}\tau, \tag{7.11}$$

$$\mathrm{d}A = -DS \frac{\mathrm{d}|A|}{\mathrm{d}x}\mathrm{d}\tau_{CH}. \tag{7.12}$$

For absorbed heat and for diluted fluid absorption, the following is valid

$$c\rho\mathrm{d}T = \frac{\partial(\mathrm{d}Q)}{S\,\partial x}, \tag{7.13}$$

$$c_k\,\mathrm{d}|A| = \frac{\partial(\mathrm{d}|A|)}{S\partial x}. \tag{7.14}$$

In Eqs. (7.11) to (7.14) there are the following quantities: temperature T, chemical activity $|A|$ of the diluted substance A, heat amount Q, time τ, surface S, perpendicular distance x to the surface, thermal conductivity λ, specific volume heat capacity $c\rho$, diffusion coefficient D, and complex stability constant c_k.

From the preceding equation, one can proceed to fundamental equations for one-dimensional propagation of heat and diluted substance

$$\frac{\lambda}{c\rho}\frac{\partial^2 T}{\partial x^2} = \frac{\partial T}{\partial \tau}, \tag{7.15}$$

$$\frac{D}{c_k}\frac{\partial^2|A|}{\partial x^2} = \frac{\partial|A|}{\partial \tau_{CH}}. \tag{7.16}$$

From these equations, the analogies between the heat flux and the diluted substance flux, between the temperature and chemical activity, between the specific heat capacity and the complex stability constant, and between the thermal conductivity and the diffusion coefficient are obvious. The chemical activity measurement consists of the solution acidity measurement. Some remarkable possibilities for electrochemical analogy application appear primarily in the solution of volume tasks.

Friction analogy

This refers to the physical analogy between hydrodynamic and temperature boundary layers. Usually, it is called the Reynolds analogy. It is applied in the fluid flow. It is based on the mathematical similarity principle between the fluid friction and heat transfer velocity through a surface unit. It was considered first by Reynolds and developed further by Prandtl and Taylor. In addition, Kármán considered the analogy between the momentum in the turbulent flow and heat flux. In the friction analogy, the following equations for a laminar sublayer are valid

$$\frac{\tau}{\rho} = \nu \frac{\mathrm{d}u}{\mathrm{d}y}, \tag{7.17}$$

$$\frac{q_A}{A} = -\lambda \frac{\mathrm{d}T}{\mathrm{d}y}, \tag{7.18}$$

for a transient sublayer

$$\frac{\tau}{\rho} = (\nu + \varepsilon) \frac{\mathrm{d}u}{\mathrm{d}y}, \tag{7.19}$$

$$\frac{q_A}{A c_p \rho} = -\left(\frac{\nu}{Pr} + \varepsilon\right) \frac{\mathrm{d}T}{\mathrm{d}y}, \tag{7.20}$$

and for a turbulent core

$$\frac{\tau}{\rho} = \varepsilon \frac{\mathrm{d}u}{\mathrm{d}y}, \tag{7.21}$$

$$\frac{q_A}{A c_p \rho} = -\varepsilon \frac{\mathrm{d}T}{\mathrm{d}y}. \tag{7.22}$$

In Eqs. (7.17) to (7.22), there are the following quantities: shear stress τ, fluid density ρ, kinematic viscosity ν, axial velocity u at a point with the coordinate y, temperature T, area A perpendicular to the heat flux q_A, fluid thermal conductivity λ, fluid specific thermal capacity c_p, vortex diffusivity ε, and Prandtl number Pr.

7.3 Electrical Analogues of Physical Circuits and Systems

Static and dynamic circuits and systems can be investigated not only on computer models but also in relatively simple equivalent electrical circuits. In such a way, not only mechanical, but even hydraulic, pneumatic and other circuits and systems can be studied.

7.3.1 Analogues of Static Circuits and Systems

These enable the solution of tasks of static mechanics and elasticity, such as, for example, force conditions on beams or in bar systems and similar static tasks in electric and other circuits. Essentially, these are problems described by algebraic equations and sets thereof.

Electric analogue of a set of two algebraic equations

Figure 7.1 shows a simple electric circuit composed of current sources I and conductivities G. The circuit is described by a set of two linear equations in the following form

$$a_{11}x_1 - a_{12}x_2 = b_1, \tag{7.23}$$

$$-a_{12}x_1 + a_{22}x_2 = b_2. \tag{7.24}$$

By expressing the set of Eqs. (7.23) and (7.24), one obtains

$$(G_1 + G_{12})U_1 - G_{12}U_2 = I_1, \tag{7.25}$$

$$-G_{12}U_1 + (G_2 + G_{12})U_2 = I_2. \tag{7.26}$$

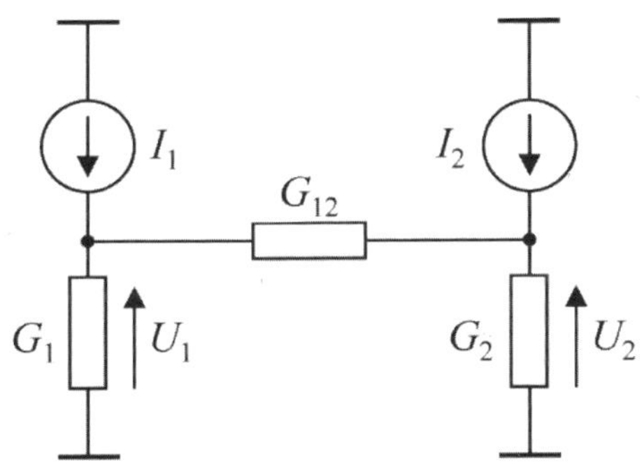

Fig. 7.1 Electrical analogue diagram for a set of two algebraic equations.

An *analogue of a beam on two supports* is an example of an electrical analogue of a two-equation set. The analogy principle is obvious from Fig. 7.2. The relation between acting forces F_1, F_2 and reactions F_A, F_B is given by the following equations

$$F_A = \frac{l_2 + l_3}{l} F_1 + \frac{l_3}{l} F_2, \tag{7.27}$$

$$F_B = \frac{l_1}{l} F_1 + \frac{l_1 + l_2}{l} F_2. \tag{7.28}$$

For the analogous electric circuit (Fig. 7.2b), the following holds

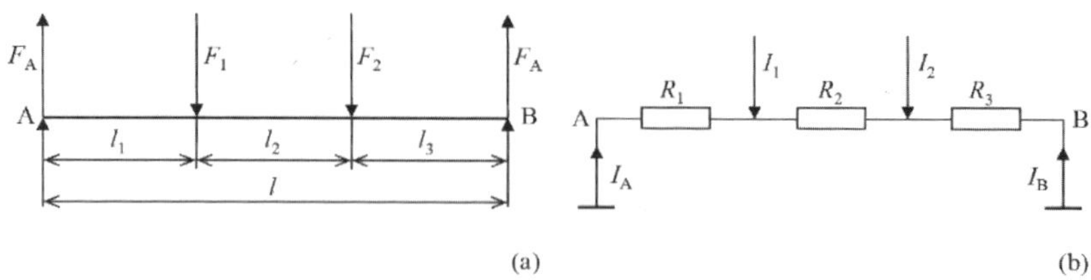

Fig. 7.2. Electrical analogue (b) of a beam on two supports (a).

$$I_A = \frac{R_2 + R_3}{R} I_1 + \frac{R_3}{R} I_2, \tag{7.29}$$

$$I_B = \frac{R_1}{R} I_1 + \frac{R_1 + R_2}{R} I_2. \tag{7.30}$$

with $R = R_1 + R_2 + R_3$. From the comparison of Eqs. (7.27) to (7.30), the mathematical similarity results, where the currents I correspond to forces F and the resistances R to beam lengths l, are as follows

$$I_i = F_i \mu_F, \quad R_i = l_i \mu_L, \tag{7.31}$$

where μ_F, μ_L are scales of forces and lengths.

Similarly, electrical analogues for a variable-cross-section beam, cantilever beams, a mechanical lever, a torsion-loaded bar, and other tasks can be constructed.

Matrix electrical analogue of a set of more linear algebraic equations

The set of linear algebraic equations describing a static mechanical system has the following form

$$\mathbf{Ax} = \mathbf{b}, \tag{7.32}$$

where the matrix of coefficients is

$$\mathbf{A} = \begin{bmatrix} a_{11}, & a_{12}, & \cdots & a_{1n} \\ a_{21}, & a_{22}, & \cdots & a_{2n} \\ \vdots & \vdots & \ddots & \vdots \\ a_{n1}, & a_{n2}, & \cdots & a_{nn} \end{bmatrix}, \tag{7.33}$$

x are unknown variables and **b** loading forces. The matrices of canonic sets of equations occurring in, for example, the static uncertainty solution of bar systems, are symmetrical.

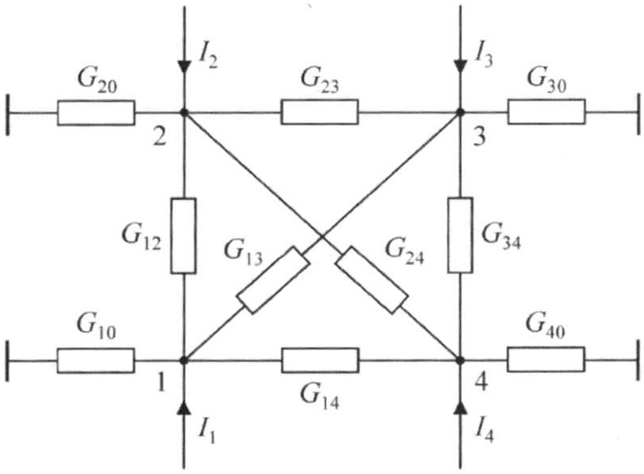

Fig. 7.3. Matrix electrical analogue of a set of algebraic equations.

It is clear that complicated bar (plate, shell, ...) systems and their stability can be solved either with a computer model or a matrix electrical analogue in some specific cases.

The matrix electrical analogue of the set of algebraic equations is presented in Fig. 7.3. The nodes are interconnected by conductivities G_{ik} and each node is joined to a reference point (the earth) by conductivity G_{k0}. The currents I_k are led to the nodes. The node voltage is described on a model by a set of equations

$$\left(G_{10}+\sum_{i=2}^{n}G_{1i}\right)U_1-G_{12}U_2-\cdots-G_{1n}U_n=I_1, \tag{7.34}$$

$$-G_{21}U_1+\left(G_{20}+\sum_{i=1,\,i\neq 2}^{n}G_{2i}\right)U_2-\cdots-G_{2n}U_n=I_2, \tag{7.35}$$

$$-G_{n1}U_1-G_{n2}U_2-\cdots+\left(G_{n0}+\sum_{i=1}^{n-1}G_{ni}\right)U_n=I_n, \tag{7.36}$$

although this is analogous to the original set (7.32), it is not general. It is a special case only, for which the following conditions are valid:

• the main coefficients (the elements laying on the diagonal) are always greater than or equal to the sum of adjacent coefficients of a matrix row

$$a_{ii}\geq\sum_{k=1,\,k\neq i}^{n}a_{ik}, \tag{7.37}$$

• the adjacent matrix coefficients are always negative

$$a_{ik}<0,\qquad i\neq k, \tag{7.38}$$

• the matrix of the set is symmetrical

$$a_{ik} = a_{ki}. \tag{7.39}$$

The first and second conditions can be eliminated by using an alternating current with positive and negative conductivities. Due to this, even the main coefficients can be entered by determining the conductivity G_{j0} as a difference between the required value G_{jj} and the sum G_{ji}

$$G_{j0} = G_{jj} - \sum_{j=1,\ j\neq i}^{n} G_{ji}. \tag{7.40}$$

To solve linear algebraic equations with an unsymmetrical matrix ($a_{ik} \neq a_{ki}$), transformer analogues are used, which are represented by a passive network with inserted ideal transformers which do not consume energy.

7.3.2 Analogues of Dynamic Circuits

These operate on the principle of the analogy between dynamic mechanical and electrical circuits. The circuits are described by analogous differential equations. Mathematical similarity is a necessary but not sufficient condition. The hydraulic and pneumatic circuits can be modelled like the mechanical circuits.

To solve complex dynamic systems, matrix models can also be applied, which are expressed by a set of linear algebraic Eqs. (7.32). The relation between acting dynamic forces F_i and displacements y_i is expressed in a canonical form ($a_{ik} = a_{ki}$) where the forces $\mathbf{F}$ correspond to $\mathbf{x}$ and the displacements $\mathbf{y}$ correspond to $\mathbf{b}$. The coefficients a_{ik} represent the displacements at the point i due to the action of unity forces at the point k. In vibration, the inertia forces are additional forces F, etc.

Basic kinds of electromechanical analogies

Figure 7.4a shows a mechanical vibrating system with one degree of freedom composed of mass m, a spring with linear yielding c and a damper with damping constant k. Dynamic force $F(\tau)$ is acting on this system.The fundamental equation describing the system has the following form

$$m\ddot{x} + k\dot{x} + cx = F(\tau), \tag{7.41}$$

or

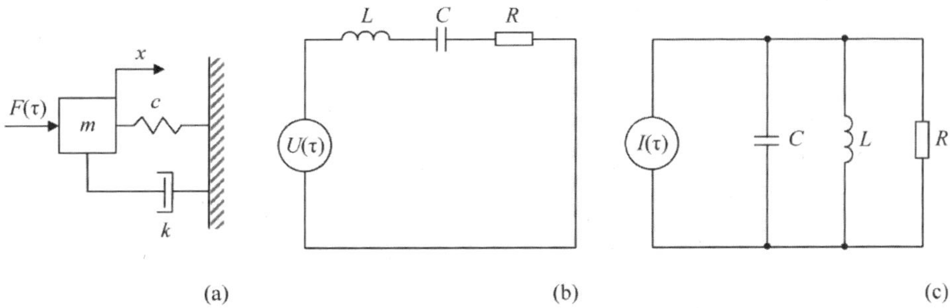

Fig. 7.4. Voltage (b) and current (c) electromechanical analogy of a dynamic system with one degree of freedom (a).

$$m\dot{v} + k v + c \int v \, \mathrm{d}\tau = F(\tau). \tag{7.42}$$

The following equation holds for an analogous resonance circuit with the elements in series connection (Fig. 7.4b),

$$C\ddot{U} + G\dot{U} + \frac{1}{L}\mathrm{U} = \dot{I}(\tau), \tag{7.43}$$

or

$$L\dot{I} + RI + \frac{1}{C}\int I \, \mathrm{d}\tau = U(\tau). \tag{7.44}$$

With the elements arranged in parallel connection in the electric circuit (Fig. 7.4c), one obtains

$$L\ddot{I} + R\dot{I} + \frac{1}{C}I = \dot{U}(\tau), \tag{7.45}$$

or

$$C\dot{U} + GU + \frac{1}{L}\int \mathrm{U} \, \mathrm{d}\tau = I(\tau). \tag{7.46}$$

Two fundamental kinds of analogies result from the comparison of corresponding equations for the three circuits mentioned. From the similarity of Eqs. (7.42) and (7.44), the *voltage analogy* is implied and so is the *current analogy* from the similarity of Eqs. (7.42) and (7.46). In the voltage analogy, the electric current I corresponds to velocity v and electric voltage U corresponds to force F. In the current analogy, it is the other way round. The survey of mutually corresponding quantities is in Table 7.1 for both kinds of electromechanical analogy. The table was completed for a dynamic mechanical rotational circuit.

The electromechanical analogy presumes idealized planar systems with one degree of freedom and the strings of elements in which each

Tab. 7.1. Analogous quantities of some dynamic circuits

Electric voltage circuit	**Electric current circuit**	**Mechanical rectilinear circuit**	**Mechanical rotational circuit**
voltage $U = L\frac{\mathrm{d}I}{\mathrm{d}\tau} = L\dot{I}$	current I	force $F = m\ddot{x}$	torque $M_k = I\ddot{\varphi}$
charge Q	magnetic flux Φ	displacement x	rotation φ
current $I = \dot{Q} = C\dot{U}$	voltage U	velocity $v = \dot{x} = \frac{F}{Z_m}$	angular velocity $\omega_r = \dot{\varphi} = \frac{M_k}{Z_r}$
electric impedance $Z_e = R + j\,\omega\,L - \frac{j}{\omega C}$	electric impedance $Z_e = R + j\,\omega\,L - \frac{j}{\omega C}$	mechanical impedance $Z_m = R_m + j\,\omega\,m - \frac{j}{\omega c}$	mechanical impedance $Z_r = R_r + j\,\omega\,I - \frac{j}{\omega c_t}$
electric resistance $R = \frac{U}{I}$	electric conductance $G = \frac{I}{U} = \frac{1}{R}$	mechanical resistance $R_m = \frac{\mathrm{F}}{v}$	mechanical resistance $R_r = \frac{M_k}{\omega_r}$
inductance L	capacitance C	mass m	moment of inertia I
capacitance $C = \frac{\mathrm{Q}}{U}$	inductance $L = \frac{\Phi}{I}$	linear yielding $c = \frac{x}{F}$	torsional yielding $c_t = \frac{\varphi}{M_k}$
electric power $P = U\,I$	electric power $P_i = U\,I$	mechanical power $P_m = F\,v$	mechanical power $P_r = M_k\,\omega_r$
kinetic energy $W = \frac{1}{2}\,L I^2$	kinetic energy $W_i = \frac{1}{2}\,C U^2$	kinetic energy $W_m = \frac{1}{2}\,m\dot{x}^2 = \frac{1}{2}\,m v^2$	kinetic energy $W_r = \frac{1}{2}\,I\dot{\varphi}^2 = \frac{1}{2}\,I\omega_r^2$
potential energy $E = \frac{1}{2}\frac{Q^2}{C} = \frac{1}{2}C U^2$	potential energy $E_i = \frac{1}{2}\frac{\Phi^2}{L} = \frac{1}{2}L I^2$	potential energy $E_m = \frac{1}{2}\frac{x^2}{c}$	potential energy $E_r = \frac{1}{2}\frac{\varphi^2}{c_t}$
power losses $P = RI^2$	power losses $P_i = \frac{U^2}{R}$	power losses $P_m = F\,\dot{x} = R_m\,\dot{x}^2$	power losses $P_r = M_k\,\dot{\varphi} = R_r\,\dot{\varphi}^2$
resonant frequency $f = \frac{1}{2\pi\sqrt{LC}}$	resonant frequency $f_i = \frac{1}{2\pi\sqrt{LC}}$	resonant frequency $f_m = \frac{1}{2\pi\sqrt{mc}}$	resonant frequency $f_r = \frac{1}{2\pi\sqrt{I c_t}}$

element is always represented by one parameter. Usually, the mentioned simplifications do not influence greatly the required accuracy.

From the kinds of analogies mentioned, the voltage analogy is more widely used, in which the electric voltages, which can be more easily and more precisely measured, correspond to acting forces or moments in a model. On the other hand, the advantage of the current analogy consists of its simplicity and clarity in designing the model diagram.

Mechanical system analysis and electrical analogue design

Detailed elaborated analytical methods for electric circuits can be applied to mechanical circuits. The Kirchhoff, Ohm and other laws are valid for these circuits analogically. The operator calculus can be applied in advantage. Sometimes, mechanical systems with distributed parameters can be replaced by mechanical circuits. In practice, this means the transition from a mathematical model expressed by a partial differential equation to a model represented by the ordinary differential equation. Mechanical systems consist of active and passive elements like electrical circuits. The active elements are: force F, torque M_k, linear velocity v or angular velocity ω. The passive elements are: mass m,momentofinertia I, linear stiffness k or torsion stiffness k_t, and mechanical damping resistance k_k. Each body is characterized, in addition to mass, by, for example, elasticity, etc. There are energy losses due to the influence of internal friction.

7.4 Electrical Analogues of Physical Fields

Elliptic, parabolic or hyperbolic 2nd-order partial differential equations, possibly even sets thereof, and 4th-order partial differential equations are the fundamental equations of the mathematical models of physical fields. Depending on the nature of the physical field to be solved, electrical models can be divided into analogues of steady or unsteady fields, analogues of motion and wave fields, analogues of coupled processes and fields and analogues of processes and fields described by 4th-order partial differential equations.

Depending on the nature of the solution, electrical models can be continuous (CS) or discrete (DS) in space, and continuous (CT) or discrete (DT) in time.

7.4.1 Analogues of Steady Fields

Steady physical fields are described by elliptic equations. The Laplace equation describes a field without internal sources. The Poisson equation describes those with internal sources or sinks.

DS analogues of R type – based on differences

For the *Laplace equation* (Fig. 7.5a,b,c), the following holds

$$\nabla^2 u(M)=0, \qquad M(x,y,z), \tag{7.47}$$

where $u(M)$ is the variable dependent on space coordinates. After converting this equation to a difference one, one obtains the following equation for the electric *circuit of the DS analogue of R type* for a cell of a one-dimensional field (Fig. 7.5c)

$$U_1 + U_2 - 2U_0 = 0, \tag{7.48}$$

and for a cell of an n-dimensional field

$$\sum_{i=1}^{2n}(U_i - U_0)=0 . \tag{7.49}$$

The resistances in a regular rectangular cell

$$R_1 = R_2 = \frac{1}{2}\rho\,\frac{\Delta x}{\Delta y \Delta z}, \tag{7.50}$$

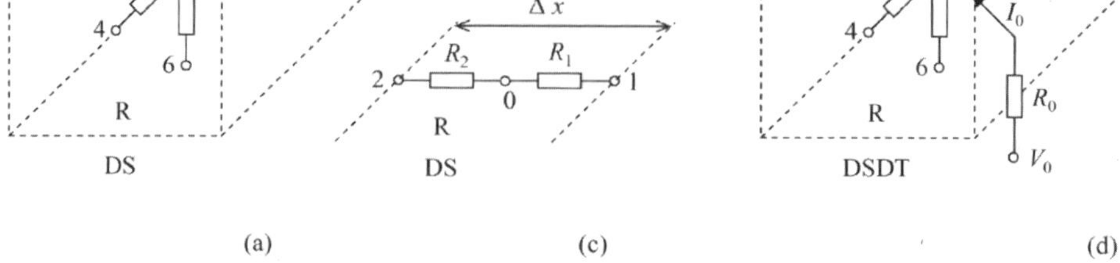

Fig. 7.5. Electrical DS analogues – a three-dimensional (a), a two-dimensional (b), and a one-dimensional (c) cell of the Laplace equation; a three-dimensional (d) cell of the Poisson equation.

$$R_3 = R_4 = \frac{1}{2}\rho\ \frac{\Delta y}{\Delta x \Delta z}, \tag{7.51}$$

$$R_5 = R_6 = \frac{1}{2}\rho\ \frac{\Delta z}{\Delta x \Delta y}, \tag{7.52}$$

where ρ is the specific electrical resistance of a conductor. For the *Poisson equation* (Fig. 7.5d), the following holds

$$\nabla^2 u(M) = F(M),\ \ M(x,y,z). \tag{7.53}$$

By converting the equation into a difference form, one obtains for the electric *circuit of the DS analogue of R type* for a cell of an n-dimensional field

$$\sum_{i=1}^{2n} \frac{U_i - U_0}{R_i} + \frac{V_0 - V_0}{R_0} = 0, \tag{7.54}$$

and the following condition holds

$$|V_0| \gg |U_0|. \tag{7.55}$$

The current flowing into a node of a square cell is

$$I_0 = \frac{V_0}{R_0} = \frac{\Delta x^2}{4R} F(M). \tag{7.56}$$

DS analogues of R type - based on elements

The finite element method, which works on the principle of total potential system energy minimization, can be applied for these analogues. Using the Galerkin method, for instance, the elliptic partial differential equations can be replaced by a set of algebraic homogenous equations (in the Laplace equation) or by a set of right-side equations (in the Poisson equation). Instead of solving the set of equations as usual with a computer, a relatively simple and inexpensive electrical network analogue can also be used, which is composed of resistances only. This has also been shown in [7.3]. This book relates to solutions of multi-dimensional tasks. Here, the possibilities of finite element implementation with an electrical circuit and proceeding to solve a set of algebraic equations in a matrix network analogue are suggested only briefly.

The principle of replacing a triangular finite element with an electrical circuit, composed of three resistors, is given in Fig. 7.6. A

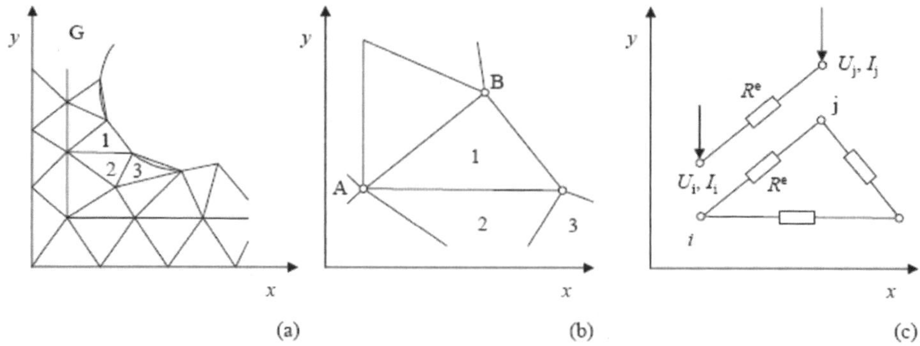

Fig. 7.6. Diagram of a triangular finite element implementation by means of an electrical analogue – the network of elements (a), the triangular element (b), and the electrical analogue of element number 1 (c).

chosen network of elements in the investigated region G (Fig. 7.6a), and the replacement of an element with an electrical circuit (Fig. 7.6c), are suggested. According to Ohm's law, the following relation is valid between current I and voltage U on the resistor R_e on the element e

$$I_i = \frac{U_i - U_j}{R^e}, \quad I_j = \frac{U_j - U_i}{R^e}, \tag{7.57}$$

which can be written in a matrix form

$$\begin{Bmatrix} I_i \\ I_j \end{Bmatrix} = \frac{1}{R^e} \begin{bmatrix} 1, & -1 \\ -1, & 1 \end{bmatrix} \begin{Bmatrix} U_i \\ U_j \end{Bmatrix}. \tag{7.58}$$

The following relation generally holds between the current and voltage of the element e,

$$\{I\}^e = [k]^e \{U\}^e. \tag{7.59}$$

On the AB side of the asymmetrical triangular network (Fig. 7.6b), the resistance R^e can be determined from the relation

$$R^e = \frac{R_{ref}\, l_{AB}}{r} = \frac{2R_{ref}}{\operatorname{ctg}\beta_1 + \operatorname{ctg}\beta_2}, \tag{7.60}$$

where R_{ref} is the reference resistance which is chosen so that the total network resistance and also individual resistances in the network are in a suitable range. Should the resistance R^e be a physically real quantity, β_1 and β_2 must be acute angles in compliance with the relation (7.60).

Then the following holds for the current passing into a network of all elements of the electrical analogue

$$\{\mathbf{I}\} = [\mathbf{K}]\ \{\mathbf{U}\}, \tag{7.61}$$

where the square stiffness submatrix for the whole region is given by the sum of stiffness matrices for individual elements

$$[\mathbf{K}] = \sum [\mathbf{k}]^{e}. \tag{7.62}$$

The practical solution procedure by the electrical matrix analogue of *R* type involves several stages. Thus, at the beginning, in choosing the element shape and whole network design, care must be taken with every triangular element that all internal angles are acute. This means that in a symmetric stiffness matrix all diagonal elements would be, for example, positive but all out-of diagonal ones negative. With obtuse angles, out-of-diagonal elements can have both signs and so an electrical analogue would contain negative resistances. In the subsequent solution stage, a mathematical matrix model will be constructed. In the symmetrical matrix construction, it is enough to determine the on-diagonal and above-diagonal elements only. The procedure goes on by entering boundary conditions into the complete matrix equation. During the actual solution using a simulation model, the currents and voltages are measured as output quantities.

CS analogues of R type

Models using continuous electrically conductive material enable one to model steady physical fields, described by the Laplace partial differential equation (7.47) and the Poisson equation (7.53), in the simplest possible way. Thus, the CS analogue of *R* type can be obtained for a planar continuous cell of the Laplace (Fig. 7.7a) and Poisson (Fig. 7.7b,c) equations.

The following resistance of a square is the fundamental quantity in an h-thick cell (Fig. 7.7) with dimensions Δx, Δy

$$R_{box} = \rho \frac{\Delta x}{\Delta y h}. \tag{7.63}$$

For unity thickness of a layer and $\Delta x = \Delta y$, $R_{box} = \rho$ is obtained, where ρ equals the specific conductor resistance. For an h-thick, b-wide, and l-long planar conductor resistance, the following holds

$$R = R_{box} \frac{l}{b} = \rho \frac{l}{hb}. \tag{7.64}$$

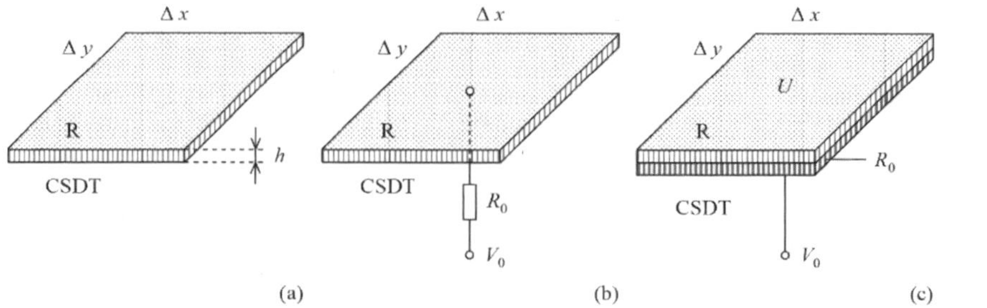

Fig. 7.7. Electrical CS analogues – a planar continuous cell of the Laplace equation (a) and the Poisson equation (b), (c).

By changing the specific resistance or thickness, e.g. by overlaying, the planar conductor resistance can be changed.

The continuous electrically conducting material can be solid, fluid or even plastic if need be. A solid planar modeling environment can consist of special electrically conducting paper or foil but, at present especially, the resistant layer coating technologies are being utilized.

To measure the voltage on solid planar models, simple direct-current *zero indicators* are usually used. The *direct gradient measurement* on planar models consists of replacing the gradient of electric potential $\mathrm{d}V/\mathrm{d}s$ by the ratio of differences $\Delta V/\Delta s$, where ΔV is the voltage between two points, distant from each other by Δs and touching the model surface, of a *double-probe.*

With the modeled quantity representing a harmonic function which fulfils the Laplace equation, the gradient field of electric potential can be modelled directly (see the subsection 7.4.6).

7.4.2 Analogues of Unsteady Fields

The simplest unsteady physical fields are described by parabolic equations. The Fourier equation describes a field without internal sources and the heat convection equation one with internal sources.

The most widely-used analogues to solve unsteady fields are *continuous-in-time CSCT and DSCT analogues of RC type* (the DSCT analogues are called *Beuken models*) and *discrete-in-time CSDT and DSDT analogues of R type* (the DSDT analogues are called *Liebmann models*). In addition to these, there are some other modifications of these basic models.

The analogues of an unsteady fields, described by more complicated equations such as the incomplete wave equation, Schrödinger equation, telegraph equation, and partial 4th-order differential equation (e.g. the biharmonic equation) are presented in [2.6].

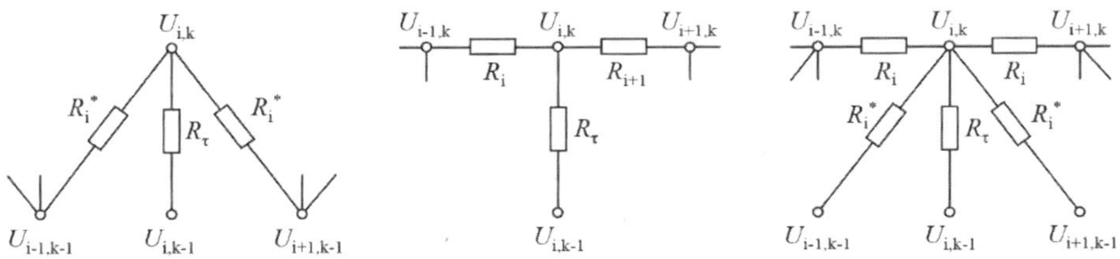

Fig. 7.8. Electrical analogue of three difference diagrams of the Fourier equation — the explicit one (a), the implicit one (b), and the combined one (c).

(A) Fourier equation

$$\nabla^2 u(M) = B(M)\frac{\partial u}{\partial \tau}, \qquad B(M) > 0, \qquad M(x,y,z). \tag{7.65}$$

The Fourier equation (7.65) can be converted into a difference form. The analogous equation is also valid even for an electrical analogue. Depending on the kind of difference diagram, three different arrangements, presented in Fig. 7.8, can be obtained for a one-dimensional cell.

For the *explicit diagram* (Fig. 7.8a) by making use of the Kirchhoff law for a nodal point, one obtains

$$\frac{U_{i,k} - U_{i,k-1}}{R_\tau} = \frac{U_{i-1,k-1} - 2U_{i,k} + U_{i+1,k-1}}{R_\tau}, \tag{7.66}$$

and hence

$$p_k\left(U_{i,k} - U_{i,k-1}\right) = U_{i-1,k-1} - 2U_{i,k-1} + U_{i+1,k-1}, \tag{7.67}$$

where

$$p_k = 2 + \frac{R_i^*}{R_\tau}. \tag{7.68}$$

The lowest value $p_k = 2$ is obtained with $R_\tau \to \infty$ and, therefore, the solution is always stable. Similarly, the *implicit diagram* (Fig. 7.8b) leads to the equation

$$p_k\left(U_{i,k} - U_{i,k-1}\right) = U_{i-1,k} - 2U_{i,k} + U_{i+1,k}, \tag{7.69}$$

where

$$p_k \frac{R_i}{R_\tau}. \tag{7.70}$$

The lowest value $p_k = 0$ is obtained with $R_\tau \to \infty$. The *combined diagram* (Fig. 7.8c) is most complicated. The implicit and explicit diagrams can be considered as special cases of the combined diagram. Of the three diagrams mentioned, the implicit one is used most frequently. The relevant electrical analogue type is called the *Liebmann model.*

DT analogues of R type

These are presented at Fig. 7.9 for solving the Fourier equation. Case (a) represents a spacial cell of the Liebmann DSDT analogue of *R* type, and case (b) is a continuous-in-space modification to solve two-dimensional fields. The time discretization converts the Fourier equation (7.65) solution to the solution of the Helmholtz–Poisson equation

$$\nabla^2 U = A\left(U - U_\tau\right), \quad A > 0. \tag{7.71}$$

For a *spacial model cell* (Fig. 7.9a), the following equation can be written

$$\sum_{i=1}^{6} \frac{U_i - U_0}{R} = \frac{U_0 - U_{\tau-\Delta\tau}}{R_\tau}, \tag{7.72}$$

where the resistance R_τ corresponds to the selected time step $\Delta\tau$ and the

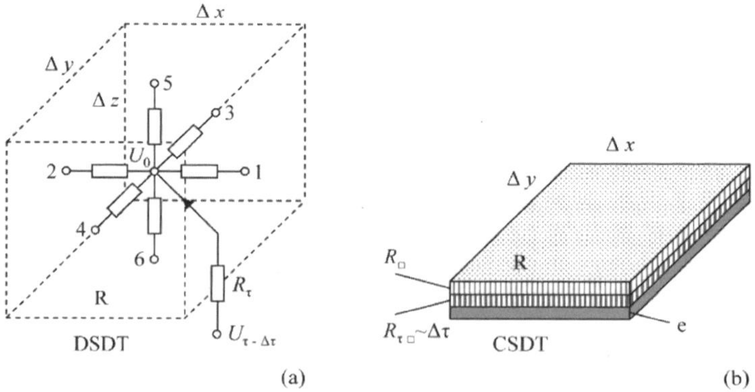

Fig. 7.9. Electrical DT analogues of the Fourier equation – the three-dimensional (a) and the planar continuous (b) cell of *R* type.

following holds

$$R_\tau = \frac{a\Delta\tau}{(\Delta x)^2} 2R = 2\Delta Fo\, Rn^2, \tag{7.73}$$

where 2 R equals the resistance between two adjacent nodes, ΔFo is the dimensionless time step, n is the number of cells related to the reference length of the object, a is the constant denoting thermal diffusivity in, for example, in theat transfer, and Δx is the geometric step.

The two-dimensional tasks can be solved approximately on a *continuous-in-space three-layer model of R type* (Fig. 7.9b). The top resistive measuring layer has the square resistance R_{box} and the middle layer has the resistance $R_{\tau,box}$. Each of the electrodes e belongs to a square planar cell with a side Δx. The cell resistance in the x-direction and the middle layer resistance are

$$R_x = \rho \frac{\Delta x}{h\Delta x} = R_{box}, \tag{7.74}$$

$$R_\tau = \rho_\tau \frac{h_\tau}{(\Delta x)^2} = R_{\tau,box} \frac{h_\tau^2}{(\Delta x)^2}. \tag{7.75}$$

After inserting (R_x =2R) into Eq. (7.73), one obtains the following relation

$$R_{\tau,box} = \frac{a\Delta\tau}{h_\tau^2} R_{box}. \tag{7.76}$$

Here ρ, ρ_τ are the resistances of the top and middle layers and h, h_τ are thicknesses thereof.

CT analogues of RC type

Figure 7.10 shows two different kinds of continuous-in-time analogues of *RC* type. Case (a) represents a discrete spacial cell of DSCT analogue of *RC* type for the classic Beuken model. Case (b) is an analogue of *RC* type with both continuously distributed parameters. On the electrode *e,* a thin dielectric layer *C* is deposited with a resistive layer *R* on it. Special self-adhesive resistive, dielectric, and conductive foils can be used as well.

For the *Beuken analogue of RC type* (Fig. 7.10a), the original Fourier equation (7.65) can be rewritten in a semidiscrete form

$$\sum_{i=1}^{2n} \frac{U_i - U_0}{R} = C\frac{\partial}{\partial\tau}(U_0 - 0), \tag{7.77}$$

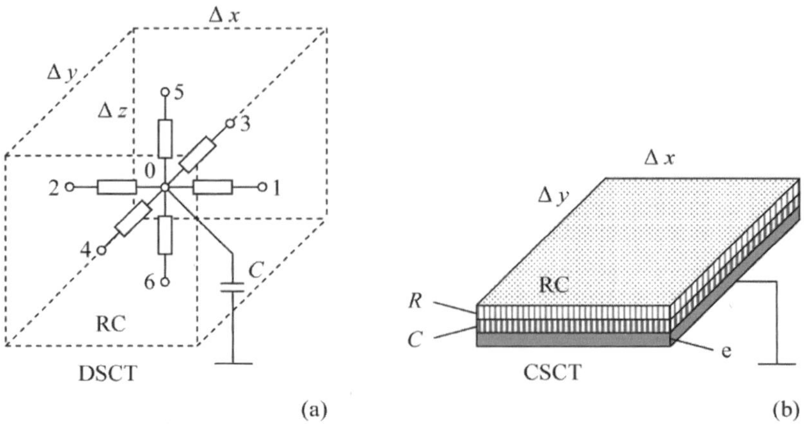

Fig. 7.10. Electrical CT analogues of the Fourier equation – three-dimensional (a) and planar continuous (b) cell of *RC* type.

where *R* and *C* are the resistance and the capacitance of the cell. These values can be determined from the equality of mutually corresponding Fourier numbers for the original and the model $Fo_O = Fo_M$.

(B) Heat conduction equation

$$\nabla^2 u = B(M)\frac{\partial u}{\partial \tau} + F(M),\ M(x,y,z). \tag{7.78}$$

The equation can be converted into a difference form and its analogous form can be found for some of the electrical analogues from Fig. 7.11.

CTDS analogues of RC type

In the case of a continuous-in-time analogue of *RC* type (Fig. 7.11a), one can obtain for one cell

$$\sum_{i=1}^{2n}\frac{U_i - U_0}{R} + \frac{V_0 - U_0}{R_0} + C\frac{\partial(0 - U_0)}{\partial \tau} = 0, \tag{7.79}$$

where

$$I_0 = \frac{U_0 - V_0}{R_0}. \tag{7.80}$$

CTCS analogues of RC type

For the continuous-in-space-and-time analogue of *RC* type of a planar

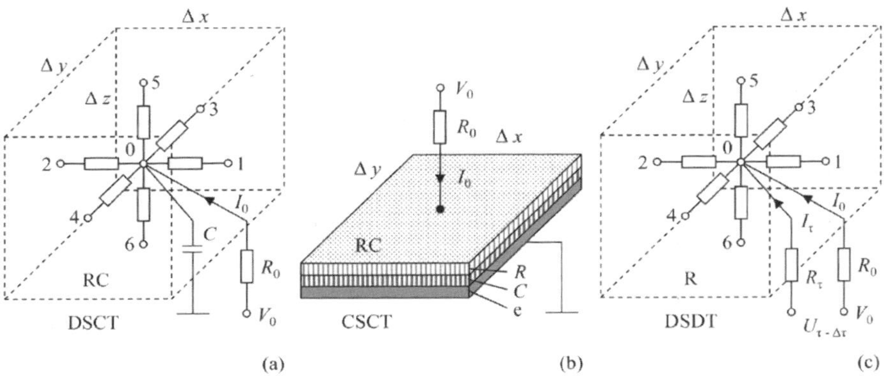

Fig. 7.11. Electrical analogues of the heat conduction equation – three-dimensional (a) and planar continuous (b) cell of *RC* type, and three-dimensional (c) cell of *R* type.

field with internal sources (Fig. 7.11b), the source of the current I_0 is connected into the centre of the planar cell.

DTDS analogues of R type

The following equation describes the discrete-in-time-and-space analogue of *R* type (Fig. 7.11c), the following

$$\sum_{i=1}^{2n} \frac{U_i - U_0}{R_i} = \frac{U_0 - V_0}{R_0} + \frac{U_0 - U_{\tau-\Delta\tau}}{R_\tau}, \tag{7.81}$$

where

$$I_0 = \frac{U_0 - V_0}{R_0}, \quad I_\tau = \frac{U_0 - U_{\tau-\Delta\tau}}{R_\tau}. \tag{7.82}$$

7.4.3 Analogues of Basic Boundary Conditions

Figue 7.12 shows diagrams of the electrical analogues of basic boundary conditions. The continuous boundary for the continuous-in-space analogues is divided into parts, called the electrodes, and the relevant voltage is led to each of them according to the boundary condition type. The width and distance of the electrodes must be chosen with respect to the nature of the task and the required accuracy. For the discrete-in-space analogues, the relevant voltage is led into the boundary nodal cells directly.

1st-kind, Dirichlet, boundary condition

The 1st-kind boundary condition is given by the known voltage

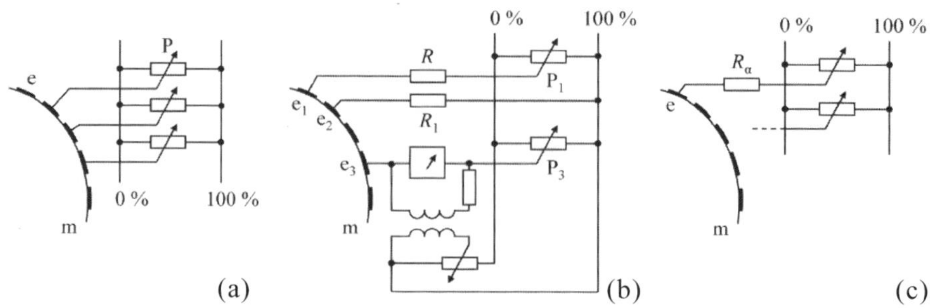

Fig. 7.12. Electrical analogues of basic boundary conditions – 1st-kind Dirichlet BC (a), 2nd-kind Neumann BC (b), and 3rd-kind Fourier BC (c).

function of the investigated region boundary (Fig. 7.12a). It can change continuously along the boundary. The continuous boundary is divided into parts and the relevant voltage from a potentiometer is connected to each of them. In a discrete-in-space network model, the voltage is connected into boundary nodal cells and in continuous-in-space analogues the voltage is connected to an electrode set. The potentiometers are connected in parallel to the source of 0 to 100 % of the voltage. To ensure the independency of the circuit, its resistance must be sufficiently small. Then the current flowing to the electrode can be neglected in comparison to the current flowing through the potentiometer.

2nd-kind, Neumann, boundary condition

The 2nd-kind boundary condition is given by electric current distribution along the model boundary. Analogously, the following current is flowing through each electrode on the boundary

$$I = \frac{h\,s}{\rho}\left(\frac{\mathrm{d}U}{\mathrm{d}n}\right)_{ave}, \tag{7.83}$$

where h is the model thickness, s is the equivalent electrode width corresponding to the distance of the electrodes, ρ is the specific conductor resistance, and $\left(\frac{\mathrm{d}U}{\mathrm{d}n}\right)_{ave}$ is the average gradient value in the normal direction. To set the relevant current value, usually one of three ways, given in Fig. 7.12b, is used. In the upper part, the method shown has a known resistance R connected between the cursor of the potentiometer P_3 and the electrode e_1 and the potentiometer is set gradually, so that the necessary voltage would be reached on the resistor. This is a very lengthy procedure because the voltages influence each other on the electrodes.

In the second method, a high-value resistor R_1 is used with the current set semi-automatically, but the voltage differences are relatively small on the model. The third method represents independent power supply by means of a circuit with a transformer. In principle, this is a zero-resistance ampere meter. Potentiometer P_3 is set to obtain the necessary current. The voltage on the electrode e_3 does not change during setting of other electrodes. With the 2nd-kind condition, the current value is determined from the equality of the Kirpitchev numbers for the object and the model.

3rd-kind, Fourier, boundary condition

The 3rd-kind boundary condition expresses the linear relation between the function U and its derivation in the normal direction

$$U - a\frac{\mathrm{d}U}{\mathrm{d}n} = b, \tag{7.84}$$

and therefore

$$U + \frac{a\rho}{hs} I = b, \tag{7.85}$$

where the dimensional constants a, b are given at each point on the area border.

The resistance value R_α mentioned in Fig. 7.12c is determined from the condition of the equality of the Biot number for the object and the model.

7.4.4 Analogues of Coupled Fields

It occurs often in physical field solutions that two or more dependent quantities appear in the process. For example, this is true for the heat transfer in a dispersion material where temperatures of both material components appear. Here, there is heat conduction in each material component and heat transfer on its boundaries. Similarly, the coupled processes represent the heat and mass transfer as it occurs, e.g. drying of material. In it, material temperature and humidity are of use, possibly also pressure if a rapid shock process occurs. The heat and thermal stress transfer in material is another very frequent case. These coupled processes can be expressed by a set of 2nd-order partial differential equations or by a single 4th-order partial differential equation in some cases.

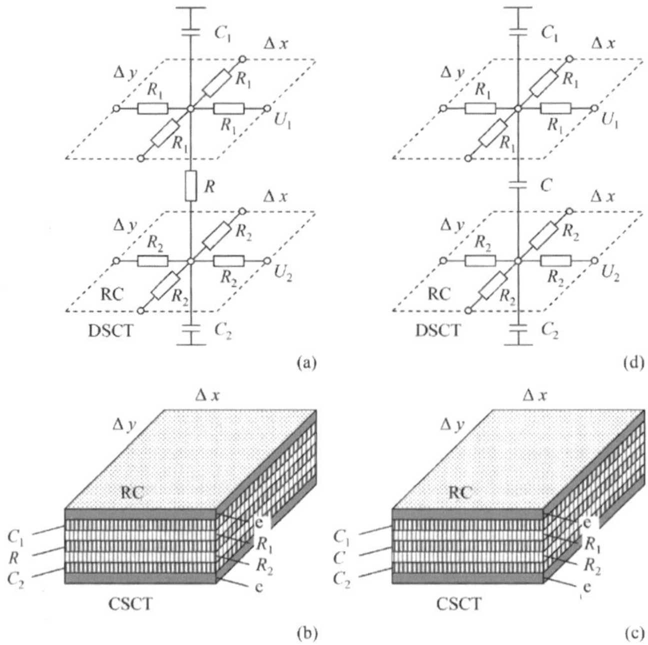

Fig. 7.13. Electrical DSCT (a, d) and CSCT (b, c) analogues of *RC* type to simulate the coupled heat transfer by conduction and convection (a, b) and the heat and mass transfer (c, d).

The following examples of coupled fields show a mathematical model and corresponding electrical analogue for coupled heat transfer processes in dispersion material (Fig. 7.13a,b) and heat and mass transfer (Fig. 7.13c,d). These analogues are expressed by means of discrete-in-space and continuous-in-time models (DSCT) and continuous-in-space-and-time models (CSCT).

Heat transfer in dispersion material

For two-component material, e.g. loose material and gas, the heat transfer process is described by the set

$$\nabla^2\Theta_1 + Bi_1\left(\Theta_2 - \Theta_1\right) = \frac{\partial\Theta_1}{\partial Fo_1}, \tag{7.86}$$

$$\nabla^2\Theta_2 + Bi_2\left(\Theta_2 - \Theta_1\right) = \frac{\partial\Theta_2}{\partial Fo_2}, \tag{7.87}$$

where the temperature of solid material is Θ_1 and that of gas is Θ_2, the Biot number related to the solid material is Bi_1 and that for gas Bi_2, the Fourier numbers are Fo_1, Fo_2, where

$$Fo_2 = \frac{Bi_2}{Bi_1} Fo_1. \tag{7.88}$$

At the beginning of an unsteady process in a dispersion material, the heat transfer is accelerated in the gas part, which acts as continuously distributed surroundings, and is decelerated in solid material particles acting as thermal sinks. However, the temperature field for higher *Fo* numbers is equalized in both phases and the process can be suitably described by a single equation with average values of physical quantities.

Analogously to Eqs. (7.86) and (7.87), a set of differential equations in the following dimensionless form can be written for an analogue of *RC* type

$$\frac{\partial \Phi_1}{\partial Fo_1} = \frac{\partial^2 \Phi_1}{\partial X_M^2} + Bi_{1M}\left(\Phi_2 - \Phi_1\right), \tag{7.89}$$

$$\frac{\partial \Phi_2}{\partial Fo_2} = \frac{\partial^2 \Phi_2}{\partial X_M^2} - Bi_{2M}\left(\Phi_2 - \Phi_1\right). \tag{7.90}$$

describing the dimensionless electrical potential distribution in an electric circuit with distributed (Fig. 7.13b) or concentrated (Fig. 7.13a) *R* and *C* parameters.

The *modeling scales* can be determined from the equality of mutually corresponding criteria of an object and a model. At the same time, with the cell being an assumed square ($\Delta x_M = \Delta y_M$), the following holds for, for example, the Fourier numbers of the model

$$Fo_{1M} = \frac{\tau_M \Delta x_M^2}{C_1\left(R_1 + R\right)L_M^2}, \qquad Fo_{2M} = \frac{Bi_2}{Bi_1} Fo_{1M}, \tag{7.91}$$

where the Biot electrical numbers B_{i1M}, B_{i2M}, are the electrical potentials corresponding to temperatures of the solid and gas phases Φ_1, Φ_2 and the following is valid for them

$$Bi_{1M} = \frac{R_1 L_M}{R_{\alpha 1}\Delta x}, \quad Bi_{2M} = \frac{R_2 L_M}{R_{\alpha 2}\Delta x}, \tag{7.92}$$

$$\Phi_1 = \frac{U_1}{U_{ref}}, \; \Phi_2 = \frac{U_2}{U_{ref}}. \tag{7.93}$$

The designation of the electrical resistances (R_1, R_2, R), capacitances (C_1, C_2), and dimensions (Δx, Δy) of the cell corresponds to the analogue of *RC* type in Fig. 7.13a. The heat transfer through solid material (1) and the gas (2) is interconnected mutually by the resistance *R*, which represents the transient thermal resistance between solid particles of the loose material and the gas.

Heat and mass transfer

With both fields influencing each other, the simultaneous heat and mass transfer in the material being considered, provided the mass (steam, water, ice) is transferred by molecular diffusion or effusion, is described by the following equation set according to [17]

$$\nabla^2\Theta + Ko^* \frac{\partial U}{\partial Fo} = \frac{\partial \Theta}{\partial Fo}, \tag{7.94}$$

$$Lu\nabla^2 U + Lu\, Pn\nabla^2\Theta = \frac{\partial U}{\partial Fo}, \tag{7.95}$$

where temperature Θ, humidity transfer potential (humidity content) U, Lykov number Lu, Posnov number Pn, Kossovitch modified number Ko^*, and Fourier number Fo are dimensionless quantities.

The equation set in the form of (7.94), (7.95) describes the heat and mass (humidity) transfer in, for example, a material during drying or hardening. With heat and mass transfer being in progress under the action of pressure mass filtration (molar flow), the set of Eqs. (7.94), (7.95) would extend to three equations, in which the pressure filtration potential (filtering pressure) would be the third dependent variable.

The equation set for the electric analogue of *RC* type is analogous to the set of Eqs. (7.94), (7.95),

$$\frac{\partial \Phi_1}{\partial Fo_M} = \nabla^2\Phi_1 + Ko_M \frac{\partial \Phi_2}{\partial Fo_M}, \tag{7.96}$$

$$\frac{\partial \Phi_2}{\partial Fo_M} = Lu_M\nabla^2\Phi_2 + Lu_M Pn_M \nabla^2\Phi_1, \tag{7.97}$$

which describes the dimensionless electrical potential distribution in an electrical circuit with distributed (Fig. 7.13c) or concentrated (Fig. 7.13d) parameters *R* and C. Heat transfer (1) and the humidity content (2) are interconnected by capacitor *C*.

Similarly, the following can be written for boundary conditions

$$\nabla\Phi_1 - Bi_{t,M}(1-\Phi_1) - (1-\varepsilon_M)Ko_M\, Lu_M\, Bi_M(1-\Phi_2) = 0, \tag{7.98}$$

$$\nabla\Phi_2 + Pn_M\nabla\Phi_1 - Bi_{m,M}\left(1-\Phi_1\right)=0\,, \tag{7.99}$$

$$\nabla\Phi_1 = -Ki_{t,M}\,, \tag{7.100}$$

$$\nabla\Phi_2 + Pn_M\nabla\Phi_1 = -Ki_{m,M}\,. \tag{7.101}$$

Similarly, one can write for the initial condition

$$\Phi_1\left(X_M,0\right)=\Phi_{1,ini}\left(X_M\right), \quad \Phi_2\left(X_M,0\right)=\Phi_{2,ini}\left(X_M\right). \tag{7.102}$$

The following also holds

$$0 \le X_M < 1, \quad 0 < Fo_M < \infty. \tag{7.103}$$

The electric potentials Φ_1, Φ_2, of which Φ_1 corresponds to temperature and Φ_2 to the humidity content, the Fourier electrical number Fo_M, the Lykov electrical number Lu_M, the Kossovitch electrical number Ko_M, the Posnov electrical number Pn_M, the Biot electrical numbers for a thermal process $Bi_{t,M}$ and for a mass process $Bi_{m,M}$, the Kirpitchev electrical numbers for a thermal process $Ki_{t,M}$ and for a mass process $Ki_{m,M}$, and the spacial model coordinate X_M are dimensionless quantities.

The modeling scales can be determined from the equality of the mutually corresponding object and model quantities.

7.4.5 Analogues of Wave-Diffusion Fields

Simultaneous action of heat transfer diffusion and wave processes appears primarily when various intensive shock energy sources are acting on a material. In this case a wave-diffusion equation is the mathematical model. The great significance of the wave-diffusion process solution is related to the introduction of new technologies which utilize intensive heat sources such as laser or electron beams. This involves very complicated processes in which the influence of relaxation and thermalization times of material apply. Figure 7.14 illustrates the regions of diffusion and wave-diffusion heat propagation for surface heat flux thermal loading and for various thermal systems in non-technical systems and technologies.

The following electrical analogues for wave-diffusion heat propagation utilize various types of *RCL* cells where discrete-in-space and continuous-in-time (DSCT) models are used.

The unsteady physical fields in which diffusion, wave-diffusion, and wave propagation take place are described by the *extended complete*

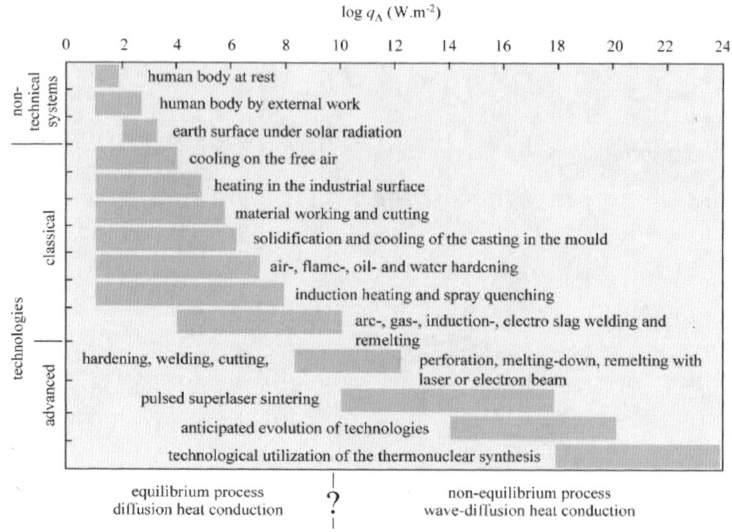

Fig. 7.14. Classification of equilibrium and non-equilibrium processes according to surface heat flux loading in various non-technical systems and technologies.

wave equation,which is among the group of hyperbolic equations. For the equation, the following is valid

$$\nabla^2 u = B(M)\frac{\partial u}{\partial \tau} + C(M)\frac{\partial^2 u}{\partial \tau^2} + F(M),\ M(x,y,z). \qquad (7.104)$$

The wave-diffusion heat propagation model originates by extending the cell of the heat conduction equation, i.e. the diffusion heat propagation model, by a coil inserted in series to the resistor, as in Fig. 7.15. The equation describing the voltage in network nodes of this model has the following form

$$U_{i-1} - 2U_i + U_{i+1} = R_L C_0 \frac{\mathrm{d}U_i}{\mathrm{d}\tau} + LC_0 \frac{\mathrm{d}^2 U_i}{\mathrm{d}\tau^2}. \qquad (7.105)$$

This model describes the relaxation phenomena in the heat propagation.

The electric circuit dimensionless criteria can be defined by an analogy to the wave-diffusion heat propagation process. The Fourier electrical number can be defined by the relation

$$Fo_{el} = \frac{1}{R_L C_0}\frac{\tau}{n^2}, \qquad (7.106)$$

and similarly the Fourier electrical relaxation number by the relation

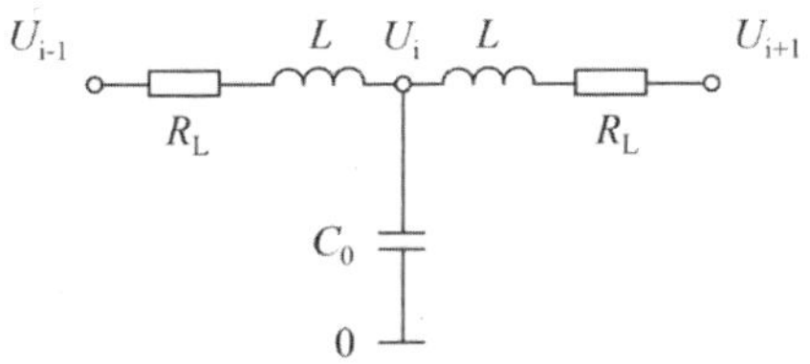

Fig. 7.15. Basic cell of the electrical analogue of *RCL* type for wave-diffusion heat propagation.

$$Fo_{q,el} = \frac{L}{R_L^2 C_0} \frac{1}{n^2}, \tag{7.107}$$

where n is the number of cells of the electrical analogue for wave-diffusion heat propagation. In this analogue, the analogy conditions have been fulfilled exactly and, in addition, the coil self-resistance problem is eliminated because the resistance value can be included in that of the total resistance R_L in the analogue parameter design.

The internal heat source can be modelled by electric current led into a relevant network node, due to which Eq. (7.105) is extended as follows

$$\left(U_{i-1} - 2U_i + U_{i+1}\right) + R_L I_V + L\frac{\mathrm{d}I_V}{\mathrm{d}\tau} = R_L C_0 \frac{\mathrm{d}U_i}{\mathrm{d}\tau} + L C_0 \frac{\mathrm{d}^2 U_i}{\mathrm{d}\tau^2} \tag{7.108}$$

and the dimensionless inner source criterion, the Pomerantsev electrical number, is defined as

$$Po_{el} = \frac{R_L n^2}{\Delta U_{ref}} I_V. \tag{7.109}$$

If necessary, the influence of material thermalization time can also be included in the model by inserting resistance R in parallel to the coil into the basic cell, due to which a general wave-diffusion heat propagation model is created as shown in Fig. 7.16.

Due to this, Eq. (7.108) will be extended to the form

$$\begin{aligned}&\left(U_{i-1} 2U_i + U_{i+1}\right) + \frac{L}{R R_{rel}} \frac{\mathrm{d}}{\mathrm{d}\tau}\left(U_{i-1} - 2U_i + U_{i+1}\right) + \\ &+ \frac{R_L}{R_{rel}} I_V + \frac{L}{R_{rel}} \frac{\mathrm{d}I_V}{\mathrm{d}\tau} = \frac{R_L C_0}{R_{rel}} \frac{\mathrm{d}U_i}{\mathrm{d}\tau} + \frac{L C_0}{R_{rel}} \frac{\mathrm{d}^2 U_i}{\mathrm{d}\tau^2}\end{aligned} \tag{7.110}$$

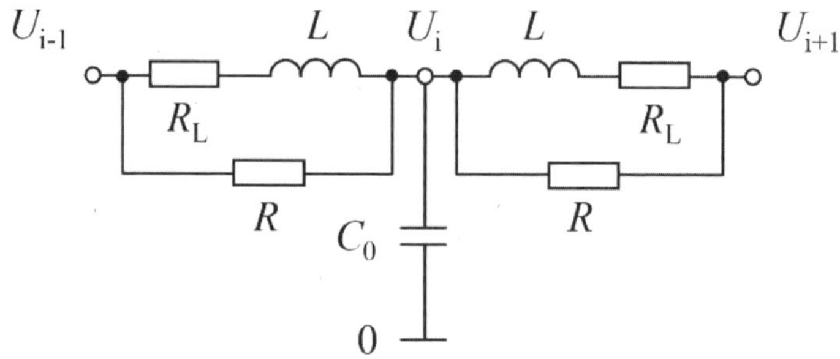

Fig. 7.16. General cell of the electrical analogue of *RCL* type for wave-diffusion heat propagation.

and for R_{rel}, the following holds

$$R_{rel} = 1 + \frac{R_L}{R}. \tag{7.111}$$

The dimensionless criteria are defined as follows

$$Fo_{el} = \frac{1}{R_L C_0} R_{rel} \frac{\tau}{n^2}, \tag{7.112}$$

$$Fo_{q,el} = \frac{L}{R_L^2 C_0} R_{rel} \frac{1}{n^2}, \tag{7.113}$$

$$Fo_{T,el} = \frac{L}{R R_L C_0} \frac{1}{n^2}, \tag{7.114}$$

$$Po_{el} = \frac{R_L n^2}{\Delta U_{ref}} R_{rel} I_V, \tag{7.115}$$

where the dimensionless criterion of the Fourier electrical thermalization number $Fo_{T,el}$ is defined for the first time. The conditions for the analogy with wave-diffusion heat propagation are always fulfilled and the wave-diffusion heat propagation condition is valid

$$K_{\tau,el} = \frac{Fo_{T,el}}{Fo_{q,el}} < 1, \tag{7.116}$$

where $K_{\tau,el}$ is the number of non-equilibrium heat propagation.

In the electrical analogues of *RCL* type mentioned previously, the analogy with heat conduction is not negatively affected and, therefore, these models seem to be suitable from the theoretical point of view. In addition, the analogy with the conduction in the material is obvious. Of course, no influence of real parameters of electrical components used in electric circuits is implied from the theoretical description. In addition, the series coil self-resistance which is involved in the

R_L value, especially the leakage resistance of capacitors or parasite capacitances of wires, can act unfavourably. The proposed values, especially those of capacitors or coils, can be outside the limits of produced components and it might appear in such cases that these models, despite of all their advantages, can be implemented only with difficulty and other *RCL,* models, as given in, for exampe [5.11], should be used instead.

7.4.6 Direct Modeling Method for Gradient Fields

The direct modeling of gradient fields concerns the electrical *CS* analogues of *R*-type for steady physical fields that are described by the Laplace equation, i.e. the physical fields without inner sources or sinks.

In many various problems described by the Laplace equation, it is far more important to determine directly the gradients of functions (e.g. gradients of temperature, flow velocity, shear stress) than the harmonic functions (temperature, hydrodynamic or stress potential). The direct modeling method exploits those properties of harmonic functions which allow equigradient fields to be modeled by means of electrical equipotential fields. The difficult process of measuring the gradients is thus replaced by a usually simple one of determining the electrical potential on the model. In this way, the gradient measurement accuracy which corresponds to the measurement accuracy of electric voltage, i.e. by about one order higher, can be reached.

Gradients of harmonic functions

The solution of Laplace's differential equation, expressed in the form

$$\nabla^2\rho(x,y)=0, \tag{7.117}$$

yields a harmonic function $\varphi(x,y)$. If the object of study is the flow of electric current, this function corresponds to the electric potential $V(x,y)$; in the case of heat transfer it corresponds to temperature $T(x,y)$. As the function $\varphi(x,y)$ is a scalar quantity, its derivation results in a gradient, i.e. in a vector defined by its modulus (length of a vector) and argument (plane angle between a vector and positive x-axis). Accordingly, the field of gradients in a certain region can be determined only after two other fields have been established for that region – equimodulus and equiargument fields.

The complex variable z comprises a real and an imaginary part, where i is the imaginary unit and x, y are real variables. Equation (7.118) is the algebraic form of a complex variable. The goniometric and exponential forms of a complex variable are in Eqs. (7.119) and (7.120)

$$z = x + iy, \tag{7.118}$$

$$z = r\,(\cos\alpha + i\sin\alpha), \tag{7.119}$$

$$z = re^{i\alpha}. \tag{7.120}$$

The value $r = |z|$ is the *modulus of the complex variable* and α is the *argument of the complex variable.*

The natural logarithm of a complex variable gives

$$\ln z = \ln r + i\,(\alpha + 2\pi n), \tag{7.121}$$

where n is a natural number. The $\alpha \in (-\pi, \pi)$ is the argument of the main logarithm value.

An analytic function is any function that is equal to its Taylor series in its domain. It implies that the analytic function is actually infinitely differentiable. There exist both real analytic functions and complex analytic functions. The complex function

$$f(z) = \varphi(x, y) + i\psi(x, y), \tag{7.122}$$

is the analytic complex function of a complex variable if and only if the partial derivatives of real functions $\varphi(x,y)$ and $\psi(x,y)$ of two real variables satisfy the Cauchy-Riemann equations

$$\frac{\partial\varphi}{\partial x}(x, y) = \frac{\partial\psi}{\partial y}(x, y), \tag{7.123}$$

$$\frac{\partial\varphi}{\partial y}(x, y) = -\frac{\partial\psi}{\partial x}(x, y). \tag{7.124}$$

The real and imaginary parts of any analytic complex function yield harmonic funtions, these are said to be a pair of conjugate harmonic functions.

The derivative of an analytic complex function is again an analytic complex function, and so is the logarithm of that derivative. Therefore, both the real and imaginary part of this logarithm are harmonic functions. *The point is that all harmonic functions can be modeled utilizing the direct modeling method for gradient fields.*

Let us consider the analytic complex function $f(z)$, see Eq. (7.122). The derivative of this function remains an analytic complex function

$$f'(z) = e^{\Lambda(x, y) + i\,\Omega(x, y)}, \tag{7.125}$$

where $\Lambda(x,y)$ and $\Omega(x,y)$ are real functions of real variables x, y.

According to Eq. (7.121), the natural logarithm of this derivative gives

$$\ln f'(z)=\Lambda(x,y)+i\,\Omega(x,y), \tag{7.126}$$

that is an analytic function too; consequently, $\Lambda(x, y)$ and $\Omega(x, y)$ are conjugate harmonic functions.

The gradient function $f'(z)$ is then, according to Eq. (7.125), defined by its own modulus $e^{\Lambda(x,y)}$ and argument $\Omega(x, y)$. The relationship between the gradient of the analytic function $f(z)$ and the gradient of the harmonic function $\varphi(x,y)$ can be determined from the theory explaining the functions of a complex variable. According to this relationship, the following is valid

$$\operatorname{mod}\left[\operatorname{grad}\varphi(x,y)\right]=\operatorname{mod}\left[\operatorname{grad} f(z)\right]=e^{\Lambda(x,y)}, \tag{7.127}$$

$$\arg\left[\operatorname{grad}\varphi(x,y)\right]=-\arg\left[\operatorname{grad} f(z)\right]=-\Omega(x,y). \tag{7.128}$$

The last two equations may be rewritten in the form

$$\Lambda(x,y)=\ln\left\{\operatorname{mod}\left[\operatorname{grad}\varphi(x,y)\right]\right\}, \tag{7.129}$$

$$\Omega(x,y)=-\arg\left[\operatorname{grad}\varphi(x,y)\right]. \tag{7.130}$$

The above equations point to the following conclusions. Given the values that the gradient of the function $\varphi(x, y)$ assumes along the border of an enclosed region $S(x,$ y), the boundary conditions of the functions $\Lambda(x, y)$ and $\Omega(x, y)$ for this region can be found from Eqs. (7.127) and (7.128). These functions can then be modeled in any of the usual ways. If the values of functions $\Lambda(x, y)$ and $\Omega(x, y)$ are determinable on the model and are known within the enclosed region $S(x, y)$, then the gradient modulus mod [grad $\varphi(x, y)$] and gradient argument arg [grad $\varphi(x, y)$] fields can be determined from Eqs. (7.129) and (7.130).

Direct modeling of the gradients of function $\varphi(x, y)$ is only possible if the values of the gradient along contour S (the border of the examined region) are known in advance. This condition is fulfilled if the relevant boundary conditions are of the second kind, but not if they are of the first or third kind. In either of the latter cases, the gradient values along the border line must be determined experimentally The resultant gradient of function $\varphi(x, y)$ on the border of the examined region is perpendicular to the equipotential line and is defined by its own modulus and argument.

It follows from the conditions of electrical analogy that if a harmonic function $\varphi(x, y)$ is to be expressed on the model, an electrical potential must be assigned to it according to the equation

$$V = \frac{\varphi - \varphi_{min}}{\varphi_{max} - \varphi_{min}} 100\,(\%), \tag{7.131}$$

where φ_{min} and φ_{max} are the minimum and maximum values of harmonic function $\varphi(x,y)$.

All the data on the temperature gradients determined from electrical analogue are relative values. The absolute values may be obtained from the equation

$$\text{grad}T = \text{grad}\,\varphi \frac{T_{max} - T_{\min}}{100}, \tag{7.132}$$

where T_{max} and T_{min} are temperatures of the gas and cooling medium, and T is the temperature at the examined point of the blade.

To give a better idea, an example of the practical solution of temperature gradient modulus and argument fields in a turbine blade with internal air cooling is shown in Fig. 7.17. At every blade point, both fields enable one to determine not only the temperature gradient vector but also the local heat flux in the blade, and to give an idea about the nature of the thermal loading of the blade. These fields point to the presence of considerable thermal stress in the blade material, particularly in all the leading- and trailing-edge regions of the profile adjacent to the cooling channel.

A more detailed description of the direct modeling method for gradient fields and applications in various areas such as thermo-mechanics, aerodynamics and elasticity are presented in [7.10] and [7.12], where the problem of the argument function discontinuity

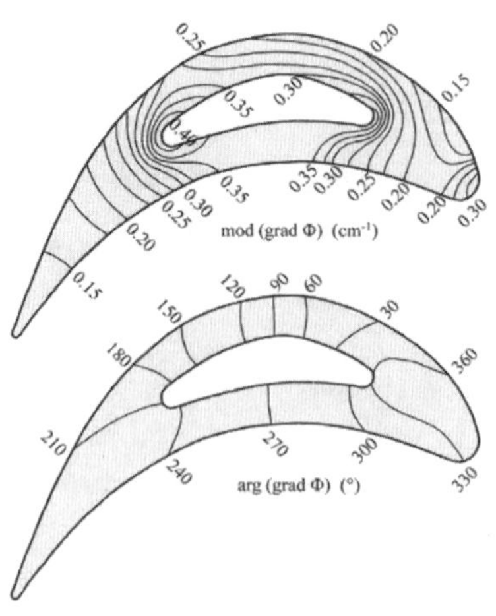

Fig. 7.17. Temperature gradient modulus and argument fields in an air-cooled blade of a gas turbine.

is solved as well (the so-called logarithmic singularity problem) A practical application of this method in aerodynamics is presented in Example 7.2

7.5 Applications

All the presented examples illustrate various ways of physical analogues utilization in different areas of mechanics of solid bodies, mechanics of fluids and thermomechanics.

7.5.1 Mechanics of Solid Bodies

The electrical analogue of a dynamic mechanical circuit is designed in the following example.

Example 7.1 *Dynamics of systems. Analogue for a mechanical differential*

An electrical analogue for a mechanical rotating system, with two degrees of freedom, as in Fig. 7.18, will be designed. Three rotating shafts are joined by a differential so that shaft 3 can rotate at a velocity which is k-times greater than the difference of angular velocities of shafts 1 and 2. The corresponding velocities of individual shafts are ω_1, ω_2, ω_3. The moments of inertia of the shafts and of the differential bevel gears set on them are I_1, I_2, I_3. The outer turning moments on the shafts are denoted as M_1, M_2 a M_3.

As a result of mechanical system analysis, the mathematical model can be composed in the following form

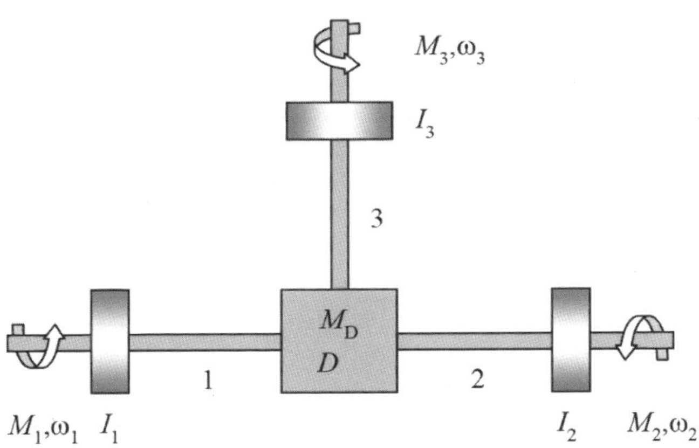

Fig. 7.18. A mechanical gear system consisting of three shafts joined by a differential,

$$\omega_3 = k(\omega_1 - \omega_2), \tag{7.133}$$

$$I_1 \frac{\mathrm{d}\omega_1}{\mathrm{d}\tau} = M_1 + M_D, \tag{7.134}$$

$$I_2 \frac{\mathrm{d}\omega_2}{\mathrm{d}\tau} = M_2 + M_D, \tag{7.135}$$

$$I_3 \frac{\mathrm{d}\omega_3}{\mathrm{d}\tau} = M_3 + \frac{M_D}{k}, \tag{7.136}$$

where M_D equals the equivalent loading moment of corresponding shafts on the differential side.

In solving this, the operator expression of the equations can be used advantageously and an equivalent diagram of the mechanical circuit can be constructed. In each block, the inputs and outputs which express the transmission are examined. Angular velocity ω_1 is considered as the basic variable in the first block, and with Eq. (7.134) being used, the following is valid for it

$$\omega_1 \frac{1}{I_1 p}(M_1 + M_D). \tag{7.137}$$

Here M_1 and M_D are the input quantities and ω_1 is the output quantity. The second block is described by Eq. (7.133), in which ω_1 and ω_2 are the input quantities and ω_3 is the output quantity. With Eq. (7.136) modified, the following holds

$$M_D = k(\omega_3 p I_3 - M_3). \tag{7.138}$$

The remaining Eq. (7.135), after being rewritten in the form

$$\omega_2 = \frac{1}{I_2 p}(M_2 + M_D), \tag{7.139}$$

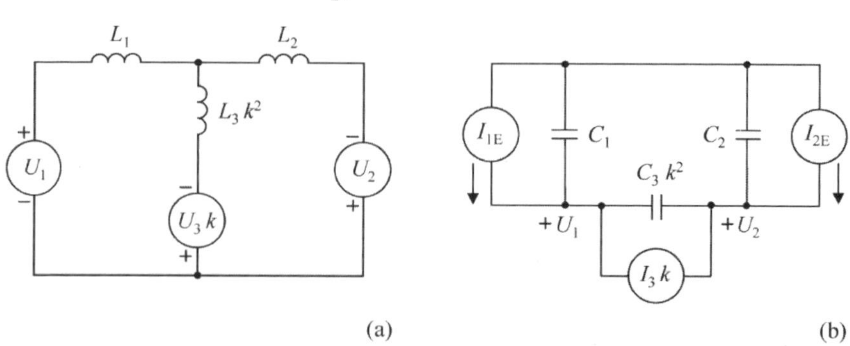

Fig. 7.19. Electric voltage (a) and current (b) analogues for a mechanical circuit with a differential.

describes the last block, the output of which is ω_2.

By replacing mechanical quantities with analogous electrical ones, the diagrams for voltage and current analogies can be obtained according to the selected analogy. The synthesis of corresponding electrical circuits follows, which can be seen in Fig. 7.19 for both types of analogies. In designing the analogue, the reference voltage polarity and reference current direction to one input must be stated and the signs thereof must be compared.

7.5.2 Mechanics of Fluids

The planar potential flow of an ideal fluid through a blade cascade is solved using an electrical analogue. The comparison of results with those obtained from aerodynamic tunnel measurement are introduced.

Example 7.2 *Aerodynamics. Modeling of potential flow through a blade cascade in an electrical analogue and a physical model*

The planar potential flow of an ideal incompressible fluid through a compressor blade cascade is considered for the given geometrical cascade parameters. The velocity distribution on the blade profile surface will be determined and the measured results in an electrical analogue will be compared with those obtained for the same blade cascade in an aerodynamic tunnel with the input flow Mach number $Ma_1 = 0.39$. The vector velocity field in a blade channel will be modeled using the results of the solution of the surface velocity with an electrical analogue. The solution with an electrical analogue involves two fundamental tasks:

1. the *indirect task* in which a set of complex conjugated equipotentials and streamlines is modeled for a given input velocity and incidence angle. From the gradient thereof on a blade profile surface, the flow velocity is determined, to which the current density corresponds in the electrical analogue.

2. the *direct task* in which the vector velocity field in a blade channel is determined from the electric current density representing a 2nd-kind boundary condition. At the same time, the harmonic function properties and the so-called logarithmic potential can be utilized to direct modeling of the fields of the velocity gradient modulus and velocity gradient argument. The accuracy corresponding to that of the voltage measurements can be reached as a result of modeling of these fields with the electric potential, .

The electrical model consists of a thin-layer resistance R analogue. More detailed information can be found in [2.6] and [7.10].

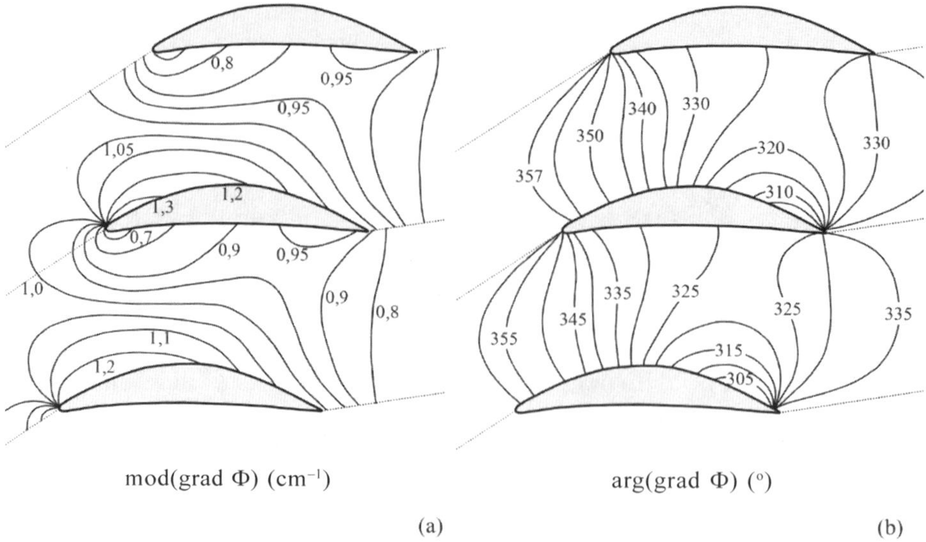

Fig. 7.20. Results of modeling of the vector velocity field in a compressor blade cascade mod(grad Φ) (a), arg(grad Φ) (b).

The modeling results for the flow velocity vector field in a compressor blade cascade channel are in Fig. 7.20. They are expressed by isolines of the constant modulus (a) and argument (b) of the hydrodynamic potential Φ or the constant value of the relative velocity W (a) and constant angle (b). With this, not only the size but also the direction of the flow velocity is determined in an arbitrary place on the blade channel.

To verify the credibility of results obtained in an electrical model, the same blade cascade is measured in an aerodynamic tunnel with the input air velocity expressed by the Mach number $Ma_I = 0.39$. In the experiment, various input angle alternatives from 64° to 42° are considered which are of optimum value. For this alternative, the velocity profile on the blade surface measured on the electrical model is compared in Fig. 7.21 with that measured on a physical model of the blade cascade in an aerodynamic tunnel. The results of mutual comparison show a relatively good agreement, with the fact that the more distinct deviation is on the overpressure blade side and is obviously related to not considering the air compressibility.

7.5.3 Thermomechanics

The following example solves temperature asymmetry in a turbo-alternator using an electrical analogue. The procedure of an electric model design of an coupled heat and mass transfer is outlined in the

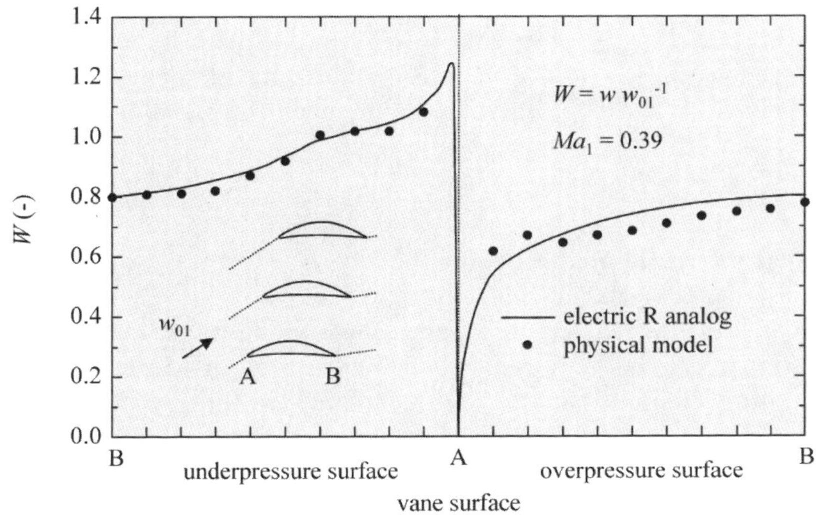

Fig. 7.21. Comparison of the velocity profiles on the compressor blade surface from measurements in the electrical analogue and the physical model in an aerodynamic tunnel.

next example. The last example utilizes an electrical and computational models for modeling of nonequilibrium heat propagation.

Example 7.3 *Temperature asymmetry in a turbo-alternator rotor*

Object of the solution	The temperature asymmetry in a turbo-alternator rotor has to be found and its influence od the dynamic instability has to be explained. The internal hydrogen cooling of conductors and the steady operational mode are considered.
Task characterization	The solution of a quasi-stationary task with irregularly distributed internal heat sources and asymmetric boundary conditions distribution. The task is characterized by considerable breaking up of the zone. A high requirement to define the temperature field precisely and non-uniformities thereof is the specific property of the task. The temperature asymmetry of 1 K between opposite surface locations appears to be able to cause vibrations of 200 approximately in the case of big rotors. The 2D-task and the non-variable temperature field in the rotor axis direction are assumed.

Applied method	With respect to the task characterization, zone breaking up, and especially the relatively high local temperature gradient, the electrical continuous-in-space (CS) analogue of R type has been chosen.

Physical principle of the problem

There are several reasons of a generator asymmetric temperature field. Essentially, all of them can be divided into four groups. These are the cooling channel function, current non-loading of the conductor, technological reasons, and different contact thermal resistances. The model solution involves 21 alternatives, some of which correspond to experiments executed on a test rotor. In this example, only one alternative is considered, with one cooling channel plugged.

The symmetric field corresponding to the nominal rotor operating mode is the fundamental relative temperature field. In it, internal heat sources with 450 $W \cdot m^{-1}$, related to 1 m of the conductor length and caused by the excitation full-load current, are acting in the conductors. The surface heat sources are acting at the intensity of 10^4 $W \cdot m^{-2}$ on the main poles and with 10^3 $W \cdot m^{-2}$ on teeth and wedges. All sources are held as mutually independent. In the solution, the possibility of superimposing the temperature fields is utilized.

Mathematical model

In the transverse cross-section, the temperature distribution is represented by a temperature field with 2nd to 4th-kind boundary conditions. In the steady state, this field is described by the following equations in dimensional and dimensionless states

$$\nabla^2 T(x,y) + \frac{q_V}{\lambda} = 0, \quad \nabla^2 \Theta(X,Y) + Po = 0, \tag{7.140}$$

and the following holds for the 4th-kind boundary condition (perfect contact of different-conductivity bodies),

$$\lambda_1 \left(\frac{\partial T}{\partial n} \right)_1 = \lambda_2 \left(\frac{\partial T}{\partial n} \right)_2, \quad Ki_1 = Ki_2. \tag{7.141}$$

The 3rd-kind boundary condition represents heat transfer by convection in cooling channels and in the gas gap

$$-\lambda\left(\frac{\partial T}{\partial n}\right)=\alpha\left(T_S-T_a\right),\ -\frac{\partial\Theta}{\partial N}=Bi\left(\Theta_S-\Theta_a\right), \tag{7.142}$$

and the 2nd-kind boundary condition primarily represents the rotor surface heat losses

$$q_A=-\lambda\frac{\partial T}{\partial n},\ K_i=-\frac{\partial\Theta}{\partial N}. \tag{7.143}$$

In the preceding equations there are *dimensional quantities:* T (K) – temperature, x, y (m) – coordinates, q_V (W m^{-3}) – volume heat source, q_A (W m^{-2}) – surface heat source, λ (W m^{-1} K^{-1}) – thermal conductivity, n (m) – normal to the surface, T_S, T_a (K) – surface and environmental (ambient) temperatures, α (W m^{-2} K^{-1}) – heat transfer coefficient; and *dimensionless quantities:* Θ (-) – temperature, X, Y (-) – coordinates, Po (-) – Pomerantsev number, Ki (-) – Kirpitchev number, Bi (-) – Biot number, N (-) – normal to the surface.

The equations, analogous to the preceeding ones for the thermal process, can also be used for an electrical model and the modeling scales can be determined from the similarity criteria.

Electrical model (Fig. 7.22)

An electrical model of R type has been chosen, as represented by a resistance layer deposited on a support pad. The layer resistance is 3.2 kΩ, the layer anisotropy and non-homogeneity is within 0.9 to 3 %, the length scale is $\mu_L = 1$, and the scale of the electric to temperature potentials ratio is $\mu_T = 10^{-1}$ V K^{-1}. The model electric current corresponds to the heat length source in a conductor $q_L = 450$ W m^{-1}. The model electric current to the cell has been determined from the model and object equality of Pomerantsev numbers

$$I=q_V V_C \mu_T \mu_\lambda \mu_L^{-2}, \tag{7.144}$$

where V_C (m^3) is the volume of the model cell, μ_λ is the scale of the electric and temperature conductivity $\mu_\lambda = \gamma\,\lambda^{-1}$ and γ (S m^{-1}) is the electric conductivity of the model.

Similarly, for the electric current, expressing the surface heat source analogicaly, the following holds from the model and object equality of Kirpitchev numbers

$$I=q_A S_C \mu_T \mu_\lambda \mu_L^{-2}, \tag{7.145}$$

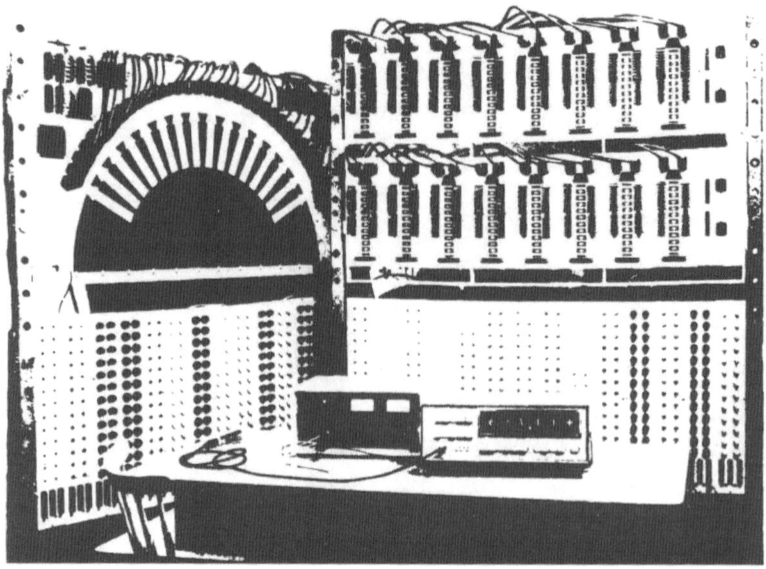

Fig. 7.22. Electrical continuous-in-space (CS) model of R type for solving temperature asymmetry in a transversal cross-section of a turbo-alternator rotor.

where qA (W m^{-2}) is the heat source expressing the surface losses and S_c (m^2) is the surface of the surface model cell.

The preceeding relation and scales enable one to design a model, but it is relatively difficult to realize. However, new technological procedures can e utilized for this, above all in creating a quality surface resistance layer for the model and measurement automatization. The mode function is verified on a task with symmetrical boundary conditions. The deviations of measured values on symmetrically located places over the whole model area are less than 0.1 K.

Solution results

Figure 7.23 shows the results for the temperature field model in a 200 MW turbo-alternator rotor under unsymmetrical conditions caused by the fact that the first conductor in a groove is not cooled. Other conductors are cooled and loading by surface sources on poles, teeth and wedge is not considered in this case. Since the conductor is not cooled, the local temperature rises by up to 25 K and the temperature in neighbouring conductors also increases. However, the maximum temperature difference also rises on the rotor diameter. This can increase if the first conductor is not cooled, depending on the kind of loading by surface forces, approximately 0.5 to 2.5 K.

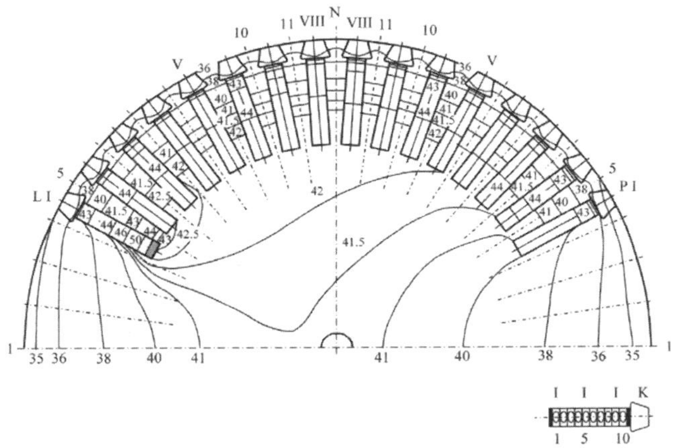

Fig. 7.23. Temperature field in a turbo-alternator rotor under symmetrical conditions and with a non-cooled first conductor in a groove.

The rotor deformation, caused by the non-uniform temperature field in the cross section, can be determined approximately from the following relation

$$\delta = k \iint_A \Delta T(x, y)\, x \,\mathrm{d}A, \tag{7.146}$$

where A is the rotor body section area, $\Delta T(x, y)$ is the temperature increment in a location with the coordinates x, y, and k is the constant which expresses the reduction of the rotor bending resistance due to slits.

The vibration amplitude due to rotor temperature asymmetry has been solved by making use of an analogue solution. The amplitude on the rotor bearing can reach values of 500 to 1000 μm, which exceeds the admissible vibration limit. To retain the vibration amplitude, the maximum difference between the left and right side of the rotor cross-section must be in the range of several tenths of Kelvin degree only. It should also be noted that the temperature asymmetry is only one of the reasons which influence the rotor dynamics.

Example 7.4 *Thermomechanics. Abridged procedure to design an electric model for coupled heat and mass transfer*

A limitless plate with finite thickness $2L$ is exposed to step temperature and humidity changes on the surface. The thermophysical properties of the plate and surroundings are expressed by dimensionless similarity criteria of the Lykov number $Lu = 0.48$, Kossovitch number $Ko = 0.10$, and Posnov number $Pn = 0.28$.

To solve this, a *continuous CSCT analogue of the RC type* is used (Fig. 7.13c). The equality of corresponding criteria for the model and the object is used to determine the electrical analogue parameters, for example, from the condition $Lu_O = Lu_M$, the basic model parameters can be obtained after inserting into the following expression

$$R_1 C_1 \left[R_2 C_2 \left(1 - \frac{C^2}{C_1 C_2} \right) \right]^{-1} = 0.45. \tag{7.147}$$

With the capacities C, C_1 and C_2 being chosen, the resistances R_1 and R_2 are determined and the following conditions must be fulfilled. From the condition $Ko_O = Ko_M$ it follows that

$$\frac{C}{C_A} \frac{U_1}{U_2} = 0.10, \tag{7.148}$$

and the voltage ratio $U_1 U_2^{-1} = 0.2$ is obtained. From the condition $Pn_O = Pn_M$, one obtains

$$\frac{R_2 C}{R_1 C_A} \frac{U_1}{U_2} = 0.28. \tag{7.149}$$

From the condition $Fo_O = Fo_M$, the following time scale is obtained

$$\mu_\tau = \frac{\tau_M}{\tau_D} = \frac{a R_1 C_A}{\Delta x_M^2} \left(\frac{L_M}{L_D} \right)^2 . \tag{7.150}$$

The solution results are presented in [1.7]. Some details about heat and mass transfer simulation in electrical models can be found in the book [7.8]

Example 7.5 *Thermomechanics of non-equilibrium heat propagation. Design of an electric and computer RCL model – a simulator*

The possibility of realizing an electrical *RCL* model, forming a part of a single-purpose hybrid system, and the properties thereof to solve the non-equilibrium wave-diffusion heat propagation, are to be verified. Further, a computer analogy for an *RCL* analogue is to be designed and mutual properties of both models compared.

Electrical RCL model

A universal electrical model for wave-diffusion non-equilibrium heat propagation has been designed as a ten-cell model in which

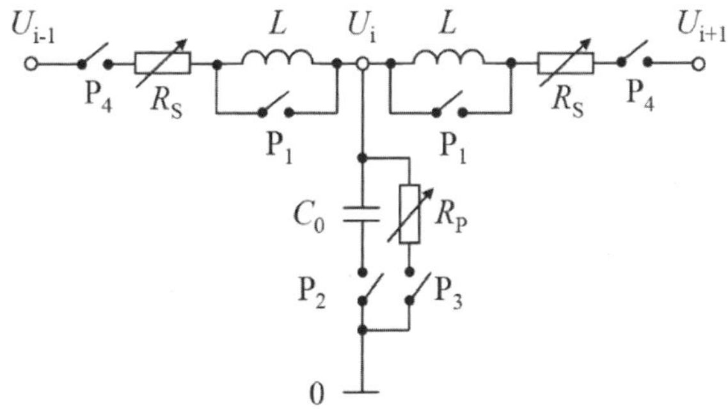

Fig. 7.24. Electric circuit of a basic *RCL* cell of the designed wave-diffusion heat propagation model.

the parameters can be changed. The electric circuit of the basic cell is shown in Fig. 7.24. The switches P_1 to P_4 enable one to change the analogue type, composed of coils with inductance $L = 4.05$ mH, capacitors with capacitance $C_0 = 10$ nF, and adjustable resistors with ranges $R_S = 1$ kΩ and $R_P = 100$ kΩ. With this, the diffusion and wave-diffusion heat propagation can be modeled by making use of various modifications of the basic model cell. The *RCL* analogue is a part of a single-purpose hybrid sampling system, operating at a sampling frequency up to 10 MHz and having several moduli for various kinds of boundary conditions, internal sources and other parameters. The *RCL* model and the hybrid sampling system are described in more detail in [5.11]

Computer RCL model – the simulator

This represents a computerized electrical *RCL* model which reflects not only the arrangement of electrical elements in basic cells but also that of other model parts such as voltage and current sources, an oscilloscope, a graphic unit, and others. The display panel of the computer simulator is in Fig. 7.25. In its left part, there is the simulator control panel, whereas in the right part the model parameters are set. Similarly, as in an electrical *RCL* model, the electric circuit properties are determined with the switches P_1 to P_4 and so the corresponding heat propagation analogue is selected. Some element values (R_S, R_P) can be set, with others being constant (C, L). The boundary conditions and inner sources are defined by voltage or current generators. On the oscilloscope, the time interval is chosen in which the voltage in a network node, stated in the graphic unit, is determined. The solver sets a discrete or continuous model, number of cells, and the parameters of numerical method and

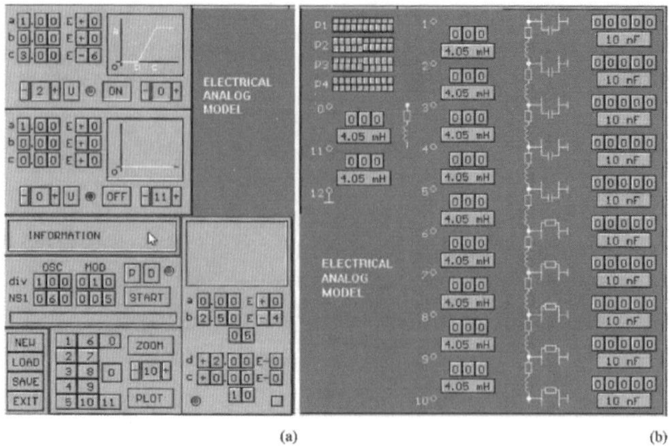

Fig. 7.25. Computer simulator for wave-diffusion heat propagation – the simulator control panel; (a) and analogue parameters setting (b).

computation. This is followed by computation, observation of results on the screen, and recording.

Results

The computer *RCL* model has many advantages, for example, easy multi-cell analogue realization and easier verification of wiring, function, and behavior of a real *RCL* analogue. The comparison of computer simulation results with those obtained with an *RCL* analogue is shown in Fig. 7.26. It expresses the time voltage dependence at the point with the coordinate $X = 0.5$. The results show good agreement between the computer *RCL* model and the electrical *RCL* analogue.

The influence of discretization, i.e. that of various numbers of cells, on the accuracy of solution results for wave-diffusion and diffusion

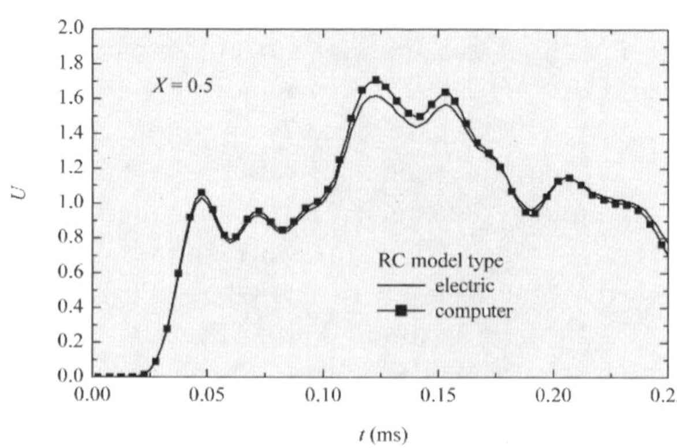

Fig. 7.26. Comparison of results for an electrical analogue and a computer simulation.

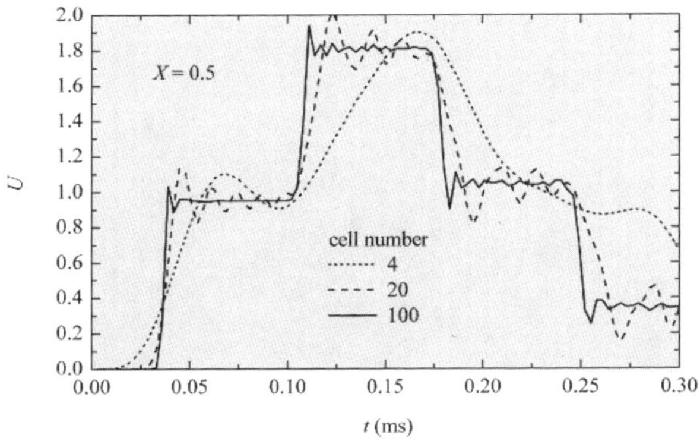

Fig. 7.27. Influence of the number of cells of a wave-diffusion propagation model on the accuracy of the simulation.

heat propagation, is another observed phenomenon. If the velocity member $\partial T/\partial\tau$ in diffusion propagation has an influence in the heat conduction equation, in the wave propagation the member $\partial^2 T/\partial\tau^2$ is decisive. This member has acceleration nature. It influences the number of discrete model cells necessary to replace continuous material credibly and, similarly, the convergence and accuracy of numerical methods. Although the discrete electric 10-cell analogue is suitable for diffusion heat propagation, it is implicitly unsuitable for wave-diffusion heat propagation. The results in Fig. 7.27 show the influence of three different numbers of cells on the voltage in the wave-diffusion heat propagation. Good agreement and accuracy of the solution are obtained with approximately 100 cells, i.e. with the number of cells which is higher by one order than was necessary for diffusion heat propagation.

"Give me a (partial) derivative and I will invert the world"

Isaac Newton *(4.1.1643–31.3.1727)*

8

Deterministic Computer Models

Deterministic models are based on the transformation of mathematical differential or difference equations of the investigated process to the form that can be realized on a computer. This form of a mathematical model is usually called an *operational model.*

According to the type of the operational model, the model is applied on an universal or single-purpose digital, hybrid or analogue computer. It provides only a file of basic operations which proceed in series or partially in series and in parallel. In a numerical operational model, these can be arithmetic operations only, while for a hybrid or analogue operational model integration is the fundamental operation.

8.1 Characterization of Deterministic Models

The deterministic models can be divided to numerical, hybrid or analogue ones. The *numerical deterministic models* transform mathematical equations to the *numerical matrix operational model,* which is a set of algebraic equations that can be expressed in a matrix form. The *hybrid* or *analogue deterministic models* convert the equations to the *hybrid* or *analogue operational model,* that is an ordinary differential equation in which a member with the derivation of a time-dependent variable occurs.

The *deterministic simulation model* is a technical means by which a deterministic operational model is realized. Because a digital or hybrid computer is the most frequent means, it is simply called a *computer model.*

In contemporary modeling, the *digital computer* represents the most significant means not only from the point of view of its universality,

J. Kuneš, *Similarity and Modeling in Science and Engineering*,
DOI: 10.1007/978-1-907343-78-0_8, © Cambridge International Science Publishing 2012

wide accessibility, power output, and operational speed, but also in terms of the development of numerical methods. However, the role of the computer, as the simulation model, differs fully from that for calculations, mass data processing and control. Especially, the difference consists of the model assignment of an investigated object or process and in the modeling solution algorithm. Because deterministic models belong to the higher-abstraction-degree models, admittedly sometimes a concrete idea is lost, e.g. about the geometric similarity, scales, etc., but this is virtual only.

Nowadays, hardly anybody is aware of the role of the physical similarity theory in mass data processing and model information compression and also in the generalization of results for other physically similar systems or processes. It must be noted that the significance of the similarity theory and criteria increases markedly with the development of computer modeling.

The *fundamental conception* of deterministic models is determined by the way the information is displayed. The methods of information processing, abilities and limitations of the models depend on this. Two different ways of displaying information are used, the digital and analogue ones or a combination thereof, if need be. In the *digital displaying,* the values of discrete or continuous quantities are expressed by a number in the selected numerical system, e.g. the binary one, which enables one to create elements with two unambiguous states in the simplest way. In *analogue displaying,* a continuous quantity in the machine unit range is assigned to the continuous or discrete quantity from the interval $<a, b>$. The machine unit is the characteristic quantity of an analogue model, e.g. 10 V, etc.

In the deterministic models, *information processing* can be parallel or serial. In *parallel processing,* the model structure corresponds better to that of the investigated model structure. Between the model and the object, the correlation in space and time can be better kept. The parallel working models contain more single operational units, digital or analogue, according to the type of model. However, they are much more complicated and expensive. The hybrid models thereof are examples. On the other hand, only one operational unit is employed in *serial processing.* The simulation speed is less and the time and space correlations are wholly affected.

Single-purpose or *specialized simulation models* are made to simulate one task type or a narrow class of tasks. The number of elements and kinds thereof is determined by the extent and complexity of the task to be solved. The memory is limited to the most necessary range. On the other hand, the *universal simulation models* can be employed on the solution of a wide range of tasks generally.

8.2 Numerical Deterministic Models

Numerical deterministic models transform mathematical model of the investigated proces to matrix operational model, as was already written. On the basis of a operational model, the simulation model is created. In Fig. 8.1, the transformation of a mathematical model into a numerical simulation one is outlined schematically. Using numerical methods, the mathematical model is transformed to a matrix numerical operational model, that involves only the basic arithmetic operations. With one of the higher programming languages used, the operational model is translated into the computer instruction code. This is the model of a certain class of tasks. Together with concrete input data, it creates the *simulation model* for an investigated process. The utilization of simulation languages or computational software systems simplifies the formation of simulation model. The computational software system can be shortly characterized as a software for the description of the investigated process, simulation model devevelopment, solution process and results evaluation.

A numerical deterministic model can ordinarily be built on a digital computer as a multi-purpose simulation means or on a single-purpose digital model or system. The simplest diagram of the serial modeling principle is denoted on Fig. 8.2. After formulating the problem, the mathematical and operational models are selected. With a computer, the instruction code involves a sequence of codes of operations. The serial processing of instructions, otherwise the serial information flow, proceeds as a sequence of fundamental arithmetical operations with binary numbers controlled by logical circuits. All information on the investigated system and processes, and als on the strategy and the course of simulation, are involved in the control program of the operational unit. Considering the already existing extensive literature on computers, we will present here only the matrix operational model principle, which is the fundamental model with which the numerical deterministic computer models work.

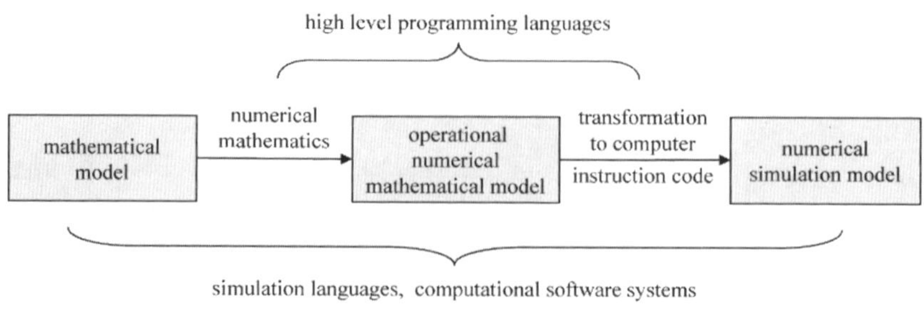

Fig. 8.1. Transformation of a mathematical model to a numerical simulation model.

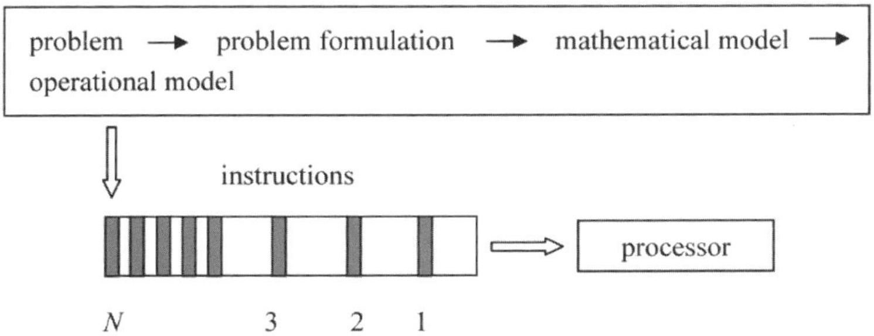

Fig. 8.2. Simplest diagram for serial numerical modeling.

According to the type of the deterministic method used, the *deterministic numerical models* are divided to the *models based on the finite difference method principle, finite element method principle, finite volume method principle* and *boundary element method principle.* All numerical methods transform the mathematical model of the investigated physical field to the set of algebraic equations for individual nodes from the zone to be solved. The solution of the original differential equation can be obtained from the solution of the sets of algebraic equations.

According to the transformation procedure, the numerical methods can be divided to the *methods approximating differential equations* and the *methods approximating the solution of differential equations.* The finite difference method and finite volume method belong to the methods approximating differential equation. These methods are based on the approximation of the differential equation in individual elements of the computational grid. The methods approximating the solution of the differential equation employ so-called shape functions for the approximation of the solution in individual elements of the computational network. The finite element method and the boundary element method belong to them.

According to the space discretization of the solution zone, the methods can be divided to three groups. *The structured network of nodes* is characteristic for the first group to which the classic finite difference method belong. In the second group of methods, *the arbitrary unstructured network of nodes* is employed. The finite element method and the finite volume method come to this group. The third group consists of boundary element method. The *discretization* is performed *at the zone boundary only,* so that the distinctively less number of nodes is generated.

The accuracy, stability, convergence and error of the solution obtained by numerical methods are the important properties for their exploitation. The solution *accuracy* characterizes the proximity of the approximated solution to the exact solution. The numerical method *stability* expresses the fact that small variations in input data cause only a small changes in the solution. The numerical method is supposed to be *convergent,* when almost arbitrary accurate solution can be obtained. It is achieved using the smaller time or space steps usually.

The numerical solution is accompanied by three main sources of the *solution errors* – the errors of model, discretization errors and rounding errors. Each of these errors lowers the accuracy of the solution. The *errors of the model* are caused by all the simplifications made during the model formation. The *discretization errors* imply from the fact that numerical methods work with certain network of nodes. This error can be lowered using the finer network of nodes, thus lowering the space and time steps. The *rounding errors* results from the fact that calculations on the computer are made with certain accuracy. It must be taken into consideration, that lowering the discretization errors due to finer network of nodes causes the increase of rounding errors due to raised number of arithmetical operations.

8.2.1 Matrix Operational Model

The solution of multidimensional physical fields described by differential equations using numerical deterministic methods such as finite difference method, finite element method, finite volume method, and boundary element method leads to the *matrix operational model* in the form of the large set of algebraic equations. Generally, the set of algebraic equations has the following form

$$\begin{array}{ccccccc} a_{11}u_1 + & a_{12}u_2 + & \cdots & + & a_{1n}u_n & = & b_1, \\ a_{21}u_1 + & a_{22}u_2 + & \cdots & + & a_{2n}u_n & = & b_2, \\ \vdots & \vdots & \ddots & + & \vdots & & \vdots \\ a_{n1}u_1 + & a_{n2}u_2 + & \cdots & + & a_{nn}u_n & = & b_n, \end{array} \tag{8.1}$$

in which a_{ij} and b_i are known constants and u_j are unknown quantities. The set can be written in the following matrix form

$$\mathbf{Au} = \mathbf{b}, \tag{8.2}$$

where **A** denotes a square matrix of $(n,\ n)$type, **u** and **b** are column vectors composed of n elements

$$\mathbf{A} = \begin{bmatrix} a_{11}, & a_{12}, & \cdots & a_{1n} \\ a_{21}, & a_{22}, & \cdots & a_{2n} \\ \vdots & \vdots & \ddots & \vdots \\ a_{n1}, & a_{n2}, & \cdots & a_{nn} \end{bmatrix}, \quad \mathbf{u} = \begin{pmatrix} u_1 \\ u_2 \\ \vdots \\ u_n \end{pmatrix}, \quad \mathbf{b} = \begin{pmatrix} b_1 \\ b_2 \\ \vdots \\ b_n \end{pmatrix}. \tag{8.3}$$

To determine **u**, the matrix **A** must be inverted, and the matrix **C** is obtained

$$\mathbf{C} = \mathbf{A}^{-1} = \begin{bmatrix} c_{11}, & c_{12}, & \cdots & c_{1n} \\ c_{21}, & c_{22}, & \cdots & c_{2n} \\ \vdots & \vdots & \ddots & \vdots \\ c_{n1}, & c_{n2}, & \cdots & c_{nn} \end{bmatrix}. \tag{8.4}$$

Consequently, the sought quantity **u** is determined by the following relation

$$\mathbf{u} = \mathbf{A}^{-1}\,\mathbf{b} = \mathbf{Cb}. \tag{8.5}$$

Because the quantities b_i are known, the physical field in the zone nodes can be determined by the inversion of the matrix **A** on a computer by using standard numerical methods.

8.2.2 Models Based on the Finite Difference Method Principle

The *finite difference method (FDM)* is based on approximations enabling one to substitute an asymptotic differential equation by a difference equation. After previous development, the models, operating on the FDM principle, are in such a state that they hold up well in comparison with the progressive finite element method (FEM), which is closely connected with the development of computers. For example, this involves the possibility to utilize arbitrary unstructured networks and the potential of both methods to solve mathematical physics equations is approximately equal.

The task can be formulated either as a local or global one. The *local formulation* represents a classic task formulation utilizing discrete approximation method with structured difference network. The *global formulation* uses the task variation formulation. The following description of the finite difference method concerns classic FDM formulation.

The solution by the finite difference method implies the following steps. After task formulation, solution zone discretization, local approximation and difference operators creation, equations set formulation and their solution, processing and evaluation of results are made.

Modeling procedure

General unsteady boundary task for the quantity u can be defined by the following differential equation at the zone Ω with boundary conditions defined at the boundary Γ

$$\mathcal{L}u(x,y,z,t) = f(x,y,z,t) \qquad \text{for} \quad (x,y,z,t) \in \Omega, \tag{8.6}$$

$$\mathcal{L}_b u(x,y,z,t) = g(x,y,z,t) \qquad \text{for} \quad (x,y,z,t) \in \Gamma, \tag{8.7}$$

where $\mathcal{L}$, $\mathcal{L}_b$ are given differential operators and $f(x,y,z,t)$, $g(x,y,z,t)$ are prescribed functions. The initial condition have to be included in these equations.

The discretization of the solution zone is approximated with a discrete network of nodes which are regular and can be of different kinds, mostly in accordance with the zone shape, so that the network boundary may agree with that of the zone. The most commonly used networks are the square and triangular ones, but polar networks are sometimes used (Fig. 8.3). The precision of the FDM method depends among others on the space and time discretization and the selected difference operator.

The discretization of the differential operator $\mathcal{L}$ in the point $P_i = (x,y,z,t)$ *is* formed by the *finite difference* (*FD*) *operator.* For the linear differential operator $\mathcal{L}$, the relevant FD operator has the following form

$$\mathcal{L}u(P_i) \approx \sum_{j=0}^{m} B_{j(i)}\, u(P_{i+j}), \qquad j = 1,\ldots,m, \tag{8.8}$$

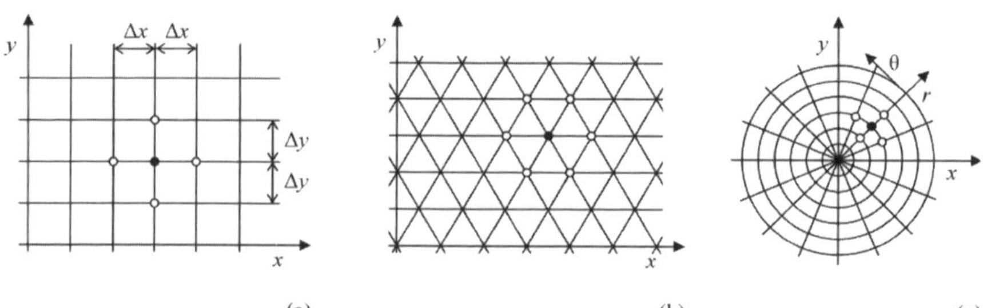

Fig. 8.3. Basic elements of the two-dimensional network of the finite difference method: the square (a), triangular (b), and polar (c) elements.

that is linear combination of function values in this point and the points selected from its surroundings, where $B_{j(i)}$ are known coefficients that follows from FD approximation. The points P_{i+j} form the so-called *finite difference star*, where P_i is the central point of the star. The arrangement of the star depends on the form and the order of the differential operator $\mathcal{L}$, the required precision of the approximation and the selected type of computational network. Because the network is structured and regular, the same FD operator is used for the inner network nodes.

FD operator is obtained as a substitution of the differential operator by means of appropriate difference ratio. For the derivatives of the function $u_{i,j,k,l} = u(x,y,z,t)$ in the regular rectangular network, the following relations can written

$$\frac{\partial u}{\partial x} = \frac{1}{h_i}\left(u_{i+1,j,k,l} - u_{i,j,k,l}\right), \tag{8.9}$$

$$\frac{\partial u}{\partial x} = \frac{1}{h_i}\left(u_{i,j,k,l} - u_{i-1,j,k,l}\right), \tag{8.10}$$

$$\frac{\partial u}{\partial x} = \frac{1}{2h_i}\left(u_{i+1,\ j,k,l} - u_{i-1,j,k,l}\right), \tag{8.11}$$

$$\frac{\partial u}{\partial t} = \frac{1}{2h_l}\left(u_{i,\ j,k,l+1} - u_{i,j,k,l-1}\right), \tag{8.12}$$

$$\frac{\partial^2 u}{\partial x^2} = \frac{1}{h_i^2}\left(u_{i-1,j,k,l} - 2u_{i,j,k,l} + u_{i+1,j,k,l}\right), \tag{8.13}$$

$$\frac{\partial^2 u}{\partial x\, \partial y} = \frac{1}{4h_i h_j}\left(u_{i-1,j-1,k,l} - u_{i-1,j+1,k,l} - u_{i+1,j+1,k,l} - u_{i+1,j-1,k,l}\right) \tag{8.14}$$

where $h_i = \Delta_x$, $h_j = \Delta y$, $h_k = \Delta z$, $h_l = \Delta t$ are space and time steps. The first relation represents *forward difference,* the second is *backward difference* and third is *central difference.* All these relations come from Taylor expansion.

The *parabolic partial differential equation of the Fourier type* that can be written in one-dimension in the following form

$$\frac{\partial^2 u}{\partial x^2} = a(x)\frac{\partial u}{\partial t}, \tag{8.15}$$

represents the heat transfer equation without inner sources. The following three basic one-layer difference schemes are used to transform this differential equation to a difference equation (Fig. 8.4) for the quantity $u_{i,j} \equiv u(x,\ t)$.

For the *explicit difference scheme* (Fig. 8.4a), the following relation is valid

$$u_{i,j+1} = u_{i,j} + p\left(u_{i-1,j} - 2u_{i,j} + u_{i+1,j}\right), \tag{8.16}$$

where $p = a\Delta t\,(\Delta x)^{-2}$. The explicit scheme is simple, the variables in the following $(j + 1)^{th}$ time level are explicitly expressed. The disadvantage is the restriction for the selection of space and time steps that relates with the stability of difference scheme. The stability condition is $0 < p < 1/2$, with optimum $p = 1/6$. The advantage of the explicit scheme is also the possibility of simple expression of non-linear system parameters, boundary conditions, etc.

The *implicit difference scheme* (Fig. 8.4b) is expressed using the following relation

$$u_{i-1,j+1}(-p) + u_{i,j+1}(1+2p) + u_{i+1,j+1}(-p) = u_{i,j}. \tag{8.17}$$

The implicit scheme requires the determination of the quantity in $(j + 1)^{th}$ time level and the solution of the set of n linear equations with n unknown variables. The advantage of this difference scheme is in the fact, that its stability does not depend on the selected space and time steps.

The *combined implicit–explicit difference scheme* (Fig. 8.4c) is described by the following relation

$$\begin{aligned} &u_{i-1,j+1}(-\gamma p) + u_{i,j+1}(1+2\gamma p) + u_{i+1,j+1}(-\gamma p) = \\ &u_{i-1,j}(1-\gamma)p + u_{i,j}\left[1-2(1-\gamma)p\right] + u_{i+1,j}(1-\gamma)p, \end{aligned} \tag{8.18}$$

where $0 \le \gamma \le 1$. For the value $\gamma = 0$, the explicit scheme is obtained, while $\gamma = 1$ gives the implicit scheme. At the $(j+1)^{th}$ time level, the implicit–explicit scheme needs to solve the set of m equations with m unknown variables, where $1 < m < n$ and the remaining values are determined directly, it means explicitly.

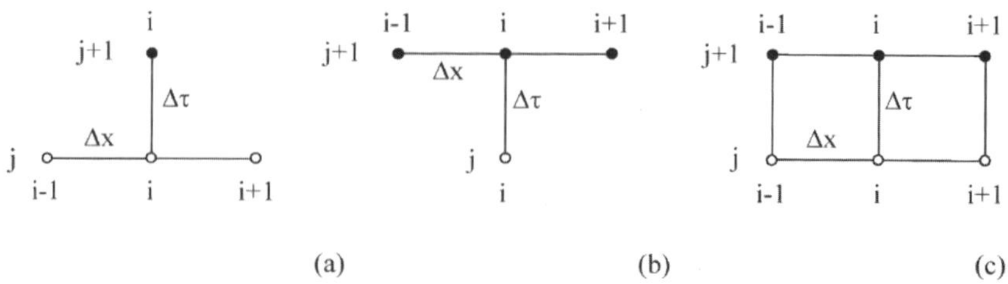

Fig. 8.4. One-layer basic difference schemes for the numerical solution of the Fourier equation – explicit (a), implicit (b), and combined (c) scheme.

Simplicity, low time requirements, and small consumption of computer memory are characteristic properties of one-layer expressions. However, except the combined scheme, they have less accuracy and convergence speed.

In addition to the mentioned and other simple schemes, sometimes multilayer and multiple schemes are used. In *multilayer schemes,* the quantity in several preceding levels must be known to calculate the quantities in the $(j + 1)^{\text{th}}$ time level. In *multiple schemes,* two or more difference expressions must be enumerated, with the average value being considered as the solution. Compared to one-layer expressions, the multilayer ones are usually more precise and stable for arbitrary relation of the space and time steps. The disadvantage of the multilayer expressions is the higher computational demands. The multiple expressions are relatively accurate but also the computational demands are higher. None of the schemes can be declared optimal and the problem of the most suitable difference substitution must be solved according to the character of the task above all.

The set of equations for the FDM method is formulated using the difference operator from (8.8) into the basic Eq. (8.6) and expanding it to all inner points of a solution zone Ω

$$\mathcal{L}u(Pi) \approx \sum_{j=0}^{m} B_{j(i)} u(P_{i+j}) = f(P_i), \qquad i = 1, \cdots, N. \tag{8.19}$$

The resulting set of equations can be written in the matrix form

$$\mathbf{Au} = \mathbf{b}, \tag{8.20}$$

where $\mathbf{u}$ is the vector of unknown values of function u in the network nodes and $\mathbf{b}$ is the known vector. This equation set is linear or nonlinear according the task considered.

The boundary conditions have to be added to the set of equations. It can be done in the step when the finite difference relations are made for boundary nodes or later to complete the final set of equations derived for inner nodes of the zone. The discretization of boundary conditions is complicated in the case all boundary nodes are not situated on the zone border. This situation is typical for zones of arbitrary irregular shape. The discretization of the differential operator on the zone border is usually performed by adding the additional points, so-called outer points, that are in the network nodes near the border but outside the zone Ω. Then, the boundary conditions are used for obtaining the unknown value of the function u in these outer points with the utilization of function vaues in the inner points of the zone. The second way is to use only the inner points inside the zone Ω. The more precise approximation of the solution is obtained using the outer points, but these additional outer points can influence the numerical solution.

The resulting matrix equation (8.20) is solved using some of the standard numerical methods. As the results of the solution, the discrete values of the function u in the nodes of the investigated zone are obtained. The gradients, fluxes, various surface or volume integrals can be evaluated from the space distribution of the quantity u.

8.2.3 Models Based on the Finite Element Method Principle

The *finite element method (FEM)* starts from the two formulations. The *Ritz* or *Rayleigh-Ritz* formulation of the FEM method is a variational method in which a task with boundary conditions is expressed by a variational term, the functional. The minimum of this functional corresponds to the solved differential equation with relevant boundary conditions. An approximate solution can be obtained by minimizing the functional with respect to variables which define the approximate solution. The *Galerkin formulation* is among the group of weighted residuals methods in which the solution is sought by weighting the residuals of the differential equation.

Both formulations mentioned use an approximation function expressed as a combination of fundamental mathematical functions which are defined for whole solution zone. This approximation function must be able to approximately express the solution and to fulfil the boundary conditions. In complicated tasks, especially two-dimensional and three-dimensional ones, an approximation function cannot be made which would satisfy in the whole solution zone. A possible way is to divide the whole zone into small sub-zones, the so-called *finite elements.* Then, the approximation functions are usually simpler because the solution does not change in a complex way on very small zones. This procedure forms the basis of the finite element method and the principle of finite element models. At present, this method and these models are among the most used ones in computer modeling for tasks of physics, mechanics, thermomechanics, electromagnetism, and other scientific and engineering fields.

In modeling, the FEM procedure consists of several steps involving the task formulation, solution zone discretization, the selection of shape functions in sub-zones, the formulation of the set of linear algebraic equations and their solution, as well as the task evaluation.

Modeling procedure

The formulation of a steady boundary task for the quantity u consists of defining the differential equation at the zone Ω with boundary conditions at the boundary Γ

$$\mathcal{L}u(x,y,z) = f(x,y,z) \qquad \text{for} \quad (x,y,z) \in \Omega, \tag{8.21}$$

$$\mathcal{L}_b u(x,y,z) = g(x,y,z) \qquad \text{for} \quad (x,y,z) \in \Gamma, \tag{8.22}$$

where $\mathcal{L}$, $\mathcal{L}_b$ are given differential operators consisting of only space derivations, and $f(x, y, z)$, $g(x, y, z)$ are prescribed functions.

The discretization of the solution zone Ω represents an important step because this influences the solution's demands and precision. The zone Ω is divided into a large number of small sub-zones, the finite elements that are denoted as Ω^e for e =1, 2, M, where M is their total number. The elements can be line segments for a one-dimensional zone, triangles or rectangles for a two-dimensional one, and tetrahedrons or three or four-wall prisms. For an unstructured computational networks, the triangles for two-dimensional zones or tetrahedrons for three-dimensional zones are used.

The task is formulated by a sought function u in the nodes joined to the elements. A linear line element has two nodes at its ends, the triangular one has three nodes, and the tetrahedron prism has four nodes in its vertices. The higher-order elements have more nodes and use higher-order shape functions. For the solution procedure, the nodes are described by the local numbering that determines their position in the element and also by global numbering that defines their position globally in the solution zone.

The next step is the selection of the *shape (interpolation) function* that approximates an unknown solution in the element. The interpolation is chosen mostly by a 1st-degree (linear) polynomial, 2nd-degree (quadratic) or a higher-degree one. Usually, the higher polynomial degrees lead to more complicated formulations of equations. Therefore, the linear interpolation is still often used. After selection of the polynomial degree, a relation for approximation of an unknown solution u^e in the element e can be written in the following form

$$u^e(x,y,z) = \sum_{j=1}^{n} N_j^e(x,y,z)\breve{u}_j^e = (\mathbf{N}^e(x,y,z))^T \breve{\mathbf{u}}^e, \tag{8.23}$$

where n is the number of nodes in the element, $\breve{u}_j^e$ is the value of u in the node j of element e and $N_j^e(x, y, z)$ is the shape function for this node. The largest degree of the polynomials $N_j^e(x, y, z)$ for the given element denotes the degree of the element. The functions $N_j^e(x, y, z)$ are non-zero only in appropriate element e.

The Ritz or Galerkin method can be used for the formulation of a set of equations. In the following text, the Ritz method is used. Under certain general assumptions, the solution of a boundary task can be obtained by the minimalization of the functional $F(u(x, y, z))$. This functional is given as the sum of functionals $F^e(u^e(x, y, z))$ for individual elements e

$$F\left(u(x,y,z)\right)=\sum_{e=1}^{M}F^e\left(u^e(x,y,z)\right), \tag{8.24}$$

where M is the element number in the solution zone. For F^e, the following relation is valid

$$F^e\left(u^e(x,y,z)\right)=\frac{1}{2}\int_{\Omega^e}u^e(x,y,z)\mathcal{L}u^e(x,y,z)\mathrm{d}\Omega-\int_{\Omega^e}f(x,y,z)u^e(x,y,z)\mathrm{d}\Omega, \tag{8.25}$$

while

$$u(x,y,z)=\sum_{e=1}^{M}u^e(x,y,z). \tag{8.26}$$

Using Eq. (8.23) in the relation (8.25), the following equation is obtained

$$F^e\left(u^e(x,y,z)\right)=\frac{1}{2}\left(\breve{\mathbf{u}}^e\right)^T\mathbf{K}^e(x,y,z)\breve{\mathbf{u}}^e-\left(\breve{\mathbf{u}}^e\right)^T\mathbf{b}^e(x,y,z), \tag{8.27}$$

where $\mathbf{K}^e(x, y, z)$ is the matrix of (n, n) type and $\mathbf{b}^e(x, y, z)$ is the column vector of $(n, 1)$ type, those elements are given by the following relations

$$K_{ij}^e(x,y,z)=\int_{\Omega^e}N_i^e(x,y,z)\mathcal{L}N_j^e(x,y,z)\mathrm{d}\Omega, \tag{8.28}$$

$$b_i^e(x,y,z)=\int_{\Omega^e}f(x,y,z)N_i^e(x,y,z)\mathrm{d}\Omega. \tag{8.29}$$

Using F^e from Eq. (8.27) in Eq. (8.24), and converting to the global numbering of the nodes, the following relation is obtained

$$F=\frac{1}{2}\left(\breve{\mathbf{u}}\right)^T\mathbf{K}\breve{\mathbf{u}}-\left(\breve{\mathbf{u}}\right)^T\mathbf{b}, \tag{8.30}$$

where $\mathbf{K}$ is the known symmetrical matrix of (N, N) type, $\mathbf{b}$ is the known column vector of $(N, 1)$ type and $\breve{\mathbf{u}}$ is the unknown column vector of $(N, 1)$ type, those elements are uknown node values. The total number of nodes N denotes the degree of freedom of the task. When determining the functional minimum, the partial derivative of F with respect to $\breve{u}_i$ is considered to equal zero

$$\frac{\partial F}{\partial \breve{u}_i} = \frac{1}{2}\sum_{j=1}^{N}\left(K_{ij} + K_{ji}\right)\breve{u}_j - b_i = 0, \qquad i = 1, 2, 3, \cdots, N. \tag{8.31}$$

Because the matrix **K** is symmetrical ($K_{ij} = K_{ji}$), the following equation is obtained

$$\mathbf{K\breve{u}} = \mathbf{b}. \tag{8.32}$$

Using Eq. (8.23) in the boundary conditions is the last step in the formulation of the set of equations. The resulting equation set can be in one of the two following forms

$$\mathbf{K\breve{u}} = \mathbf{b}, \quad \text{or} \quad \mathbf{A\breve{u}} = \lambda\mathbf{B\breve{u}}. \tag{8.33}$$

The left equation follows from the non-homogeneous differential equation or non-homogeneous boundary conditions while the right equation is obtained from both homogeneous differential equation and boundary conditions. The right equation is the certain case of the left equation under the condition of $\mathbf{b} = \mathbf{0}$, Then the matrix **K** can be written in the form $\mathbf{A} - \lambda\mathbf{B}$, where λ is the eigenvalue. This type of the equation can have more solutions, so-called eigenvalues λ and corresponding eigenvectors $\mathbf{\breve{u}}$.

The set of algebraic equations can be solved using standard direct of iterative numerical procedures. The results are the discrete values of the function u in the computational network nodes. From these values, the various other quantities can be evaluated, including gradients, fluxes, integrals, etc.

For the solution of unsteady tasks, in the case of the time independent solution zone, the following procedure can be used. The relation for approximation of an unknown solution u^e in the element e can be written in the separated form, so that space dependence is in the shape function $\mathbf{N}^e(x, y, z)$ and time dependence in the basic quantity $\mathbf{\breve{u}}^e(t)$. For element e can be written the following equation

$$u^e(x, y, z, t) = \sum_{j=1}^{n} N_j^e(x, y, z)\breve{u}_j^e(t) = \left(\mathbf{N}^e(x, y, z)\right)^T \mathbf{\breve{u}}^e(t). \tag{8.34}$$

The following discretization proceeds in two steps. At first, the partial discretization is implemented with respect to space coordinates. The matrix differential equation with the time-derivatives is obtained. After that, the equation is solved using time discretization.

In the general case, the differential equation (8.21) can be written in the following form

$$\mathcal{L}u(x, y, z, t)+p(x, y, z)\frac{\partial u(x, y, z, t)}{\partial t}+q(x, y, z)\frac{\partial^2 u(x, y, z, t)}{\partial t^2}=f(x, y, z). \tag{8.35}$$

For the constant time, the time derivatives of u and all parameters can be considered as prescribed functions of space coordinates and are the loading functions. Performing the space discretization of this equation by FEM and using the approximation of u^e from the Eq. (8.34), the following relation is obtained

$$\begin{aligned}\mathbf{K}\breve{\mathbf{u}} = \int_{\Omega^e}\mathbf{N}^\mathbf{e} f\,\mathrm{d}\Omega-\left(\int_{\Omega^e}\left(\mathbf{N}^\mathbf{e}\right)^T p\,\mathbf{N}^\mathbf{e}\,\mathrm{d}\Omega\right)\frac{\mathrm{d}\breve{\mathbf{u}}^\mathbf{e}}{\mathrm{d}t}\\ -\left(\int_{\Omega^e}\left(\mathbf{N}^\mathbf{e}\right)^T q\,\mathbf{N}^\mathbf{e}\,\mathrm{d}\Omega\right)\frac{\mathrm{d}^2\breve{\mathbf{u}}^\mathbf{e}}{\mathrm{d}t^2}.\end{aligned} \tag{8.36}$$

This equation comes to the *matrix differential equation* in the form

$$\mathbf{M}\frac{\mathrm{d}^2\breve{\mathbf{u}}}{\mathrm{d}t^2}+\mathbf{C}\frac{\mathrm{d}\breve{\mathbf{u}}}{\mathrm{d}t}\mathbf{K}\,\breve{\mathbf{u}}=\mathbf{b}, \tag{8.37}$$

where the matrices and vectors are compiled in a standard way from the matrices $\mathbf{K}^\mathbf{e}$ and vectors $\mathbf{b}^\mathbf{e}$ from the Eqs. (8.28) and (8.29), and from the matrices $\mathbf{M}^\mathbf{e}$ and $\mathbf{C}^\mathbf{e}$ that are determined by the relations

$$M_{ij}^e = \int_{\Omega^e} N_i^e\,\rho\,N_j^e\,\mathrm{d}\Omega,\;\; C_{ij}^e = \int_{\Omega^e} N_i^e\,\mu\,N_j^e\,\mathrm{d}\Omega. \tag{8.38}$$

The matrices **M, C** and **K** are known as *mass, damping* and *stiffness matrices.* The boundary conditions are processed in the same way as basic differential equations. Time discretization follows. The solution is performed gradually over individual time levels, and the initial conditions are used for the first time level values.

8.2.4 Models Based on the Finite Volume Method Principle

The finite volume method (FVM) starts from integral equations, in contrast to the finite difference method based on differential equations. In discontinuous solution tasks, the failure of the FDM method can be presumed near the discontinuity, and on the contrary, a good approximation of the solution can be expected with the FVM method.

Like the FDM method, with the FVM method the values of the quantities are determined in discrete network of nodes. The concept of the *finite volume* refers to a small volume around each network node. For a 2D zone, the volumes (cells) can have a triangular or tetragonal shape. For a 3D zone, they can have a tetrahedral or hexahedral shape. In the finite volume method, the volume integral, involving the divergence, is transferred to a surface integral using the divergence theorem. Afterwards, these members are evaluated as fluxes on surfaces of every finite volume. Because the flux entering the given volume equals the flux going out of it, the FVM method is incorporated into conservative methods. The advantage of the FVM method is that it can be easily formulated for unstructured networks. This method is used frequently, e.g. in fluid dynamics modeling.

With the finite volume method, tasks are solved in several steps involving the task formulation, solution zone discretization, linearization of discretized equations, equations set solution, and task evaluation.

Modeling procedure

The steady state equation of a conservation law for a scalar quantity transfer is considered. In the integral form for arbitrary test volume V, the equation for quantity u has the following form

$$\oint \rho u \mathbf{w} \cdot \mathrm{d}\mathbf{A} = \oint D_u \nabla_u \cdot \mathrm{d}\mathbf{A} + \int_V S_u \mathrm{d}V, \tag{8.39}$$

where ρ is the density, $\mathbf{w}$ is the velocity vector, $\mathbf{A}$ is the surface area vector, D_u is the diffusion coefficient of **u,** Δu is the gradient of u and S_u is the source of u in the unit volume. The left-side member of Eq. (8.39) expresses the quantity u transfer by the mass flux. The first right-side member of the equation represents the diffusion of the quantity u due to its gradient. The second right-side member expresses the source or sink of the quantity u from the volume source or the sink inside the test volume V.

The integral in Eq. (8.39) is used for every finite volume in the solution zone. Discretizing this equation, one obtains

$$\sum_{f=1}^{N} \rho_f \, \mathbf{w}_f u_f \cdot \mathbf{A}_f = \sum_{f=1}^{N} D_u \left(\nabla u_f\right)_n \cdot \mathbf{A}_f + S_u \, V, \tag{8.40}$$

where N is the number of walls enclosing the cell, u_f is the value of u passing through the wall f, $\rho_f \, \mathbf{w}_f \cdot \mathbf{A}_f$ is the mass flux through the wall f, $\mathbf{A}_f$ is the surface vector of the wall f, $(\Delta u_f)_n$ is the component of the gradient of u perpendicular to the wall f and V is the cell volume.

In the FVM method, the discrete values of the scalar quantity u are preserved in the cell centers. The values on the walls u_f are used for diffusion and flow members in Eq. (8.40) and must be interpolated from the cell center values. This is accomplished for the flow member by the so called upwind scheme in which the wall value is calculated from the values u in the cells located against flow with respect to the normal component of the velocity $\mathbf{w}_f$ in Eq. (8.40). The diffusion member is approximated by the central difference scheme.

The discretized scalar transport described by Eq. (8.40) contains an unknown scalar variable u in the cell center and unknown values in surrounding cells as well. This equation is non-linear with respect to these quantities in general. The linearized form of Eq. (8.40) can be written for the cell c in the form

$$a_c u = \sum_n a_n u_n + b, \tag{8.41}$$

where the index n refers to neighbouring cells, a_c, a_n and b are linearized coefficients for u and u_n. The number of the neighbouring cells for every cell depends on the network topology, but in practice it equals the number of walls enclosing the cell excepting the boundary cells. Similar equations can be written for every cell in the network. Then, a set of algebraic equations results with a thin matrix of coefficients

$$\mathbf{Au} = \mathbf{b}, \tag{8.42}$$

where **A** is the matrix of linearized coefficients, u is the vector of center values of quantity u in the cells of computational network and b is the vector of linearized coefficients.

For example, for the fluid dynamic tasks, the original integral Eqs. (8.39) for the conservation of mass (continuity equation), momentum (x-,y-, and z-momentum equations), and when appropriate for energy (energy equation), turbulence (turbulence equations for turbulent kinetic energy k and its rate of dissipation ε), etc. One matrix Eq. (8.42) is constructed for every conservation equation. The resulting matrix equations for the discrete dependent variables such as pressure, velocity components, temperature, etc., are mutually coupled. Therefore, an iterative procedure must be used in order to obtain a satisfactorily converged solution. The convergence criterion for quantity u can be a residue R^u, that is the normalized sum of the differences between left and right sides of Eq. (8.41) over all cells in the solution zone. The discrete values of the quantities such as pressure p, velocity components w_x, w_y, w_z, temperature T, turbulent kinetic energy k and its rate of dissipation e in the network cell centers are the result of the solution of the task.

In solving unsteady tasks, the governing equations must be discretized in space and time. The space discretization is the same as that in steady task solution. The time discretization involves the integration of each member in a differential equations over the time step Δt. The time change of the variable u is given by the following equation

$$\frac{\partial u}{\partial t} = F(u), \tag{8.43}$$

where the function F contains an arbitrary space discretization. The discretization of the time derivative can be expressed using the first-order backward difference in the following form

$$\frac{u^{n+1} - u^{n}}{\Delta t} = F(u) \tag{8.44}$$

or using the second-order backward difference in the form

$$\frac{3u^{n+1} - 4u^{n} + u^{n-1}}{2\Delta t} = F(u), \tag{8.45}$$

where the index $n+1$ denotes the value in the time level $t + \Delta t$, the index n denotes the value in the time level t and the index $n - 1$ denotes the value in the time level $t - \Delta t$. After time discretization, the form of the function F remains to be chosen.

With using $F(u^{n+1})$ in the Eqs. (8.44) or (8.45), it is called *implicit time integration*.With using $F(u^{n})$ in these equations, *explicit time integration* is obtained.

8.2.5 Models Based on the Boundary Element Method Principle

These models share the advantages of the classic integral equations with the finite element method. The task dimension reduction is among the most important advantages of the *boundary element method* (*BEM*), due to which the number of equations necessary to determine the unknown variables is limited significantly. In the BEM method, a smaller number of unknown variables can be used to analyze equivalent tasks because there are no unknown variables inside the zone. This property is very significant for solving tasks with boundary conditions, especially three-dimensional ones, particularly where the outer surface ratio to the volume is small.

Characterization and properties

For the user, the input data for the BEM are considerably simpler than those for the FEM. In general, the accuracy and numerical stability are better than for the FEM. With the BEM, the approximation accuracy of the derivations of the sought functions is much higher, too. With the BEM, some tasks on areas, the dimension of which increases to infinity, are solved more easily and precisely. Some tasks for which the FEM appear ineffective can be solved more effectively as well. This includes, foe example, stress concentration tasks for which the FEM is not suitable if considering the shape functions possibilities. The BEM can be combined with the FEM in case some part of the body needs to be discretized by finite elements due to anisotropy and inhomogeneity, with remaining part needing to be discretized by the BEM due to zone infinity, large gradients, etc.

The BEM leads to a non-symmetric set of equations. The matrix of the set of equations is often dense and is, therefore, not a sparse or band type matrix. However, this disadvantage can be compensated for by the relatively small number of equations. The BEM can be used only for tasks in which a fundamental solution of a differential equation is known. Such solutions are not known or they do not have a closed form for non-homogenous and anisotropic regions, therefore the BEM is not able to solve these types of tasks. The BEM is also not effective for regions with one or two dimensions which are much smaller than others, e.g. shells.

The procedure of tasks modeling using the BEM involves the formulation of the task, the representation formula and the boundary integral equation, then the zone boundary discretization, the shape functions selection in sub-zones, the formulation and solution of the set of linear algebraic equations in the form of matrix equation, and solution evaluation.

Modeling procedure

The deduction of equations for the boundary element method (BEM) depends on the differential operator form. The following procedure is performed for the Laplace equation

$$\frac{\partial^2 u}{\partial x_i^2} = 0 \tag{8.46}$$

on the region Ω, where the general variable u denotes potential and q is the potential flux

$$q = \frac{\partial u}{\partial x_i} n_i, \tag{8.47}$$

where n_i is the vector normal to the boundary Γ.

The differential equation can be transformed to an integral one in the first step by means of the weighted deviations method. By weighting Eq. (8.46) with the testing function w, one obtains

$$\int_{\Omega} \frac{\partial^2 u}{\partial x_i^2} w \, d\Omega = 0. \tag{8.48}$$

After it, the partial derivative of the potential function u is removed from the integral over the zone using the integration by parts. The following relation is obtained

$$-\int_{\Omega} u \frac{\partial^2 w}{\partial x_i^2} d\Omega = \int_{\Gamma} \left(\frac{\partial u}{\partial x_i} w - u \frac{\partial w}{\partial x_i} \right) n_i d\Gamma. \tag{8.49}$$

Then, the remaining integral over the zone Ω needs to be removed, so that the following discretization is accomplished on the boundary Γ of the zone only, but not inside the Ω zone. This is performed utilizing the so-called *fundamental solution* u^*.

For the fundamental solution u^* of the Laplace equation, the following holds

$$\frac{\partial^2 u^*}{\partial x_i^2} = -\delta(x, \xi_i), \tag{8.50}$$

where $\delta(x, \xi_i)$ is the Dirac distribution. Thus, for testing function w, that is the fundamental solution of the Laplace equation ($w = u^*$), the following holds

$$-\int_{\Omega} u \frac{\partial^2 w}{\partial x_i^2} d\Omega = u(\xi_i). \tag{8.51}$$

Choosing the testing function $w = u^*$ in Eq. (8.49), the following relation is obtained

$$u(\xi_i) = \int_{\Gamma} \left(q(x_i) u^*(x_i, \xi_i) - u(x_i) q^*(x_i, \xi_i) \right) d\Gamma, \tag{8.52}$$

where q is the flux and $q^* = \frac{\partial u^*}{\partial x_i} n_i$ is the fundamental solution for the flux. Equation (8.52) is the so-called *representative formula* and it is valid for two- and three-dimensional potential tasks. The representative formula enables one to determine the unknown values of the potential u inside the zone ($\xi \in \Omega$) from the known boundary solution of the task (potential u and its flux q).

To obtain the equation which contains the data only from the zone boundary is the next step in the modeling procedure. For this, the load point ξ_i must be shifted to the boundary which is not a simple task from the mathematical point of view. The resulting equation is called the *boundary integral equation,* which forms the basis for the following discretization process of the boundary element method.

To approximate the geometry, the boundary Γ is divided into boundary elements containing several nodes. Inside element e with local coordinates s, the potential u^e and flux q^e are expressed with the utitilization of the *shape functions* $N_j^e(s)$and node values $\breve{u}_j^e$ and $\breve{q}_j^e$

$$u^e(s) = \sum_{j=1}^{n} N_j^e(s)\breve{u}_j^e, \quad q^e(s) = \sum_{j=1}^{n} N_j^e(s)\breve{q}_j^e, \tag{8.53}$$

where n is the number of nodes in the element. Discretization is applied on the boundary integral equation and its discretized form is obtained. Then, the local numbering of nodes in individual elements is converted to the global nodes numbering $j = 1,...,N$ on the whole boundary Γ.

The loading point ξ_i is subsequently located to all boundary nodes and the matrix equation is obtained

$$\mathbf{H}\breve{\mathbf{u}} = \mathbf{G}\breve{\mathbf{q}}, \tag{8.54}$$

where the matrices **H** and **G** are of the (N, N) type and column vectors $\breve{\mathbf{u}}$ and $\breve{\mathbf{q}}$ are of the $(N, 1)$ type. Because vectors of node values $\breve{\mathbf{u}}$ and $\breve{\mathbf{q}}$ in Eq. (8.54) contain both known and unknown boundary values, the equation is modified to the following form

$$\mathbf{A}\breve{\mathbf{y}} = \mathbf{f}, \tag{8.55}$$

in which, in the left part, in the vector $\breve{\mathbf{y}}$ there are unknown boundary nodal values of the potential u and its flux q, and in the right part of the equation, **f** represents the multiple of the known boundary nodal values with corresponding matrix elements.

The set of Eqs. (8.55) is solved by standard direct or iterative procedures for the solution of sets of algebraic equations. After obtaining the sought boundary values of the potential u and the flux q, the values of the potential u inside the zone Q are determined using the representative formula (8.52).

8.3 Hybrid Deterministic Models and Systems

The *hybrid deterministic models* are a certain combination of numerical and analogue models in the effort to utilize the advantages of both of them. In the course of the solution, the information is transferred unilaterally or bilaterally between them and is converted from the digital to analogue form and vice versa by a connecting device.

Hybridization means the infiltration of digital elements into analogue ones and vice versa (Fig. 8.5). With both subsystems, i.e. the analogue and digital ones, balanced mutually in terms of capacity, one can speak about a *balanced hybrid simulation system.* This type of hybrid simulation system represents the crowning of the hybridization process. Each of the subsystems can operate individually. The system must be equipped by a connecting device for information transfer from one system to the other.

The *hybrid simulation system* represents a universal means enabling one to combine digital subsystems, operating in series, with parallel analogue ones and to control them by a program on the one hand and by a logical controlling network on the other. The *analogue subsystem* allows fast parallel information processing where the continuity of quantities can be preserved, modification of the simulation process and change the settings of the operating elements and the structure of the computing and controlling logical networks. The *digital subsystem* provides accurate processing of discrete or discretized quantities especially, the storage of their values and simulation control.

The structural diagram of a hybrid simulation system is in Fig. 8.6. Besides the input and output devices usually used in digital computers, in a hybrid system a direct input from another system is used which can be, e.g. a physical model, other analogue or digital system. The digital subsystem has its own operating system and software adequate to the capacity and speed required. The hybrid system preparation starts by deciding which simulation part will be executed on the analogue system part and which of it on the digital one. Automatic connection

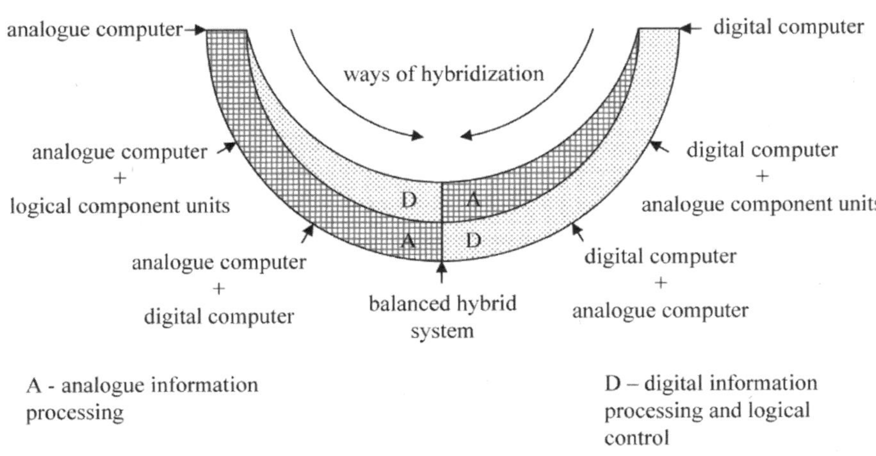

Fig. 8.5. Diagram of the hybridization procedure in computer modeling.

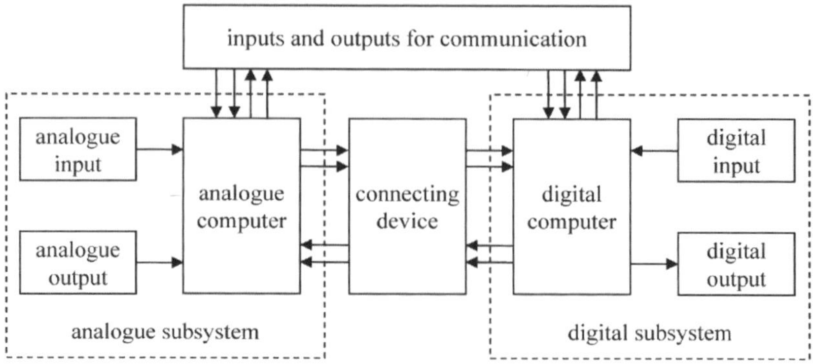

Fig. 8.6. Structural diagram of a hybrid simulation system.

of the elements of analogue subsystem represents an important part of a hybrid simulation.

In a hybrid simulation system there are *analogue, logical*, and *digital* signals. To transfer the information between these three types of signals, six basic kinds of elements are used. The *comparator* enables one to intervene into a logical network depending on the course of the analogue solution. The *switch* is used in analogue model controlling depending on the logical signals. The *reading line* provides the transmission of the control signals from the logical network to the digital subsystem and the *command line* transfers the commands from the digital subsystem to the logical network. The processed information is converted from the digital subsystem into the analogue one by means of *DA converter*, for reversal conversion an *AD converter* is used. There are exist two other elements in a hybrid system. The channels for various analogue quantities are switched over with the *multiplexer*. The *digitally controlled function converter* enables one to express various functional dependences of analogue quantities.

The *analogue models and computers* enable one to realize fundamental mathematical operations of adding, multiplying by a constant, and time integration very easily and also to utilize the functional converter, e.g., for multiplication or division or to set arbitrary change. Despite being used only rarely at present, the analogue computer is the technical means of building multi-purpose analogue models. The single-purpose analogue models are used as controlling models which operate in real time together with an object. Their speed can be high and is provided by parallel information processing on the one hand or by an applied technical solution on the other. Usually, the accuracy is less and the complexity and expenses are higher.

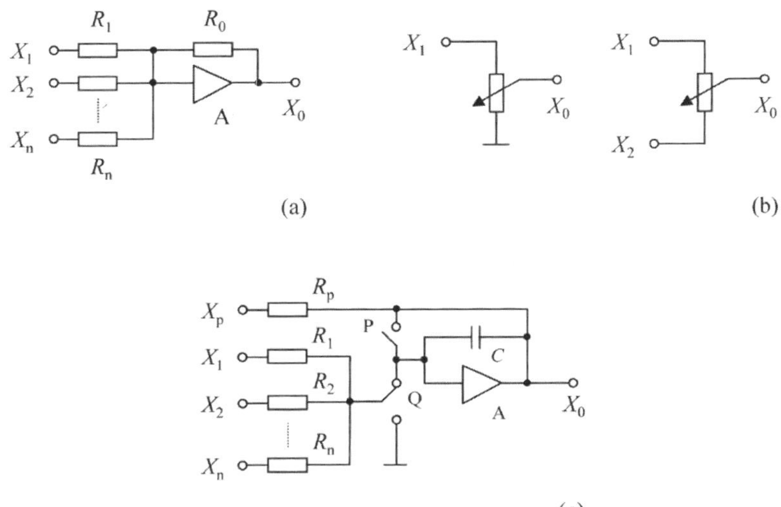

Fig. 8.7. Fundamental analogue operations – addition (a), multiplication by a constant (b) and time integration (c).

The three fundamental analogue operations are described in Fig. 8.7. The addition is performed according the following relation

$$X_0 = -\sum_{j=1}^{n} k_j X_j, \quad \text{where} \quad k_j = \frac{R_0}{R_j}, \tag{8.56}$$

multiplication by a constant in compliance with the relation

$$X_0 = kX_1, \text{ or } X_0 = k X_1 + (1 - k) X_2, \tag{8.57}$$

and the time integral is determined by the relation

$$X_0 = -\left[\int_0^t \sum_{j=1}^{n} k_j X_j \, \mathrm{d}t + X_p\right], \quad \text{where} \quad k_j = \frac{1}{R_j C}, \tag{8.58}$$

and P, Q are the controlling inputs. For the case of functional changes

$$X_0 = f(X_1), \tag{8.59}$$

the connection of the elements depends on the kind of function realized and can be fixed (limiter, quadrator, etc.) or adjustable (converters or functions).

8.4 Applications

The following examples involve the application of computer deterministic models to solve tasks of solid mechanics, thermomechanics, magneto-thermomechanics, bio-thermomechanics, physical technology, mechanical engineering, and technical cybernetics. The diversity of various types of tasks and the application of diverse deterministic numerical models, that are based on the finite difference, finite element, and volume element methods, illustrate various specific approaches in deterministic computer modeling. The examples involve relatively simple as well as complex tasks.

8.4.1 Mechanics of Solid Bodies

The beam bending during mechanical loading is solved in the following example of a beam of partially constant strength. The determination of calibration coefficients of the hole drilling method for residual stress evaluation is the task of the next example.

Example 8.1 *Linear elasticity. Bend stresses in a beam* [8.27]

Object of the solution The results of a direct experiment on a physical model of a beam of partially constant strength have to bd compared with the results of an analytical and numerical models.

Task characterization The given force F is loaded on the free end of a fixed beam. The stresses on the beam surface are determined from the solution of a direct three-dimensional steady task.

Applied method The results are obtained using analytical method, numerical FEM method and strain gauge measurement on a physical model of a beam.

Analytical model

The beam diagram in Fig. 8.8. At the fixation point it has the width b_1 and at the point with the acting force F its width is b_2. The part l_1 is designed as the constant-strength beam and the part l_2 is the constant-width beam. The origin of the coordinate system is at the free end of the beam. In compliance with the Hook law, the stress σ_x for a pure plane bend in a rectangular-cross-section beam, is

$$\sigma_x(x,y) = E\,a\,y = \frac{M_o}{J_z}\,y, \tag{8.60}$$

where E is the modulus of stress elasticity, $a = M_a (EJ_z)^{-1}$ is the proportionality constant, $M_a = F_x$ is the moment with respect to the neutral axis, and $J_z = bh^3(12)^{-1}$ is the quadratic inertia moment of the rectangular-cross-section beam. For the constant-strength beam part, $x^{b-1} = L_2 b_2^{-1}$ must hold, and stress σ_x is given by the following equation

$$\sigma_x(x,y) = \begin{cases} \dfrac{12F}{b_2 h^3} x\, y & \text{for} \quad x \in \langle 0, l_2 \rangle, \\ \dfrac{12F}{b\, h^3} x\, y & \text{for} \quad x \in \langle l_2, l_2 + l_1 \rangle. \end{cases} \tag{8.61}$$

The beam deflection can be solved by integration of the equation

$$y_p'' = \frac{M_o}{E J_z}. \tag{8.62}$$

For the total beam deflection, the following is valid

$$y_p = \frac{2F}{E b_2 h^3}\left(x^3 - \left(6l_1 l_2 + 3l_2^2\right)x + 3l_1^2 l_2 + 6l_1 l_2^2 + 2l_2^3\right) \text{ for } x \in \langle 0, l_2 \rangle, \tag{8.63}$$

$$y_p = \frac{12F l_2}{E b_2 h^3}\left(\frac{1}{2}x^2 - (l_1 + l_2)x + \frac{1}{2}(l_1 + l_2)^2\right) \text{ for } x \in \langle l_2, l_2 + l_1 \rangle. \tag{8.64}$$

Numerical model

Cosmos/M software, the linear statics solver, was used for the numerical model. The task is defined for the following values of the steel beam: $F = 100$ N, $E = 2.1 \times 10^{11}$ Pa, $\nu = 0.28$, $L_1 = 210$ mm, $L_2 = 60$ mm, $h = 5$ mm, $b_1 = 90$ mm, $b_2 = 20$ mm. The stress and deflection in the longitudinal beam axis at the upper beam surface are compared with the

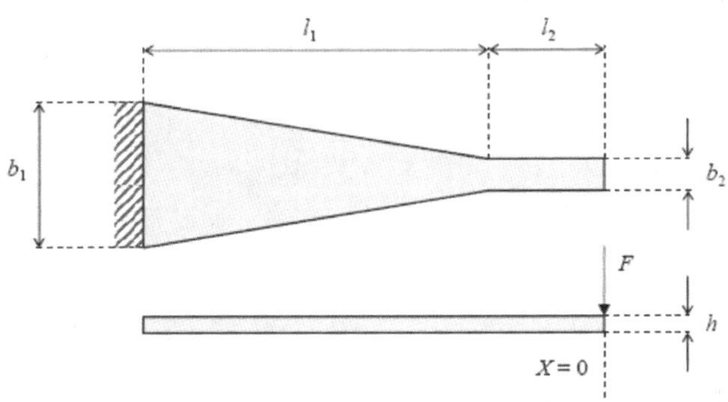

Fig. 8.8. Diagram of a beam with partially constant strength.

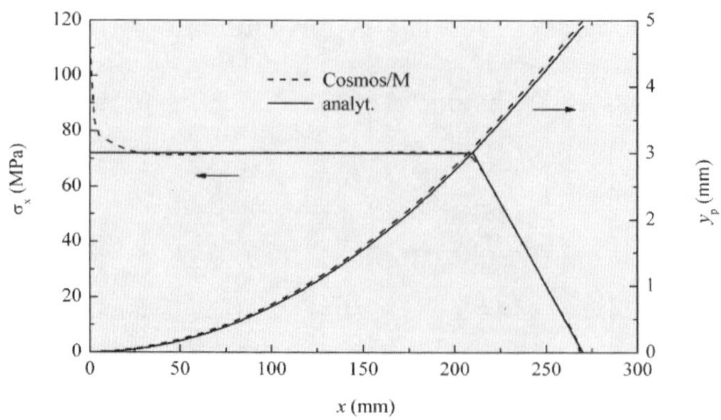

Fig. 8.9. Surface stress σ_x in the longitudinal beam axis and beam deflection *yp* dependent on the distance from fixation point ($x = 0$ mm). The force acting point is at $x = 270$ mm.

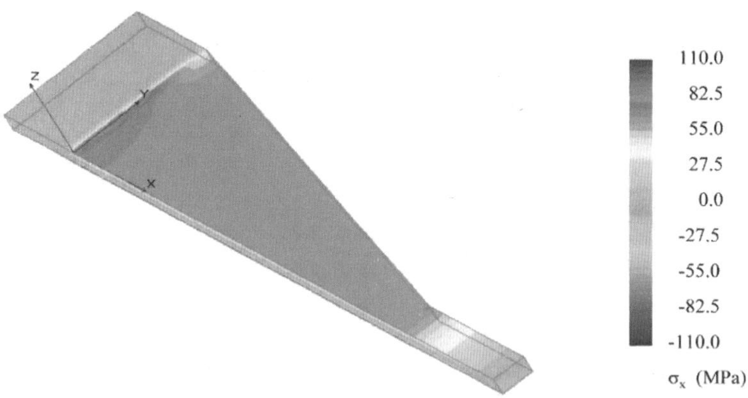

Fig. 8.10. The surface stress .eld σx on the beam.

results of the analytical solution in Fig. 8.9. From these, good agreement between analytical and numerical models can be seen. However, the solution for the numerical model show that the stress field regularity is affected due to edge effects near the fixation point and near the points with sharp changes of the beam shape. This is obvious from the nature of the surface stress field (Fig. 8.10). This fact must be kept in mind in strain gauge measurements especially.

Physical model

On the top and bottom surfaces of the constant-strength part of the steel beard model, miniature resistance strain gauges are glued, which are connected to a strain gauge measuring system. The beam has been put into a loading device (Fig. 8.11). In the considered case, the loading force F is in the range of $F \in \langle 6; 81\text{ N}\rangle$. The stresses are evaluated from the measured strain values. The analytical and numerical solutions are found for the same values of loading.

The physical model experiment confirmed the results of the numerical and analytical models. The results comparison of the analytical, numerical, and physical models show good agreement, Fig. 8.12. The differences between analytical land physical models are in the range of 0.15 to 0.35 MPa, which is essentially at the precision boundary of the strain gauge measurement.

(a) (b)

Fig. 8.11. Side (a) and front (b) views of the beam and loading device physical models.

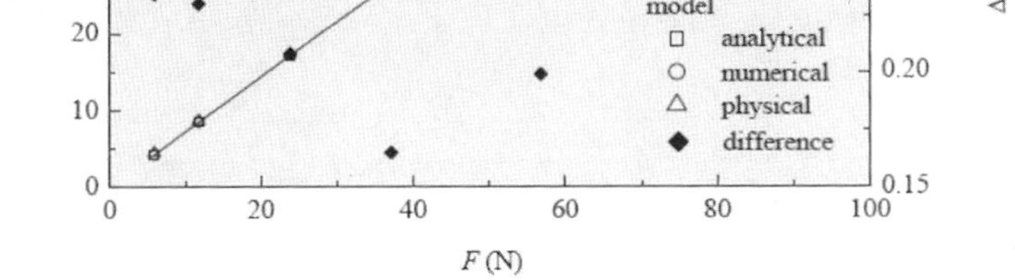

Fig. 8.12. Comparison of beam stresses from the analytical, numerical, and physical models – absolute values (a) and differences between the analytical and physical models (b).

Example 8.2 *Elasticity and strength. The hole drilling method for residual stress determination* [8.27]

Object of the solution Computer numerical model to be created for determination of calibration coefficients of the hole drilling method for residual stress measurements. The knowledge of calibration coefficients is the necessary condition .fox. residual .stresses evaluation.

Task characterization The two- and three-dimensional unsteady tasks utilize dimensional analysis for the modification of model geometry. The tasks are based on the gradual change of geometry during unsteady solution to include increasing depth of the hole within the drilling process.

Applied method A numerical finite element method (FEM) is used for the solution.

Principle of the hole drilling method

The semi-destructive hole drilling method for residual stress measurement is based on the measurement of strains relieved after drilling a hole into the material surface. The strains are measured by strain-gauge rosettes glued on the material surface. The detail of a miniature strain-gauge rosette is given in Fig. 8.13a. The principle of residual stress measurement is presented in Fig. 8.13b,c where the material before and after hole drilling is shown schematically.

The relation between residual stress relaxation and the resulting relieved surface strain is generally valid

$$\varepsilon_r = A\left(\sigma_x + \sigma_y\right) + B\left(\sigma_x - \sigma_y\right)\cos\left(2\theta\right), \tag{8.65}$$

where ε_r is the induced radial strain in the direction of angle θ, σ_x and σ_y are the residual stress components in principal directions, and A, B are dimensional calibration coefficients. For a *through-hole,* the following relations are valid

$$A = -\frac{1+\nu}{2E}\frac{1}{R^2}, \tag{8.66}$$

$$B = -\frac{1+\nu}{2E}\left(\frac{4}{1+\nu}\frac{1}{R^2} - \frac{3}{R^4}\right). \tag{8.67}$$

However, the majority of the residual stress measurements is performed using a *finite-depth* hole, where the coefficients A, B cannot

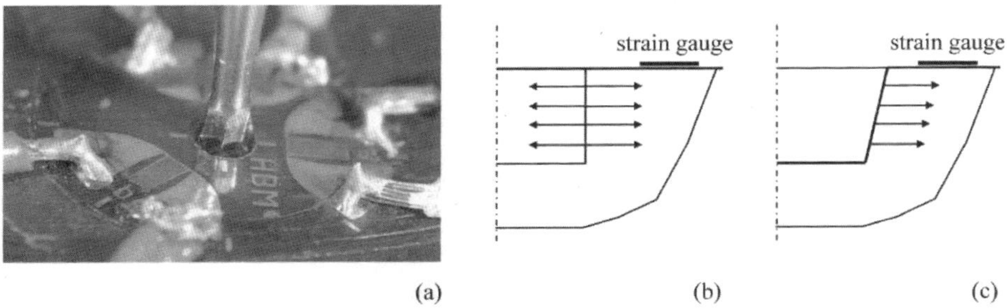

Fig. 8.13. The hole drilling method – photo of an experimental detail (a), nature of residual stresses before (b) and after drilling (c).

be used. In these cases, the coefficients $\overline{A},\overline{B}$ are used, and their values can be determined by experimental or numerical calibration. Because experimental calibration has limited possibilities, primarily the numerical calibration is used. More often, these coefficients are used in the dimensionless forms $\overline{a},\overline{b}$ that are material-independent

$$\overline{a}=\frac{2E\overline{A}}{1+\nu},\ \overline{b}=2E\overline{B}. \tag{8.68}$$

In the preceding relations, ν is the Poisson number, E is the modulus of elasticity, $R = r/r_a$ is the dimensionless distance from the hole center, r_a is the hole radius, r is the distance from the hole center and the finite length of the strain gauges is not considered.

Modeling procedure

The numerical model for determination of calibration coefficients is created in the Cosmos/M computational system. Generally, for a finite-depth hole, this is a complete three-dimensional task (Fig. 8.14b); for a symmetrical-in-space task from the geometry and boundary conditions points of view, an axially symmetrical two-dimensional model (Fig. 8.14a) can be used. The size of elements near the hole is required to be approximately $r_a/10$ to obtain precise results. With the circular-hole size in the mm-order, it is difficult to use the necessary network element size in the Cosmos/M system. Therefore, a dimensional analysis has been used to determine the scales between the modified model (M) and the original model (O), following from the equality of the similarity criteria between them. For the modified quantities, the following relations are valid

$$L_M = L_O\mu_L,\ \sigma_M = \sigma_O\mu_\sigma,\ \ E_M = E_O\mu_E,\ \rho_M = \rho_O\mu_\rho. \tag{8.69}$$

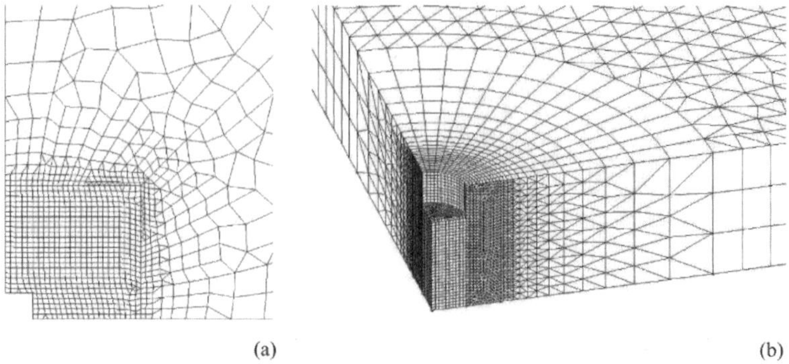

Fig. 8.14. The computational network for the model of relieved strains in the vinicity of a finite-depth hole – 2D network for the axially symmetrical model (a), 3D network for a general model (b).

Selecting the length scale μ_L the stress scale $\mu_\sigma = \mu_L^{-1}$, the modulus of elasticity scale $\mu_E = \mu_L^{-1}$ and the density scale $\mu_\rho = \mu_L^{-3}$ are obtained. Other quantities, with no length dimension, remain without change.

In the case of an unsymmetrical task, the calibration coefficients $\overline{a}$ and $\overline{b}$ have to be considered, while in the case of symmetrical task, the calibration coefficient $\overline{b}$ does not take effect.

Figure 8.15 shows the hole deepening process (a) using the hole drilling method and its simulation procedure (b) in the determination of the calibration coefficients, specifically for constant along-the-depth stress and a symmetrical task. It is considered that with each

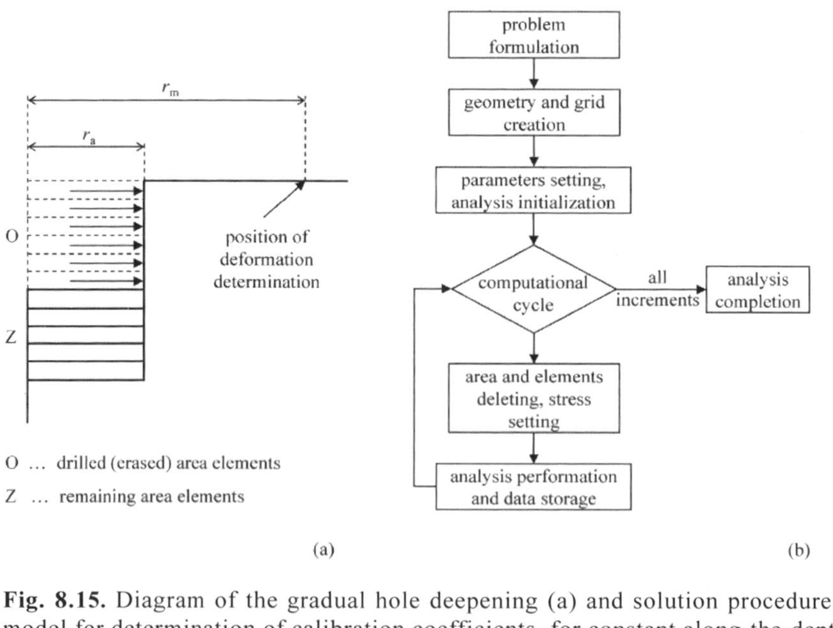

Fig. 8.15. Diagram of the gradual hole deepening (a) and solution procedure (b) in a model for determination of calibration coefficients for constant along-the-depth stress.

drilling-off increment a part of the residual stress is relieved. With the symmetrical task, the influence of the coefficient $\overline{b}$ does not appear and the coefficient $\overline{a}$ is determined only. The relation between the relieved stress, hole depth and measured strain is given by the coefficient $\overline{a}$, namely as the function of the hole depth h, hole radius r_a and strain-gauge rosette radius r_m. On the basis of task parameters, the geometry, computational network, boundary conditions and solution parameters are set up. The computational cycle is performed, those every step includes deletion of actual material element under the hole bottom, setting the boundary condition in the form of residual stress a on the new hole surface, executing the computation and saving the value of radial strain ε_r on the upper material surface at the distance r_m from hole axis. The stress is acting on the whole, already drilled-off, part of the hole. The vector of the values of the coefficients $\overline{a}$ is obtained from the simulation, where the rows i correspond to the hole depths.

In the case of variable along-the-depth stress, the procedure is more complex and several techniques exists for evaluation of residual stresses. The integral technique is based on the fact that every drilled-off increment influences the relieved strain in strain-gauge position separately. The influence of this element is dependent non only on its depth H, but also on the total hole depth h. Simultaneously, the coefficient $\overline{b}$ is considered in an unsymmetrical task. The simulation procedure is complicated, because the loading stress is set up gradually on all increments from the surface up to depth H_i and the computations are performed separately for individual loading depths H_i from H_1 up to $H_n = h$. Due to the unsymmetrical task, two computations are made, first for normal stress loading, second for shear stress loading. The stored computed values of radial strains ε_r and loading stresses are used for determination of calibration coefficients $\overline{a}_{i,j}$ and $\overline{b}_{i,j}$ in the form of triangle matrices, where rows i correspond to the hole depth and columns j to the depth of acting stresses.

Results

As an example of results from three-dimensional unsymmetrical task with variable residual stress along the hole depth, the dimensionless calibration coefficients $\overline{a}$, $\overline{b}$ are shown in Fig. 8.16. They represent the matrices of coefficients as functions of the total hole depth h and stress acting depth H related to the radius r_m and determined for the hole radius $r_a = 0.353\ r_m$, the strain-gauge length $l = 0.588\ r_m$, and the strain-gauge rosette radius $r_m = 2.55$ mm. In contrast to the experimental calibration, the numerical computer model application for numerical calibration is imore general and wide-spread. The obtained calibration coefficients

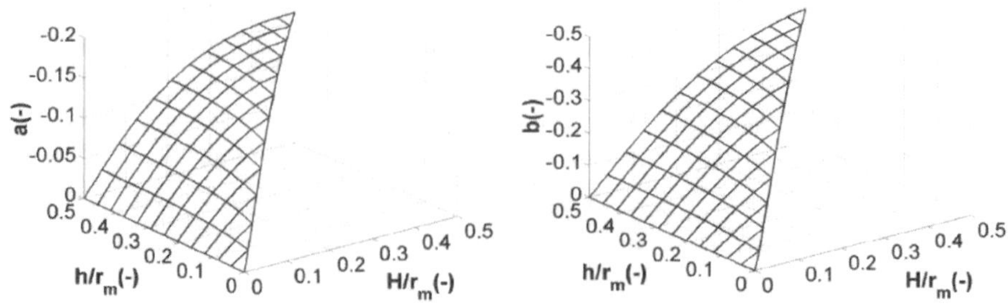

Fig. 8.16. Dimensionless calibration coefficients a, *b* determined by the numerical calibration for variable along-the-depth stress and unsymmetrical task.

are used fori the evaluation of residual stresses in the material from the strain measurement. Thus, the created computer model forms a numerical support for residual stress measurement by the hole drilling method. More information on the task solved is described in [8.27].

8.4.2 Thermomechanics

In the following example, the three-dimensional computer model of a continuous push furnace is created to solve combustion products flow and heat transfer. The thermal box-barrier is designed in the next example including the comparison of numerical and natural physical models.

Example 8.3 *Aero-thermomechanics. Combustion products flow and heat transfer in a continuous furnace [9.18]*

Object of the solution A computer model of the combustion products flow and heat transfer in a continuous push furnace has to be created, which respects the complex inner furnace geometry. The solution has to be made for a maximum burner power output. The turbulent flow, the heat transfer by convection, conduction and radiation are assumed.

Task characterization The three dimensional direct task is steady, but the model enables it to be extended to an unsteady one. The simulation results are followed up by an optimization control system for batch heating in a furnace.

Applied method To create a numerical computer model, the finite volume method (FVM) has been used.

Modeling procedure

A view of the internal surface of the whole modeled furnace space is in Fig. 8.17. The furnace is approximately 14 m long, 3 m wide, and 4 m high. In the figure, the places important for assembling the numerical mathematical model are marked. Among them are nine high-velocity burners which are arranged non-symmetrically so that the turbulent flow in the furnace can be formed. The reference batch in the furnace is prism-shaped with the dimensions 90 × 90 × 1250 mm. The discrete computational network is unstructured, composed of irregular tetrahedrons and has 154 285 cells in total (Fig. 8.18).

In formulating a mathematical model, important role is played by the boundary conditions and physical properties. With respect to the boundary conditions, the surface emissivity of the walls and loss heat fluxes are considered. The temperature of combustion products going

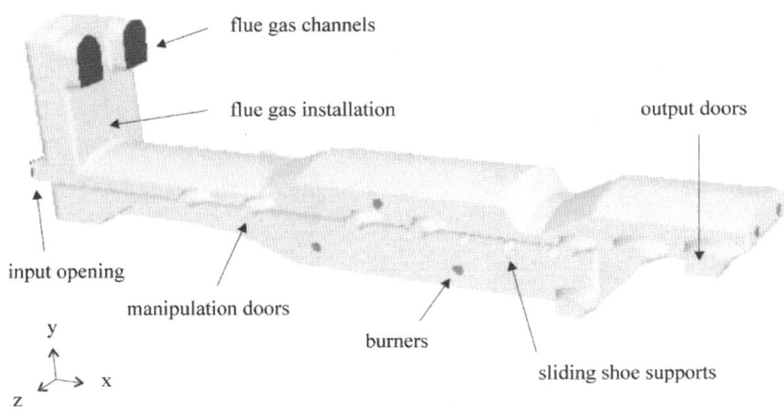

Fig. 8.17. Modeled continuous push furnace.

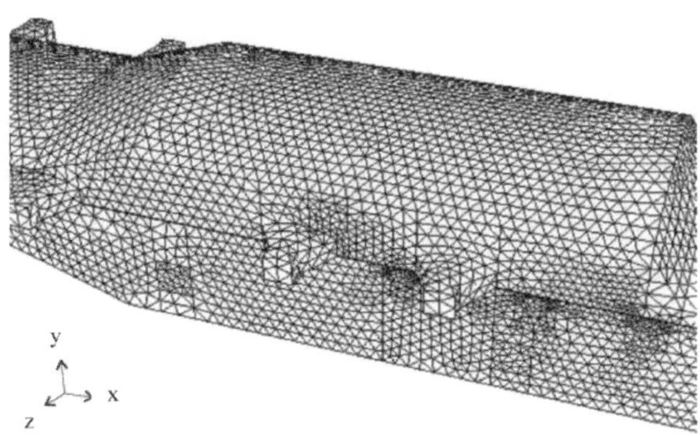

Fig. 8.18. Computational network of the furnace model.

from the burner into the furnace space is 1700°C, and the combustion products velocity is 60 m s^{-1} in the burner orifice. The batch enters the furnace with the surface temperature of 100°C and linearly increases up to 1200°C, and in the rear part of the furnace is already constant. The physical properties of the combustion products are regarded as temperature dependent.

The solution is executed with a commercial system for computational fluid dynamics FLUENT, which is based on the FVM method. Because the set of solved mathematical model equations is non-linear, the change of individual quantities between two subsequent iterations must be controlled. The computation convergence is monitored by the residue values and surface integrals of selected quantities.

Results

The velocity field in the transversal cross-section passing through the axis of the 5th side-wall burner is given in Fig. 8.19 and shows the nature

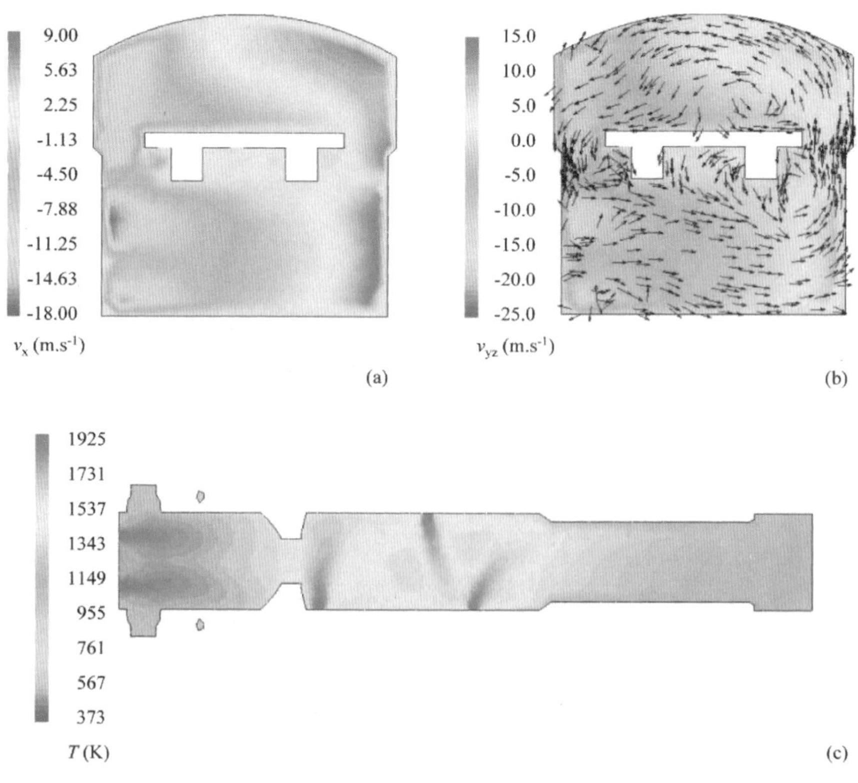

Fig. 8.19. Fields of x-axis velocity (a, b) and direction of velocity in the transversal cross-section (b), temperature field (c) in the horizontal cross-section of the furnace led by burners' axes.

of the flow in this furnace part. In the left part (a), the velocity vector component v_x is shown and in the right part (b), besides the velocity field v_x, also the flow velocity direction v_{yz} in the furnace transversal cross-section is presented and expressed by arrows. In the lower part (c) of this figure, the temperature field is shown in the longitudinal horizontal cross-section of the furnace that is led through the axes of the burners above the batch. In this way taking advantage of selected cross-sections of the furnace, the fields of selected quantities can be observed in detail.

From the values found of heat fluxes and the temperature differences between the batch surface and the furnace atmosphere at the distance of 150 mm, the convection heat flux and the coefficient of convection heat transfer have been evaluated. The convection heat transfer coefficient depending on the distance from the furnace input opening is shown in Fig. 8.20. The low convection heat transfer values, approximately 20 W m^{-2} K^{-1}, are due to the fact that there are no burners in the front part of the furnace. The highest values are reached near the burners' position and are up to 70 W m^{-2} K^{-1}.

The results of the modeling of the aero-thermomechanic process in the furnace, as presented in this example, have been verified by direct furnace measurement under approximately equal conditions. This confirmed the good agreement of the computer model with reality and enabled the partial modification of the mathematical model. It follows from the modeling results how much heat coming from burners into the furnace is led into the batch (50.2 %), into the furnace walls (1.7 %), and out of the furnace (47.6 %), with the rest (0.5 %) being the solution error. The large values of heat flux going out of the furnace through combustion products outputs is given by the condition of burners maximum power output. During the real operational conditions,

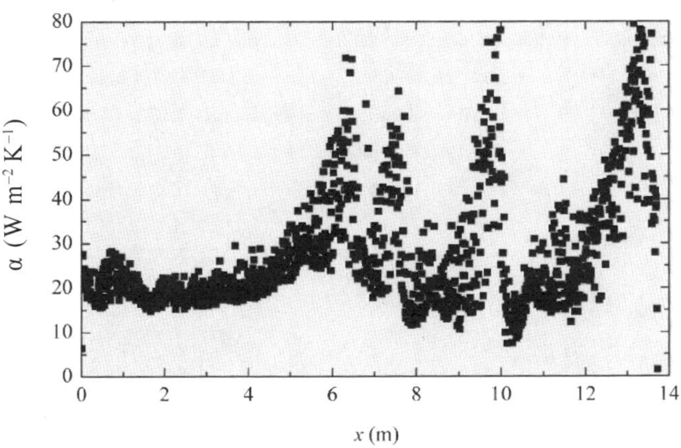

Fig. 8.20. Distribution of the coefficient of convection heat transfer into the upper batch surface.

the burners power output is approximately 70 % of their maxima. It is interesting that of the heat entering the batch, as much as 92.5 % has the character of radiation heat flux, i.e. the convection heat fluxes are low under the high-temperature conditions in the continuous furnace. Detailed results are given in [9.18].

The disadvantage of this model are the prescribed values of the batch surface temperature inside the furnace. The generalized model of the continuous furnace includes the unsteady process of the batch movement in the furnace and the computation of the temperature field in the batch. The batch movement is realized by the enhancement of the FLUENT model using own user-defined function. These types of functions are automatically started by the computational system in each iteration of the calculation. The basic idea of this enhancement is to modify the temperature values in network nodes in the batch zone to simulate the batch movement inside the furnace.

Example 8.4 *Measuring systems. Design of a protective thermal box-barrier [4.3]*

Object of the solution To create a two-dimensional computer model of a thermal box-barrier (TBB) which protect a measuring system from high temperatures for several hours during its passage through a furnace at a temperature of about 1200 °C. The model serves to find temperature field in TBB. The thickness of walls of the four-layer thermal box-barrier has to be optimized by the one-dimensional task. The model solution results have to be verified experimentally.

Task characterization These are a direct unsteady one- and two-dimensional task and optimization unsteady one-dimensional task that respect the phase changes of TBB materials. The maximum internal temperature is the restrictive condition which limits the function of a multi-channel measuring system storing the measured data which passes through the furnace at high temperature. The optimization task is solved at a large number of direct tasks.

Applied methiod The unsteady tasks are solved by the finite element method (FEM) and the finite volume method (FVM).

Thermal box-barrier characterization

The thermal box-barrier (TBB) function principle and its limiting requirements are given in Fig. 8.21. The TBB wholly surrounds the protected device (measuring system, etc.) in the internal space and lowers the heat penetrating the barrier. The temperature of the internal environment have not to exceed the maximum permitted temperature during the TBB exposion to the high temperature external environment. The box-barrier must have low thermal conductivity, in order to decrease the heat penetration into the TBB, but high capacity to slow down the temperature increase in its inner space.

The limiting requirements are determined by the protected device dimensions, maximum allowed temperature of the internal TBB space, the maximum permitted external dimensions of TBB, the external temperature and the time of protection. In the considered case, the maximum TBB thickness is 125 mm, and the maximum allowed temperature of internal space is 70°C. The time required for its remaining in the furnace is 2 to 4 hours.

In order to meet these requirements, the TBB consists of a multilayer structure composed of materials with different properties. The typical high-temperature TBB structure consists of four layers. The *outer insulating layer* of high-temperature insulating material has to reduce the absorption of the heat flux, impacting by radiation from the outer environment and to reduce the heat flux in other layers. The *evaporation layer* consists of a fluid (water). The fluid delays, by its thermal capacity involving the heat of evaporation, the temperature increase by keeping it at the boiling point of water for most of the

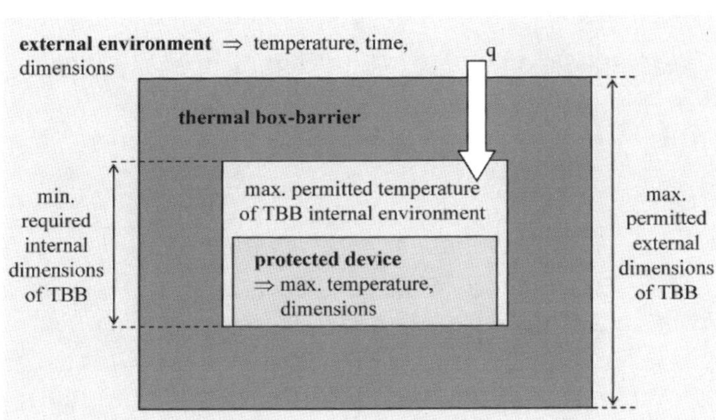

Fig. 8.21. Principle of the thermal box-barrier (TBB) operation.

time and leads the heat back into the outer environments as steam. The *inner insulating layer* of low-temperature insulating material further reduces the heat flux into the *inner capacity layer* having large thermal capacity (steel).

Mathematical phase change models of TBB

The mathematical model of heat-transfer in a TBB represents heat propagation in a heterogeneous thermal system. In the solid-material layers, the heat conduction is considered without inner sources. In the case of the evaporating layer, the heat conduction is combined with the fluid phase change expressed by a strong evaporating-heat influence. In solving this problem, three different mathematical models of the phase change process are taken into account. These are *capacity, enthalpy* and *balance models.* The thermophysical properties must be held as temperature dependent.

The phase change *capacity model* expresses the latent heat by means of the material thermal capacity dependence

$$c_m\left(T_m\right)=\mathrm{c}_{lat}\left(T_m\right) \text{ for } \mathrm{T}_L \leq T_m \leq T_G, \tag{8.70}$$

where T_L and T_G are temperatures of the beginning and end of the change, and T_m is the temperature in the evaporation layer dependent on the space and time. The phase change thermal capacity $c_{lat}(T_m)$ is defined so that it equals the total latent heat of the change l according to the following relation

$$\int_{T_L}^{T_G} c_{lat}\left(T\right)\mathrm{d}T=l. \tag{8.71}$$

The *enthalpy model* operates with a latent heat increment h_{lat} and it holds for the total enthalpy of the layer m

$$h\left(T_m\right)\int_{T_{ref}}^{T_m} c_p\left(T\right)\mathrm{d}T+h_{lat}\left(T_m\right). \tag{8.72}$$

The phase change is considered in the temperature interval (T_L, T_G) in which h_{lat} reaches the values from zero to l.

In the phase change *balance model*, the convective heat transfer is considered and the space-constant fluid temperature and its time change are presumed in the whole volume of the layer according the balance of heat fluxes on the layer boundary. The volume source or sink of the layer heat, determined by the thermal fluxes on its boundary, is

$$q_{V,m} = -\lambda_{m-1}\left(T_{m-1}\right)\frac{S_{m-1}}{V_m}\frac{\partial T_{m-1}}{\partial x_i}n_i\Bigg|_{\Gamma_{m-1}} + \lambda_{m+1}\left(T_{m+1}\right)\frac{S_m}{V_m}\frac{\partial T_{m+1}}{\partial x_i}n_i\Bigg|_{\Gamma_m} . \quad (8.73)$$

It is consumed for heating and evaporating the fluid. λ_m is the thermal conductivity of the layer m, S_m is the interface area of the layers m and $m+1$, V_m is the volume of the layer m, n_i is the component of the unity vector which is normal to the area, and Γ_m is the interface of the layers m and $m+1$. For the temperature above the evaporation point T_G, the internal temperature of the evaporation layer is not defined by the balance model. Instead, the heat transfer between the layer boundaries is described by the boundary condition of radiation heat transfer.

Modeling procedure

The limiting requirements for the TBB application being considered, where high-temperature protection is required for the measuring system, have already been mentioned. To satisfy these conditions, the optimized structure of layers and of the total arrangement are required.

Figure 8.22a shows a diagram of the arrangement of the 1D model of the TBB wall transverse cross-section. The model serves to compare the different phase changes models and to optimize the four-layer TBB. The balance model was chosen for optimization and different variants of the structure and barrier parameters are solved using the Cosmos/M computational system that uses the FEM method. The TBB structure involves an outer insulating layer 1, a water evaporating layer 2, an inner insulating layer 3, and a capacity layer 4 (steel plate). In adition, d denotes the thicknesses of individual layers, T_P is the outer environment temperature, T_S is the inner TBB wall temperature, q is the heat flux,

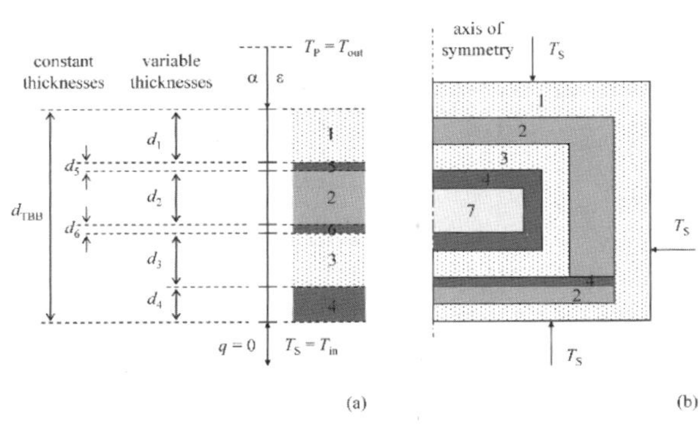

Fig. 8.22. Diagram of the 1D model for four-layer TBB optimization (a) and the 2D model of the resulting optimized TBB (b).

T_{out}, T_{in} are the temperatures of the outer and inner environment, α is the heat transfer coefficient to the external environment, and ε is relative TBB surface emissivity. Figure 8.22b displays the 2D model of the optimized TBB, in which the inner TBB space is denoted as 7. The condition of a perfect thermal contact between individual TBB layers is assumed. The 2D model of the TBB cross-section is solved using the FLUENT computational system that uses the FVM method.

Results

The typical thermal behavior of the four-layer TBB, exposed to a thermal shock, is shown in Fig. 8.23. It is valid for the outer environmental temperature T_{out} = 1200°C and the TBB surface emissivity ε = 0.7. In the figure, three time–temperature courses are presented in characteristic places of the TBB. The results are given for the capacity and balance phase change models. The outer TBB surface temperature is denoted as T_1, the outer and inner surface temperatures of the water layer are denoted as T_2 and T_3, and the inner surface temperature of the capacity layer is T_4 . The surface TBB temperature quickly increases to the value of the external environment. The water temperature reaches the phase change temperature after 30 minutes and remains constant until the water is evaporated. After that, the temperature of the walls of the water layer increases. The temperature of the inner TBB environment slowly incrases, the rapid change comes after water evaporation. The differences between the models are observed mostly in the water layer zone. The difference of the results between them influences only slightly the inner TBB environment temperature.

Figure 8.24 shows the temperature field in the cross-section of the optimalized TBB. Compared to the 2D model of the original TBB,

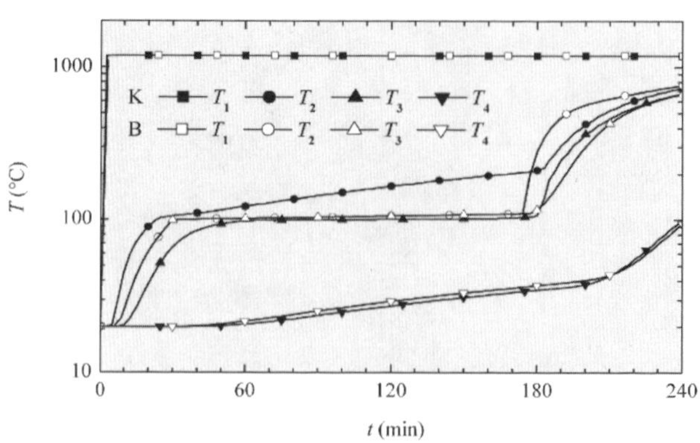

Fig. 8.23. Temperature–time courses in characteristic places of TBB for capacity (C) and balance (B) models.

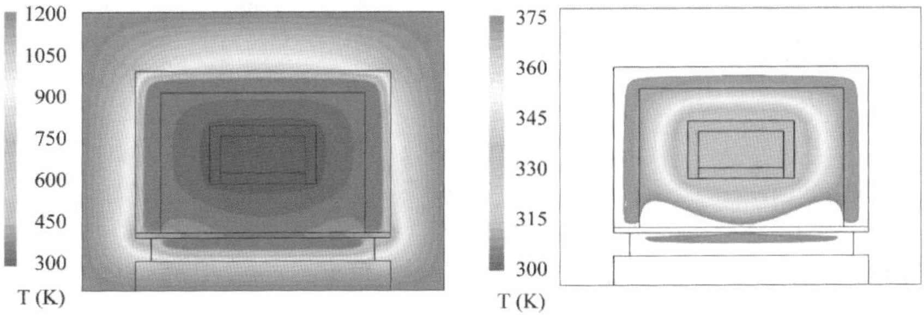

Fig. 8.24. The temperature field in the cross-section of the optimalized TBB after 4 hours of heating.

the heat transfer to the inner environment is strongly reduced in the optimalized TBB. The temperature increase of the inner space is only half of the value for original TBB. The measuring system, located in the internal TBB environment and passing through the furnace with a temperature of about 1200°C, can work reliably for more than 4 hours. This was confirmed by the operation measurements. The model parameters and detailed results can be found in [4.3].

8.4.3 Physical Technology

In the following example, the computer model of a multilayer sample heating with continuous laser is created for the evaluation of thermal properties from the measured surface temperature field. The next example concerns the laser treatment of the material surface. The three-dimensional computer model of heating of samples is created involving the complicated time-dependent boundary condition of a moving continuous laser.

Example 8.5 *Laser technology. Determination of the thermal properties of thin layers and coatings*

Object of the solution A numerical model for surface heating of a multilayer material specimen with continuous laser will be designed and used to determine the thermal properties of coatings and thin layers from the measured data of the surface temperature field. An analytical model for the same task will be designed and compared with the numerical model.

Task characterization This is an indirect unsteady two-dimensional task that is rotationally symmetric. For modeling of the thin layers on the substrate surface, the dimensional analysis is used in an interesting way to modify only a part of the model.

Applied method A numerical FEM method has been used which utilized the experimental data obtained by the laser thermographic method LQT (Laser Quasistatic Thermography).

Principle of the LQT method

The non-contact laser thermographic method LQT is developed for the determination of the thermal conductivity of layers [8.18]. Continuous laser radiation impacting the defined specimen surface is the thermal source. At the same time, the temperature field is measured in this area by a thermography camera. First, a measurement is taken of a substrate whose thermophysical properties are well known. Consequently, the measurement with the same parameters is done for the substrate with the examined layer. Substrates of both samples must consist of the same material and have the same shape and dimensions. To obtain equal parameters, it is suitable to coat the examined surfaces of both specimens with a thin layer, for example, of a paint which provides equal absorption of laser radiation and has equally high surface emissivity for the infrared radiation band for the thermography measurement.

Due to laser radiation acting on the surface, first the temperature increases fast at the place of radiation impact. After a certain time, the quasistatic mode sets in, having a nearly constant or very slowly varying temperature (Fig. 8.25). In the quasistatic mode region, the measured difference of the surface temperature for the substrate and the substrate with the layer is caused mainly by different thermal conductivity of the layer as compared to that of the substrate.

The first step in determining the thermal conductivity of the layer is to find the boundary conditions. This involves the amount of the absorbed radiation and of heat led away into the specimen holders. The boundary conditions are determined by solving the indirect task for the substrate with known thermal parameters. The second step consists of utilizing the obtained boundary conditions for solving the indirect task for the substrate with layer. The thermal conductivity of the layer is determined from the fact that the calculated temperature course in a defined place on the specimen surface would be identical with the measured temperature in the quasistatic mode region.

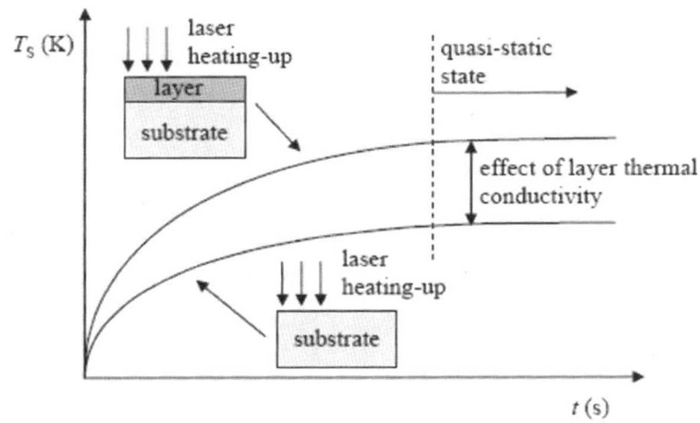

Fig. 8.25. Laser quasistatic thermography method – the influence of the thermal conductivity of the layer on surface temperature.

Numerical and analytical models

A diagram of the model is given in Fig. 8.26. The mathematical model is based on the presumption of a cylindrical sample with the diameter R and thickness d_s which is coated with two layers with thicknesses d_1 and d_2. The thermophysical properties of the substrate and layer materials are expressed by specific thermal capacity c_p, density ρ, and thermal conductivity λ, for which the direction dependence can be distinguished, i.e. perpendicular $\lambda_\perp$ to the specimen plane or parallel $\lambda_\parallel$ with it.

The imperfect contact between the layers and between the layer and the substrate is expressed by thermal resistances R_{h2l} and R_{h1s}. The heat conducted away to the environment is described by the heat transfer coefficients a_h and the environment temperature T_e. A laser with a circular shaped beam with the radius R_l, impacting in the center of the specimen, is the heat source. The space distribution of the impacting energy is described by the dimensionless function $\varphi(r)$ and the time distribution by the function $\Theta(t,t_l)$, where t_l is the total pulse length.

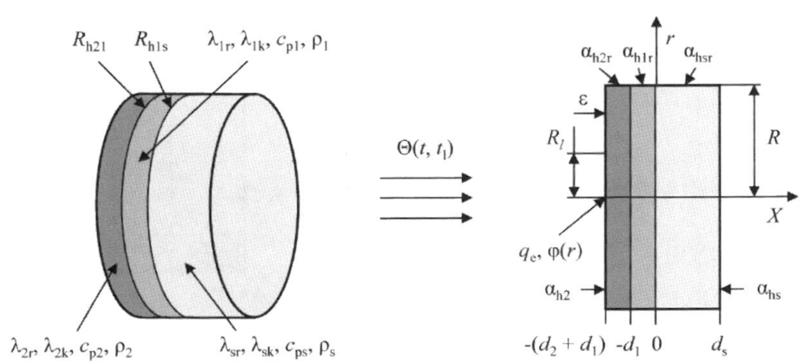

Fig. 8.26. Diagram of the model for the LQT method.

The laser pulse is absorbed in the thin surface layer with thickness $\varepsilon \ll d_2$ and the surface density of energy converted to heat is q_e.

The *numerical model* is created utilizing the computational system Cosmos/M that uses the FEM method. The task is solved under certain preconditions and simplifications. Among them are a perfect contact between the layers and the layer and the substrate; thermal conductivity equal in all directions; the equal initial temperature in all layers and the substrate; the laser beam energy absorbed in the specimen surface; the energy density constant over the cross section of the beam; the rectangular laser pulse; and the non-zero heat transfer on side surfaces. The task is solved as unsteady without internal sources and is axially symmetrical in the x axis.

It is convenient to modify the numerical model of thin layers on the substrate surface by the similarity theory and dimensional analysis. In this way, the layer thickness can be increased together with the modification of other parameters, so that the similarity criteria are equal for the object and the model. Above all, this involves the Fourier number *Fo* for time and the Biot number *Bi* for the boundary condition. The diagram for the model of 'layer stretching' with the application of dimensional analysis is shown in Fig. 8.27. Modification is performed on the layer zone only, the substrate zone remains without change. This modification is important for numerical solution in accordance with the lowering of the number of finite elements while keeping the maximum allowed ratio of the element sides.

From the equality of the similarity criteria for the model (M) and the object (O) one obtains

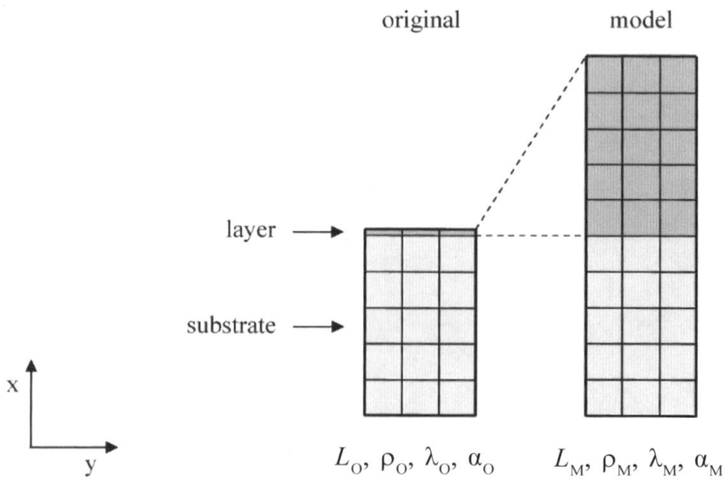

Fig. 8.27. Diagram of the model modification called 'layer stretching' using dimensional analysis.

$$Fo_{x,M} = Fo_{x,O},\ Fo_{y,M} = Fo_{y,O}, \tag{8.74}$$

$$Bi_{x,M} = Bi_{x,O},\ Bi_{y,M} = Bi_{y,O}. \tag{8.75}$$

The similarity scale indicators corresponding to these equalities are

$$\frac{\mu_{\lambda_x}\mu_t}{\mu_{c_p}\mu_\rho\mu_{L_x}^2} = 1, \quad \frac{\mu_{\lambda_y}\mu_t}{\mu_{c_p}\mu_\rho\mu_{L_y}^2} = 1, \tag{8.76}$$

$$\frac{\mu_{\alpha_x}\mu_{L_x}}{\mu_{\lambda_x}} = 1, \quad \frac{\mu_{\alpha_y}\mu_{L_y}}{\mu_{\lambda_y}} = 1. \tag{8.77}$$

The modeling scales which express the ratio of the model quantity to the object quantity can be determined. Modifying only the layer zone of the model and maintaining the y-axis layer zone dimension without change, the following scales have to be set

$$\mu_{L_x} = \mu,\ \mu_{L_y} = 1,\ \mu_t = 1, \mu_{c_p} = 1,\ \mu_{\alpha_x} = 1. \tag{8.78}$$

The scale for heat transfer coefficient in the x-axis direction is equal to one because it express the heat transfer in the x-axis direction that remains unmodified. The following scales are dependent on the *length scale in the x-axis direction*

$$\mu_{\lambda_x} = \frac{\lambda_{x,M}}{\lambda_{x,O}} = \mu, \qquad \mu_{\lambda_y} = \frac{\lambda_{y,M}}{\lambda_{y,O}} = \mu^{-1}, \tag{8.79}$$

$$\mu_\rho = \frac{\rho_M}{\rho_O} = \mu^{-1}, \qquad \mu_{\alpha_y} = \frac{\alpha_{y,M}}{\alpha_{y,O}} = \mu^{-1}. \tag{8.80}$$

The *analytical model* of the LQT method, based on the same geometry and boundary conditions, is created for comparison with the numerical model. It enables the thermal resistance between layers and the layer and the substrate to be included, but does not involve heat transfer on the side walls of the sample.

Results

A model using the FEM enables one to include temperature dependences of individual parameters of thin layers and coatings (thermal conductivity, specific thermal capacity, density, etc.). Various boundary conditions, the influence of convection, radiation, and other parameters can be expressed easily. Solutions found using the analytical model are used for comparison with the results from the numerical model. FEM solution time for a numerical task is 5 to 10 times shorter than

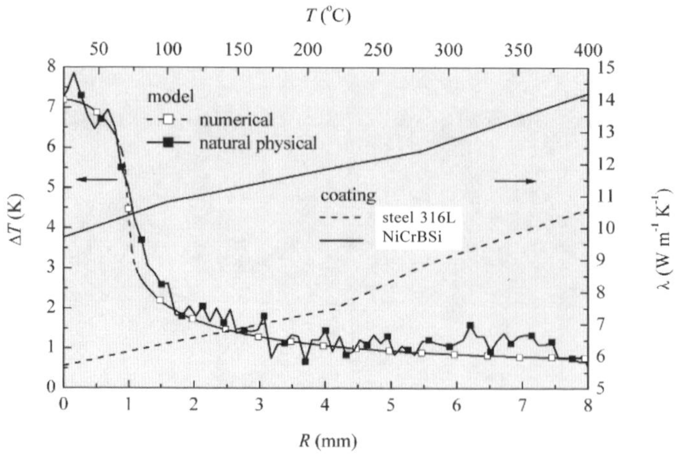

Fig. 8.28. Comparison of the substrate surface temperature increase from numerical and natural models and evaluated thermal conductivity of two coatings depending on temperature.

that by the analytical method. However, the analytical solution enables one to determine the temperature directly in an arbitrary place and time and it is more stable for very short μs-order times. It is less influenced by the ratio of the substrate thickness to that of the layers. With the FEM, this problem can be eliminated by the application of dimensional analysis with 'layer stretching'. On the other hand, the analytical model is portable into arbitrary programming language and can be connected with an optimization algorithm for the determination of layer thermal conductivity.

The results of the thermal conductivity model solution for two coatings with temperature dependence are given in Fig. 8.28. The measured (natural physical model) surface temperature increase is compared with the values obtained from the numerical model. For both coatings, the temperature dependence is approximately linear. The low thermal conductivity of the coating of steel 316L is the result of the fact that this material is less compact than the coating of NiCrBSi.

Example 8.6 *Laser technology. Model of material surface laser treatment [4.4], [5.6]*

Object of the solution A numerical model of the sample heating treatment under the action of a moving continuous laser beam with consideration of non-homogeneous beam profile.

Task characterization These are a direct and indirect unsteady three-dimensional tasks involving a complicated time-dependent boundary condition on the heated surface of the sample.

Applied method The finite element method (FEM) and an experiment on a physical natural model are used to solve the direct and indirect boundary tasks.

Principle of heating treatment

With the surface heating treatment of the material, a laser beam with a certain diameter is scanning the surface area by crossings over the sample using individual linear tracks in order to influence the whole area of the given region. Usually, the laser beam is circular, with the maximum power density in the beam center. The power density drops from the center of the beam to the sides, usually according to the Gauss curve. In order to generally treat the surface in the most homogenous way, a certain overlapping of two adjacent laser beam tracks is necessary.

After the impact, the laser beam is reflected partially from the material surface, with the rest being absorbed to a certain depth depending on the material absorption coefficient. This coefficient is high for metals and the beam is absorbed to a depth of 20 to 50 nm under the surface. Therefore, the surface absorption of the laser power can be considered for metals in a model in most cases.

Modeling procedure

The numerical model of the three-dimensional unsteady heat transfer is created using the Cosmos/M computational system. The sample material is considered to be homogeneous and isotropic. The initial sample temperature is equal in its whole volume. A boundary condition is used for the upper thermally loaded surface which expresses the thermal effect of a moving laser beam and the cooling of the sample due to radiation and convection. The side walls of the sample are thermally isolated. A boundary condition which expresses cooling due to radiation and convection is considered on the rear specimen side.

Figure 8.29 shows the diagram of the arrangement for a test specimen with built-in miniature thermocouple probes below the thermally loaded surface and geometric and boundary conditions. The laser beam moves over the specimen surface on the tracks situated in the x-axis direction between the turning points. When scanning the whole specimen area, the beam moves on individual laser beam tracks that are arranged with

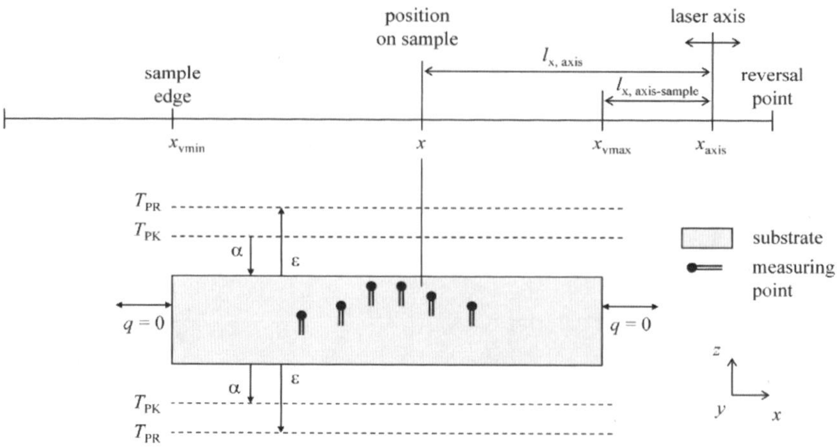

Fig. 8.29. Diagram of the arrangement of surface heating treatment of a sample.

some mutual distance in the y axis so that they cover up the whole sample surface. Over the material surface, the laser beam moves only in the x-axis direction, the motion in the y-axis direction is carried out at turning points located outside the sample surface.

The thermal effect of the laser beam is expressed as the time- and space-dependent variable boundary condition of the 3rd-kind on the sample surface. The actual value of the heat transfer coefficient depends on the distance from the laser beam axis. The temperature of the laser beam that occurs in this boundary condition is considered to be constant. The value of the heat transfer coefficient does not depend on the sample surface temperature.

The following quantities needs to be known to calculate the time dependence of the *total heat transfer coefficient* α_T on the heated sample surface: actual distance of the laser axis from the investigated point on the surface $l_{x,axis}$ in the x-axis direction, actual distance of the laser axis from the sample edge $l_{x,axis\text{-}sample}$ in the x-axis direction, the dependence of the *basic heat transfer coefficient* α_B on the distance from the laser beam axis, the dependence of the *reduction coefficient* k_α on the distance of the laser axis from the sample edge in the x-axis direction, actual distance of the laser axis from the investigated point on the surface $l_{y,axis}$ in the y-axis direction, and the dependence of the *normalized heat transfer coefficient* α_N on the distance from the laser beam axis. The time and space dependences of the total heat transfer coefficient are schematically described in Fig. 8.30.

The value of the basic heat transfer coefficient α_B dependent on the distance from the laser beam axis $l_{x,axis}$ represents the actual space

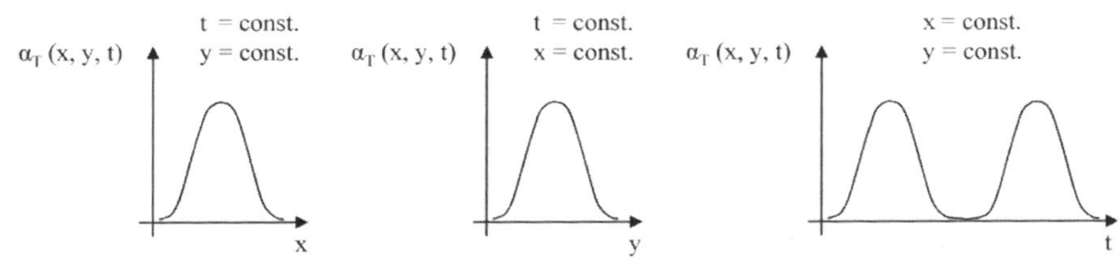

Fig. 8.30. Diagram of the heat transfer coefficient expressing the time-dependent boundary condition of the 3rd-kind.

distribution of the incident beam power intensity. This space distribution is considered to be axially symmetric. The value of the normalized heat transfer coefficient α_N dependent on the distance from the laser beam axis $l_{y,axis}$ is utilized when the boundary condition is expanded to the y-axis. The reduction coefficient k_α dependent on the distance of the beam axis from the sample edge $l_{x,axis\text{-}sample}$ takes into account the situation when the laser axis is outside the sample surface.

At first, the distribution of the total heat transfer coefficient $\alpha_T(x,t)$ on the sample surface in the x-axis direction is determined. It is then expanded to the plane distribution $\alpha_T(x, y, t)$ on the sample surface in the directions of both axes. This procedure can be expressed using the following relations

$$\alpha_T(x,t) = \alpha_B\left(l_{x,axis}(x,t)\right)k_\alpha\left(l_{x,axis\text{-}sample}(t)\right), \tag{8.81}$$

$$\alpha_T(x,y,t) = \alpha_T(x,t)\alpha_N\left(l_{y,axis}(y,t)\right), \tag{8.82}$$

$$l_{x,axis}(x,t) = \left|x(t) - x_{axis}(t)\right|, \tag{8.83}$$

$$l_{y,axis}(y,t) = \left|y(t) - y_{axis}(t)\right|, \tag{8.84}$$

$$l_{x,axis-sample}(t) = 1 \quad \cdots \quad x_{vmin} < x_{axis}(t) < x_{vmax}, \tag{8.85}$$

$$\left|x_{vmin} - x_{axis}(t)\right| \quad \cdots \quad x_{axis}(t)\langle x_{vmin}, \tag{8.86}$$

$$\left|x_{vmax} - x_{axis}(t)\right| \quad \cdots \quad x_{axis}(t)\rangle x_{vmax}, \tag{8.87}$$

$$\alpha_N\left(l_{y,axis}\right) = \alpha_B\left(l_{x,axis}\right)\alpha_{B,max}^{-1}. \tag{8.88}$$

Results

The heat treatment of a steel sample using a CO_2 laser with a power output of P = 1.125 kW, wavelength of 10.6 μm, and the beam radius of r_l = 10 mm is considered.

The sample dimensions are 100 × 70 × 20 mm, the turning points are 10 mm from the sample edge. The laser beam velocity is 24 mm s^{-1}. The laser beam track goes through the center of the front surface. The laser beam motion begins in the left turning point and continues over this track for several times. The sample absorptivity and emissivity are equal ($\varepsilon = 0.7$). The space distribution of the heat transfer coefficient α_B (r) dependent of the distance r from the beam axis, describing the laser beam, is considered, with the maximum determined by the following equation

$$P = \varepsilon S_{\alpha B,max}\left(T_l - T_S\right), \tag{8.89}$$

where S is the reduced laser beam area $S = \pi\ (r_l/2)^2$ and $T_S = 20°C$ is considered to be the specimen surface temperature for α_B evaluation. The imaginary temperature of the acting laser T_l is assumed to be 10^4°C. The profile of α_B (r) is described by the values of α_B $(r) = \alpha_{B,max}$ for $r = 0$, α_B $(r) = \alpha_{B,max}/2$ for $r = r_l/2$ and, finally, α_B $(r) = 0$ for $r = r_l$. The profile shape of the function α_B (r) or α_N (r), is determined using the indirect task. The sample temperatures during laser treatment are measured using built-up thermocouple probes in the sample (heat treatment measurement is considered to be physical natural modeling). The profile shapes of the functions α_B (r) and α_N (r) are then found so that the computed temperature values satisfy those measured by thermocouples.

Figure 8.31 shows the temperature fields for the surface and longitudinal and transversal cross-sections of the sample at the time

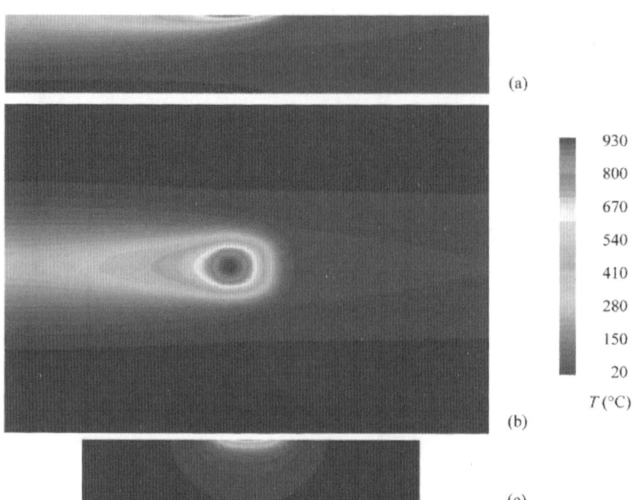

Fig. 8.31. Temperature field on the sample surface (b) and the sample longitudinal (a) and transversal (c) cross-sections at the time of 7.5 s.

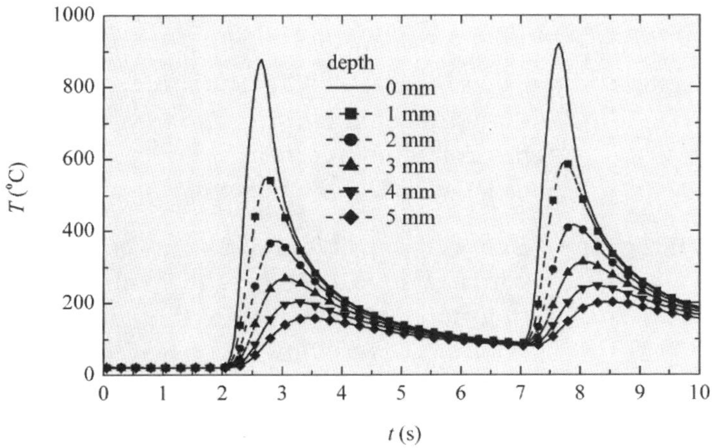

Fig. 8.32. Time-dependent temperature on the sample surface and at five different depths below.

when the laser beam crosses the surface center on its second way for the given track. The maximum surface temperature is approximately 930°C. The time-dependent temperature at a place on the surface center and at five different depths below the surface under the center of the front surface are presented in Fig. 8.32 for the first two passes of the laser beam over the sample. The depth of the heat treatment of the material can be evaluated from these temperature profiles.

The numerical model of a moving laser beam during heat treatment of the material can be used for determination of the time-dependent temperature field in the sample during the process. The resulting temperature field can give the prediction of the depth profile of the material that undergoes heat treatment, for example subsurface hardening.

8.4.4 Mechanical Engineering

The computer model of a tool during machining is created in the following example. The heat transfer and thermal deformation processes are solved to ensure the required machine parameters. The task of an inductively hardened toothed wheel is solved in the next example involving coupled electromagnetic field propagation with subsequent heat transfer and thermal induced stress. The dynamics of the thermal process and the formation of shrink holes during solidifiction in casting is simulated in the following example. The next example concerns the optimal heating of the lid preform of a reactor vessel. It is an optimalization taks for the determination of the temperature-time course of the furnace.

Example 8.7 *Machinery. Thermal deformation process in a tool during machining*

Object of the solution The thermal deformation of a cutting edge during machining, the resulting value, and the horizontal and vertical components thereof are to be stated.

Task characterization Between a workpiece and a tool, heat arises which causes tool deformation and machine thermal instability affecting the machining accuracy. The knowledge of the temperature field in the tool and the thermal deformation due to it enable one to limit the thermal influence and to ensure the required machine parameters. Also, to state the residual thermal stresses influencing the machine's lifetime is very important. The thermal task is direct, with linear 1st to 3rd-kind boundary conditions. The mathematical model is the moving temperature field and the quasi-stationary field of thermal deformations.

Applied method With respect to the thermal and stress nature of the task, the finite element method(FEM) and a numerical computer model are used for solving it.

Physical principle

In Fig. 8.33, the heat fluxes which originate during machining in the *workpiece–tool system* are presented schematically. The majority of the heat goes into the cuttings (Q_5), whereas the remaining part (Q_3) is

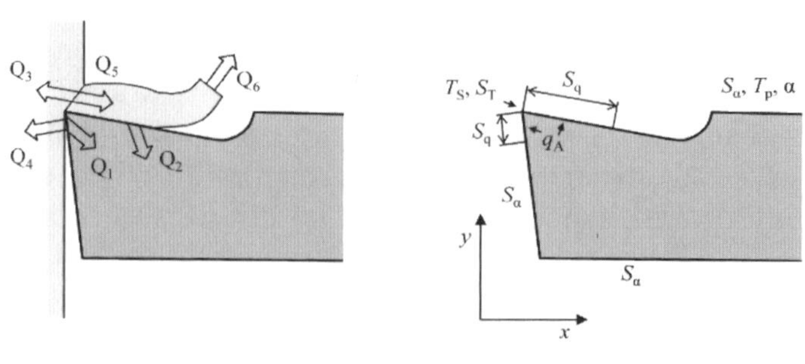

Fig. 8.33. Heat fluxes (a) and boundary conditions (b) in a tool during machining.

given over to the workpiece 1. The heat Q_1 rising on the front friction area goes into the tool 2, whereas the heat Q_4 passes into the workpiece. The part of heat in the cuttings goes into the tool together with the heat originated by the friction of cuttings, giving the heat Q_2. The heats Q_1 and Q_2, expressed by the heat flux density q_A represent the 2nd-kind boundary condition. The heat transferred by convection and characterized by the heat transfer coefficient α and the environmental temperature T_p expresses the 3rd-kind boundary condition. Sometimes, the 1st-kind boundary condition is known, e.g. temperature T_S of the tool cutting edge.

Mathematical model

For a two-dimensional temperature field, an unsteady thermal process in the workpiece–tool system is expressed by the Fourier-Kirchhoff equation

$$\nabla^2 T(x,y) - \frac{1}{a}\left(w_x \frac{\partial T}{\partial x} + w_y \frac{\partial T}{\partial y} \right) + \frac{q_V}{\lambda} = \frac{1}{a}\frac{\partial T}{\partial \tau}, \tag{8.90}$$

and for given boundary conditions, the following is valid

$$T = T_S \quad \cdots \ 1^{\text{st}}\text{-kind condition on the surface } S_T, \tag{8.91}$$

$$q_A = -\lambda \frac{\partial T}{\partial n} \quad \cdots \ 2^{\text{nd}}\text{-kind condition on the surface } S_q, \tag{8.92}$$

$$-\lambda \frac{\partial T}{\partial n} = \alpha\left(T - T_p\right) \quad \cdots \ 3^{\text{rd}}\text{-kind condition on the surface } S_\alpha. \tag{8.93}$$

The basic Eq. (8.90), representing the equilibrium equation of an energetic system, can be converted into the following dimensionless form

$$\nabla^2 \Theta - Pe_x \frac{\partial \Theta}{\partial X} - Pe_y \frac{\partial \Theta}{\partial Y} + Po = \frac{\partial \Theta}{\partial Fo}, \tag{8.94}$$

and similarly, also the equations for the boundary conditions

$$\Theta = \Theta_S, \tag{8.95}$$

$$Ki = \frac{q_A L}{\lambda T_{ref}} = -\frac{\partial \Theta}{\partial N}, \tag{8.96}$$

$$-\frac{\partial \Theta}{\partial N} = Bi\left(\Theta - \Theta_p\right). \tag{8.97}$$

In the preceding equations, there are dimensional and dimensionless quantities: T, Θ - temperatures in the system; T_p,Θ_p - environmental temperatures; x, y, X, Y - coordinates; τ, Fo - time and Fourier number;

α - thermal diffusivity; λ - thermal conductivity; q_V, Po - volume density of the heat flux and the Pomerantsev number; w_x, w_y, Pe_x, P_y - cutting velocity components and the Peclét numbers; q_A, Ki - surface density of heat flux and the Kirpitchev number; α, Bi - heat transfer coefficient and the Biot number; S_T, S_q, S_α - surface areas limited by the given temperature T_S, by heat flux density q_A, and by heat transfer α.

The preceding heat process mathematical model is solved by a variation calculus method, specifically by minimizing the equation functional with regard to temperature. A more detailed description is given in [1.7]. The examined tool zone is divided into final elements, see Fig. 8.34. The shape of these elements and the number of nodes therein is in harmony with the chosen approximation degree in the given zone. By solving the linear equations set, the deformation due to the temperature field in all element nodes can be determined.

By keeping the conditions for planar deformation of the tool, one obtains the following equation for mechanical and thermal strains

$$\varepsilon = \begin{pmatrix} \varepsilon_x \\ \varepsilon_y \\ \varepsilon_{xy} \end{pmatrix} = \begin{pmatrix} \dfrac{\partial u}{\partial x} \\ \dfrac{\partial v}{\partial y} \\ \dfrac{\partial u}{\partial y} + \dfrac{\partial v}{\partial x} \end{pmatrix}, \quad \varepsilon^t = \begin{pmatrix} \alpha_t T \\ \alpha_t T \\ 0 \end{pmatrix}, \tag{8.98}$$

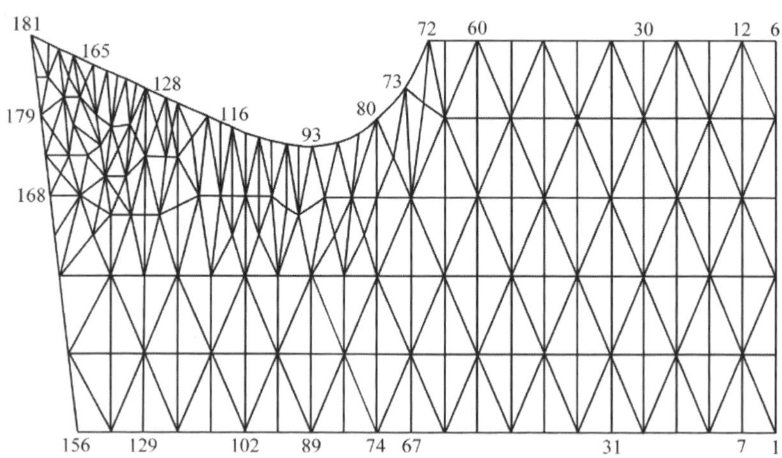

Fig. 8.34. Finite elements network used to solve the thermal deformation field in a tool.

where α_t is the linear thermal expansion coefficient, u, v are displacements into axes x and y, and T^e is the temperature of the element.

The thermal stress can also be expressed in the matrix form, and one obtains the following by applying the Hooke law

$$\boldsymbol{\sigma} = \begin{pmatrix} \sigma_x \\ \sigma_y \\ \tau_{xy} \end{pmatrix} = \mathbf{D}\left(\varepsilon - \varepsilon^t\right), \tag{8.99}$$

where **D** is the matrix expressing the stress-to-strain relation. With elastic stresses, it is called the elasticity matrix and similarly for plastic stresses, it is called the plasticity matrix.

Results

From the knowledge of elastic and elastically plastic stresses, the internal residual thermal stresses in the tool can be determined as well. From the results of solution of temperatures, thermo-elastic stresses, and tool deformations, the field of curves of constant horizontal (u) and vertical (v) tool displacements can be determined. In Fig. 8.35, this field is presented for the time τ = 5.3 min., in which the maximum thermal deformation occurs in the tool. The cutting edge displacement, caused by the temperature field, have the components u = 75 μm and v = 16 μm, can significantly influence the machining accuracy.

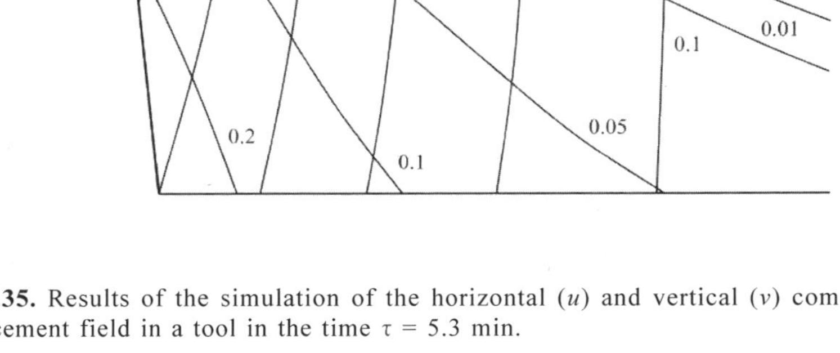

Fig. 8.35. Results of the simulation of the horizontal (u) and vertical (v) components of displacement field in a tool in the time τ = 5.3 min.

Example 8.8 *Magneto-thermomechanics. Thermal deformation process in an inductively hardened toothed wheel*

Object of the solution The cause for micro-cracks on the surface of a tooth which has been inductively heated and then cooled by a water spray has to be explained. The micro-cracks are located mostly at the base of the tooth, at the middle height of the side of the tooth, and sometimes at its top edge. With subsequent dynamic operational loading, the cracks can cause wheel damage.

Task characterization The problem consists of solving a two-dimensional coupled task of electromagnetic field propagation with subsequent heat transfer and thermal stress of the tooth. The inductive hardening influence is a specific property only in a thin sub-surface layer at the depth of 1 to 2 mm. Therefore, the process characterization needs a very fine discretization with regard to a great temperature gradient in the thin layer.

Applied method Because of the above-mentioned reason, an electric laminated *RC* analogue with both distributed parameters is chosen as one of the solution alternatives, which represents a processor in a single-purpose hybrid sampling system [1.7]. At the same time, a numerical model solution is performed with a finite element method (FEM) application, specifically on a course network (software STEP) and on a fine network (software SYSTUS). Above all, the main target of the numerical model is to explain the dynamics of the thermal deformations and the development of stresses in the tooth.

Mathematical model and similarity criteria

The thermal process during inductive hardening can be described by 17 physical dimensional quantities in the form of the following equation

$$\varphi\left(x, y, z, L, T, T_a, T_S, T_i, \lambda, \lambda_{ref}, c_V, c_{V,ref}, w, q_V, \alpha, \sigma, \tau\right) = 0, \quad (8.100)$$

where x, y, z, L are three geometric coordinates and relative length, T, T_α, T_S, T_i are the temperatures of the tooth, surroundings, surface, and the initial temperature, λ, λ_{ref} is thermal conductivity and its reference value, c_v, $c_{V,ref}$ is specific volume heat capacity and its reference value (J m^{-3} K^{-1}), w is the inductor movement velocity in the thermal system (m s^{-1}), q_V is the volume density of the heat source (W m^{-3}), α is the heat transfer coefficient (W m^{-2} K^{-1}), and σ is surface radiativity or absorptivity (W m^{-2} K^{-4}).

With the total number of dimensional quantities $N = 17$, the number of dimensionally different quantities $n = 9$ and the number of basic dimensions $r = 4$ (L, M, T, Θ), the number of dimensionless criteria for the thermal process can be determined by the Pi theorem as $k = N - r = 13$, of which the simple ones are $k_s = N - n = 8$, and the composed ones $k_k = n - r = 5$.

In the mathematical model for the unsteady non-linear heat transfer in a moving thermal system with internal heat sources, the fundamental equation has the following form

$$\mathrm{div}\left[\lambda(T)\nabla T\right] + c_V(T)w\nabla T + q_V = c_V(T)\frac{\partial T}{\partial \tau}. \tag{8.101}$$

By means of the reference quantities λ_{ref} and $c_{V,.ref}$ this equation can be converted into a dimensionless form

$$\mathrm{div}\left[\Lambda(\Theta)\nabla\Theta\right] + C_V(\Theta)Pe\nabla\Theta + Po = C_V(\Theta)\frac{\partial\Theta}{\partial Fo}. \tag{8.102}$$

The following is valid for the boundary condition on the tooth surface and in the case when it is cooled by convection and radiation,

$$-\lambda(T)\frac{\partial T}{\partial n} = \alpha(T_S - T_a) + \sigma(T_S^4 - T_a^4), \tag{8.103}$$

having the following form if expressed dimensionlessly

$$-\frac{\partial\Theta}{\partial N} = Bi(\Theta_S - \Theta_a) + Sf(\Theta_S^4 - \Theta_a^4). \tag{8.104}$$

The initial condition for inductive tooth heating is

$$T(x,y,z,0) = T_i, \qquad \Theta(X,Y,Z,0) = \Theta_i. \tag{8.105}$$

The thermal process in a toothed inductively hardened wheel is then described by the following functional relation of 13 dimensionless quantities

$$\Phi(X,\ Y,\ Z,\ \Theta,\ \Theta_S,\ \Theta_a,\ \Lambda,\ C,\ Bi,\ Fo,\ Pe,\ Po,\ Sf)=0, \quad (8.106)$$

in which there are eight simple quantities and five composed ones.

The following holds for the dimensionless quantities: the coordinates are $X = xL^{-1}$, $Y = yL^{-1}$, $Z = zL^{-1}$, the temperatures are $\Theta = TT_{ref}^{-1}$, $\Theta_S = T_S T_{ref}^{-1}$, $\Theta_a = T_\alpha T_{ref}^{-1}$, $\Lambda = \lambda\lambda_{ref}^{-1}$ is thermal conductivity, $C_V = c_V c_{V,ref}^{-1}$ is the volume specific heat capacity, $Bi = \alpha L \lambda^{-1}$ is the Biot number, $Pe = wLa^{-1}$ is the Peclét number, $Po = qL^2 \lambda^{-1} T_{ref}^{-1}$ is the Pomerantsev number, $Sf = T^3 L\lambda^{-1}$ is the Stefan number, $Fo = \dfrac{a\tau}{L^2}$ is the Fourier number.

For the inner source q_V, representing Joule heat in Eq. (8.101), one obtains

$$q_V(M,\tau) \rightarrow J_V(M,\tau). \quad (8.107)$$

In the model, the equation for the heat source in the dimensionless form is

$$Po_M(M,Fo) = \frac{J_V L^2}{\gamma U_{ref}}, \quad (8.108)$$

where $M(X,\ Y,\ Z)$ is a point with coordinates $X,\ Y,\ Z$, J_V is the volume density of current (A m^{-3}), ρ is specific electric resistance (Ω m), γ is electric conductance (Ω^{-1} m^{-1}), and U_{ref} is the relative voltage value (V). Current density J_A (A m^{-2}) can be determined from the Maxwell equation

$$\text{rot}\ \mathbf{H} = \mathbf{J}_A, \quad (8.109)$$

and the magnetic field intensity in an arbitrary place can be obtained by solving the following equation

$$\text{div}\left[\rho(M,\tau)\ \text{rot}\ \mathbf{H}\right] = \frac{\partial \mathbf{H}}{\partial \tau_{em}}, \quad (8.110)$$

which is obtained in the dimensionless form

$$\text{div}\left[R(M,Fo)\ \text{rot}\ \bar{\mathbf{H}}\right] = \frac{\partial \bar{\mathbf{H}}}{\partial Fo_{em}}. \quad (8.111)$$

Here, the dimensionless quantity $R = \rho\rho_{ref}^{-1}$ is specific electric resistance, $\bar{\mathbf{H}} = \mathbf{H} H_{ref}^{-1}$ the magnetic field intensity, and $Fo_{em} = \nu_m \tau_{em} L^{-2}$ is the Fourier electromagnetic number, μ is magnetic permeability (H m^{-1}), ν_m is magnetic viscosity (m^2 s^{-1}), L is characteristic length (m), τ_{em} is electromagnetic wave propagation time (s).

In solving inductive heating, the current distribution in the subsurface layer influences strongly the thermal and thermal stress process in the tooth. The exact solution of the coupled task, expressed by the Maxwell equation and the heat conduction equation, is almost impossible if the boundary conditions and variable electromagnetic and thermophysical parameters are not known. However, the problem can be transformed to a simple thermal task provided the maximum tooth surface temperature is found experimentally and the depth of electromagnetic field penetration below the tooth surface is stated accurately by micrographic analysis. In the considered case, the hardening temperature of 730°C corresponds to the maximum penetration depth and the maximum surface tooth temperature is 930°C for the object. In addition, the inductor power output is known and the current density distribution in the subsurface layer has been determined.

Simulation model

The single-purpose simulation system is represented by an analogue processor composed of a two-layer electric *RC* model with both continuously distributed parameters (CSCT type). A special sampling system operating at the sampling frequency up to 10 MHz, impulse generators, boundary condition modules, internal sources, and an output recording device are part of it. The sampling density can be changed according to the set-up program. The hybrid system is described in in more detail.

The *time scale* μ_T will be determined from the equality of the Fourier numbers: for the object Fo and for the model Fo_M, and the following holds for it

$$\mu_\tau = \left(\frac{L_M}{L}\right)^2 \frac{aRC}{L_M^2}, \tag{8.112}$$

where L, L_M (m) are characteristic lengths of the object and the model, a ($m^2\ s^{-1}$) is thermal tooth diffusivity, and $\mu_\tau = 8.43 \times 10^{-5}$ is the time scale for the considered case.

In Fig. 8.36, there is a view of the CSCT type *RC* model of a gear wheel tooth. Here 1 denotes the tooth body and adjacent parts of the wheel rim, 2 is the boundary condition model on the inductor side, 3 is the boundary condition model on the opposite side, 4 is the plexiglass protective housing, 5 is the measuring probe, and 6 is a digitally adjustable resistor.

Results

A brief outline of hybrid simulation results only is presented, which is focused on temperature and deformation fields dynamics in a tooth

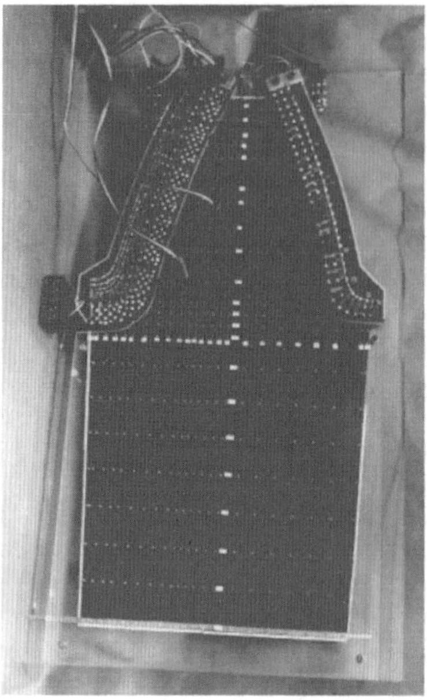

Fig. 8.36. *RC* model of a gear wheel tooth of CSCT type.

during inductive heating land subsequent hardening with a water spray. More detailed results for time responses of temperature at various tooth points, during heating and hardening, are presented for thermal stress field in [8.26].

Figure 8.37 shows the results of temperature fields in the tooth at four time levels, the first two of which (0.5 and 2 s) involve heating the tooth with an inductor, whereas the other two (5 and 12 s) involve the beginning and end of the following spray cooling. The temperatures during heating indicate a considerable temperature gradient between the tooth surface and the core. The start of cooling changes the nature of the temperature field in the tooth, the temperature gradient drops and the temperature field becomes equalized. Due to this, the character of the thermal deformation also changes, as presented in Fig. 8.38 in an enlarged scale. The character of tooth deformation for the time of 3 sec (that is 1 sec before the end of inductive heating) and 16 sec (4 sec after the end of cooling-down), when the tooth deformation becomes re-equalized, are presented. The results of the detailed model study point out the dominant influence of the extent of heat transfer on the formation of microcracks in spray hardening. The replacement of water with a suitable water solution helps to decrease the heat transfer into the tooth surface and to eliminate the nucleation of cracks.

The example shows how a single-purpose electric *RC* analogue with both distributed parameters can be utilized in a case of a short-time but extraordinarily high surface heat fluxes on the order of

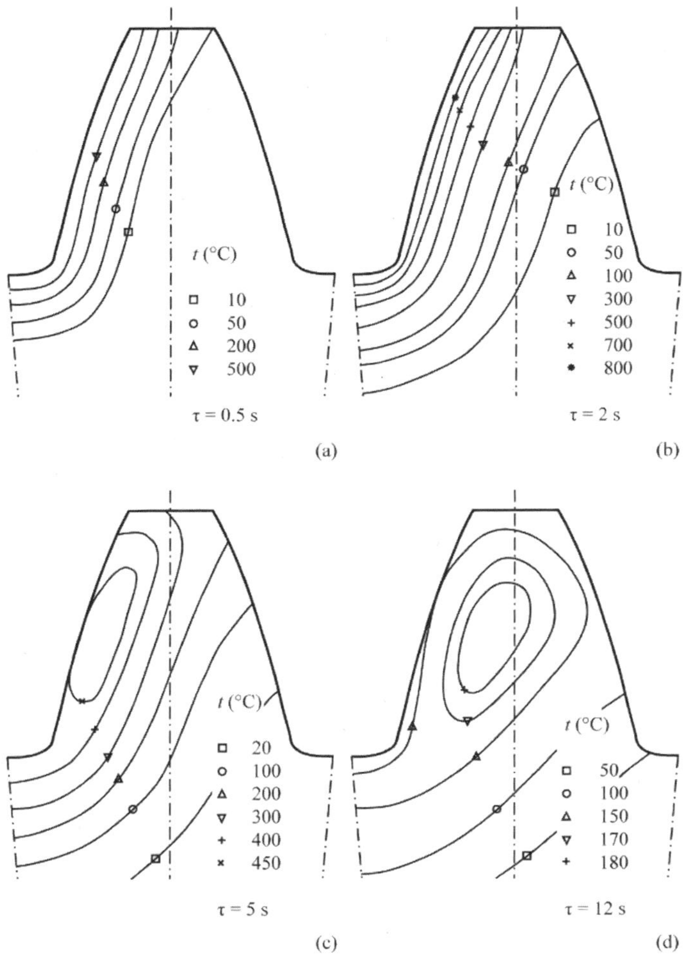

Fig. 8.37. Temperature field in the tooth at the times of 0.5 sec (a), 2 sec (b), 5 sec (c) and 12 sec (d).

10^6 to 10^7 W m^{-2}. Their action is limited to a thin sub-surface zone with a considerable temperature gradient. The *RC* model with both distributed parameters can be made by the deposition of both layers on an insulating pad with a copper coating or by using a special conducting paper and a very thin self-adhesive dielectric foil.

In the considered task, the application of the hybrid method as well as the numerical finite element method shows the unique properties of both modeling methods. The elimination of discretization and the objective approach to the solution are advantages of the hybrid method. Its disadvantage is that the electrical model is single-purpose and that there are some special requirements for workshop facilities. Above all, the numerical solution (FEM method) is favourable due

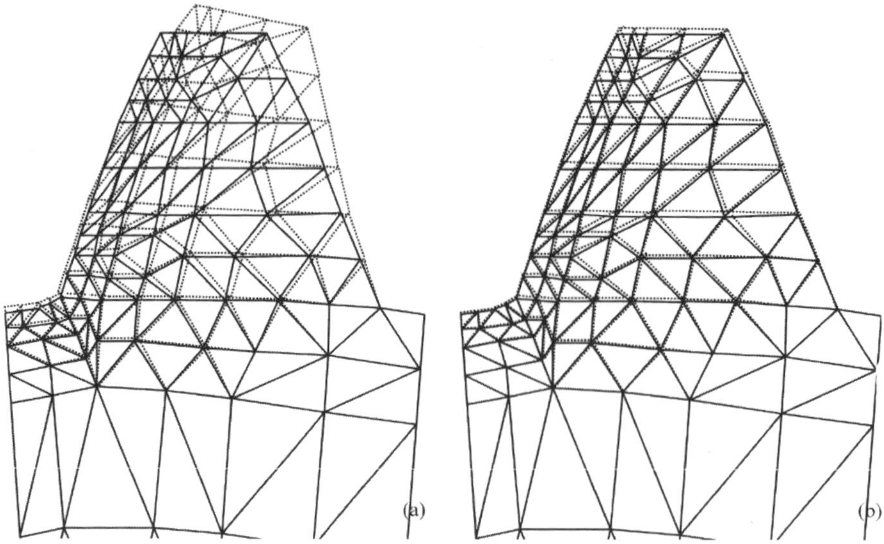

Fig. 8.38. Deformation of the tooth at the times of 3 sec (a) and of 16 sec (b).

to its universality and complexity, which make it capable of solving the thermal and thermal induced stress tasks simultaneously. Its disadvantage is less illustrative solution procedure. The procedure and results of elastic and elastoplastic modeling of thermal stresses in a tooth by the FEM method and a continuing modeling study of the influences acting in inductive hardening and its optimization is given in [8.26].

Example 8.9 *Foundry. Simulation of thermal process dynamics and shrink hole formation during casting solidification*

Object of the solution The influence of two different casting technologies, both for a shape-complicated casting in a thermal process with the creation of shrink holes, have to be clarifled on a model. The classic single-phase casting and the proposed two-phase one are considered.

Task characterization The casting consists of a rotor wheel blade of a water turbine. From the thermomechanical point of view, it represents a thermal system with the mass of 28 tons with two sub-systems created by a massive pivot with a riser and with a relatively thin blade leaf having warped areas. High thermal stress concentration is assumed, especially in the transition between the 'heat-thick' pivot with the riser and the 'heat-thin' leaf. A sand mould forms an outer thermally insulating subsystem. The casting mass flux of the melting material into the mould is 360 kg s^{-1} and the casting time is 18 sec.

Applied method The task is solved using the finite difference method (FDM).

Casting-and-mould thermal system

The given basic parameters are casting temperature of the melting material T_{ini} = 1542°C, liquidus temperature T_l = 1438 °C, solidus temperature T_s = 1407°C, specific solidification latent heat $l_{l,s}$ = 267.9 kJ kg^{-1}. Also, the thermophysical properties (thermal conductivity, specific heat capacity and density) are given for the metal and the sand mould.

Mathematical model

The mathematical model is given in the form of a non-linear unsteady heat conduction equation with a non-linear internal heat source and a 4th-kind boundary condition. The general blade leaf shape is replaced by a plate with the same surface area and mass equal to that of the leaf casting. The casting time is considered a single-step of the total solidification time. The pivot and the riser form a rotationally symmetrical body.

The solidification process can be described by the following equation for heat conduction

$$\frac{\partial}{\partial x}\left(\lambda\frac{\partial T}{\partial x}\right)+\frac{\lambda}{x}\frac{\partial T}{\partial x}+\frac{\partial}{\partial y}\left(\lambda\frac{\partial T}{\partial y}\right)=c(T)\;\rho(T)\frac{\partial T}{\partial \tau}, \qquad (8.113)$$

or by the equation with internal sources

$$\operatorname{div}(\lambda\,\operatorname{grad}T)+q_V(T)=c(T)\;\rho(T)\frac{\partial T}{\partial \tau}, \qquad (8.114)$$

which can be described in a dimensionless form

$$\mathrm{div}\left(\Lambda\,\mathrm{grad}\,\Theta\right)+Po\left(\Theta\right)=C\left(\Theta\right)\,R\left(\Theta\right)\frac{\partial\Theta}{\partial Fo}. \tag{8.115}$$

In the preceding equations, there are the following quantities: T, $\Theta = TT_{ref}^{-1}$, T_{ref} - dimensional, dimensionless and reference temperature; x, $X = xL^{-1}$, L - dimensional, dimensionless and reference length; τ, $Fo = a_{ref}\,\tau L^{-2} = \lambda_{\mathrm{ref}}\tau\; c_{ref}^{-1}\;\rho_{ref}^{-1}\;L^{-2}$ - time and the Fourier number; λ, $\Lambda = \lambda\;\lambda_{ref}^{-1}, \lambda_{\mathrm{ref}}$ - dimensional, dimensionless and reference thermal conductivity; c (J kg^{-1}, K^{-1}), $C = cc_{ref}^{-1}$, c_{ref} - dimensional, dimensionless and reference specific heat capacity; ρ, $R = \rho\rho_{ref}^{-1}$, ρ_{ref} - dimensional, dimensionless and reference density; q_V, $Po = q_V\;L^2\lambda_{ref}^{-1}\,T_{ref}^{-1}$ - volume heat source and the Pomerantsev number.

The non-linear mathematical model, as described by Eqs. (8.113) to (8.115), can be solved in several ways, e.g. by linearizing these equations with the Kirchhoff transformation application as mentioned in the subsequent example. In another way, the influence of the dimensional non-linear internal heat source, caused by phase changes, is converted to dimensional specific heat capacity change $c\,(T)$. Then, the following equation is valid

$$c\left(T\right)=\left(\frac{l_{l,s}}{2\Delta T}+\frac{c_s+c_l}{2}\right) \tag{8.116}$$
$$\text{for}\quad T_{l,s}-\Delta T\;\le\;T\;\le\;T_{l,s}+\Delta T,$$

$$c\left(T\right)=c_s\;\text{for}\quad T<T_{l,s}-\Delta T, \tag{8.117}$$

$$c\left(T\right)=c_l\;\text{for}\quad T>T_{l,\,s}+\Delta T, \tag{8.118}$$

where $T_{l,S} = (T_l + T_s)/2$ and the subscripts mean: s - the solidus, l - the liquidus, l, s - the liquidus–solidus interface. The quantity $l_{l,S}$ is the latent solidification heat.

The Eqs. (8.116) to (8.118) are obtained by introducing the Dirac function for the finite heat capacity value $c(T)$ during a small temperature interval $2\Delta T$ across the solidification temperature (Fig. 8.39). In numerical modeling, this temperature interval must be carefully chosen. It should be small enough to suppress the non-linearity due to the influence of solidification latent heat.

The following equation is obtained by converting the differential Eq. (8.113) to a difference one

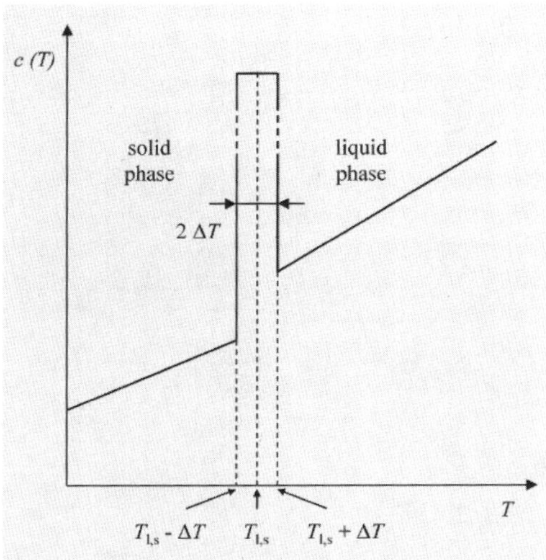

Fig. 8.39. Non-linear internal heat source of the phase change expressed by pulse change of the specific heat capacity.

$$\frac{\lambda_{i+1,j}-\lambda_{i-1,j}}{2\Delta x}\frac{T_{i+1,j}-T_{i-1,j}}{2\Delta x}+\lambda_{i,j}\frac{T_{i+1,j}-2T_{i,j}+T_{i-1,j}}{(\Delta x)^2}+$$
$$+\frac{\lambda_{i,j}}{x_i}\frac{T_{i+1,j}-T_{i-1,j}}{2\Delta x}+\frac{\lambda_{i,j+1}-\lambda_{i,j-1}}{2\Delta y}\frac{T_{i,j+1}-T_{i,j-1}}{2\Delta y}+ \tag{8.119}$$
$$+\lambda_{i,j}\frac{T_{i,j+1}-2T_{i,j}+T_{i,j-1}}{(\Delta y)^2}=c(T_{i,j})\rho(T_{i,j})\frac{T_{i,j}^{k+1}-T_{i,j}^{k}}{\Delta\tau},$$

where i and j are the coordinates of the geometric step and k is that of the time step. The number of geometric network nodes is about 10^3.

Simulation procedure

For two solution alternatives with and without supplementary pouring of the casting, the solving procedure for the difference model, Eq. (8.119), involves the determination of three characteristic fields - temperature, temperature gradient, and solidification rate fields. The temperature gradient and the solidification rate are, as vector quantities, determined by a modulus and an argument. In comparison to temperature, they deliver the information about the thermal process dynamics during solidification and about the non-compactness in the casting. Critical pressure is important for shrink hole formation and depends on the empirically verified ratio of the temperature gradient in the solidifying casting grad T (K m^{-1}) to the square root of the solidification rate of casting w_T (K s^{-1})

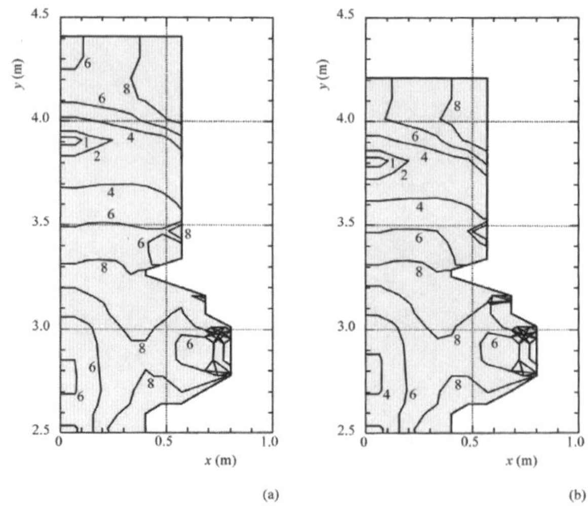

Fig. 8.40. Field of Chvorinov shrink formation number Ch_{sh} for the prediction of the shrink hole formation in a casting for single-phase casting without supplementary pouring (a) and for two-phase casting with supplementary pouring (b).

$$\frac{\mathrm{grad}\,T}{\sqrt{w_T}} = 775\ \mathrm{K}^{\frac{1}{2}}\ \mathrm{s}^{\frac{1}{2}}\ \mathrm{m}^{-1}, \tag{8.120}$$

which can be expressed in the form of the dimensionless Chvorinov shrink formation number Ch_{sh} that can be obtained by dividing the value found on the model by the reference state according to Eq. (8.120). The following is then valid

$$Ch_{sh} = \frac{\mathrm{grad}\,T}{\sqrt{w_T}}\left(\frac{\mathrm{grad}\,T}{\sqrt{w_T}}\right)_{ref}^{-1}, \tag{8.121}$$

where subscript *ref* denotes the limiting value for shrink hole formation. The shrink hole does not form if $Ch_{sh} \geq 1$, but arises if $Ch_{sh} < 1$.

Results

From the detailed analysis of the thermal process dynamics in a casting, based on the fields of unsteady temperature, those of temperature gradients, and those of solidification rate of casting, only the resulting criteria for predicting the shrink hole formation for the two solution alternatives considered are presented in Fig. 8.40. In both alternatives, casting without and with supplementary pouring, this criterion $Ch_{sh} < 1$ is located in the riser axis but does not stretch into the casting head part. In the other casting part, $Ch_{sh} \geq 1$ is valid. Both alternatives are distinguished by the position change in the riser with $Ch_{sh} < 1$.

Example 8.10 *Technical cybernetics. Optimally controlled heating of a lid preform of a reactor vessel*

Object of the solution The temperature-time course in a furnace has to be stated as a controlling function in the optimum heating of the lid preform of a reactor vessel.

Task characterization Besides the boundary and initial conditions, also the restrictive condition appears for the maximum admissible stress in a circular preform of a large lid during heating. With the solution, a temperaturetime course in a furnace must be determined in which the allowed stress limit in the material are not exceeded, the furnace power output is fully utilized, the heating time is reduced, and the energy consumption is minimized. Among thermal tasks, this one represents a qualitatively higher degree and its solution efficiency depends strongly on the selection of a suitable algorithm and means of simulation.

Applied method The task is solved using the hybrid deterministic method realized on a universal hybrid system.

Mathematical model

The mathematical model of the thermal process of the preform of a large thick-wall circular disk, heated in a gas furnace, has the following form in cylindrical coordinates with the consideration of internal parametric and source non-linearities representing heat sources during phase conversions

$$c(T)\rho(T)\frac{\partial T}{\partial \tau}=\frac{1}{r}\frac{\partial}{\partial r}\left[r\lambda(T)\frac{\partial T}{\partial r}\right]+\frac{\partial}{\partial z}\left[\lambda(T)\frac{\partial T}{\partial z}\right]-q_V(T). \quad (8.122)$$

The boundary condition expresses the composed heat transfer by convection and radiation

$$-\lambda(T)\frac{\partial T}{\partial n}=\alpha(T_P-T_S)+\sigma\left(T_P^4-T_S^4\right). \quad (8.123)$$

The initial condition

$$T(r,\varphi,z,\tau=0)=293°\mathrm{C}=\mathrm{const.} \quad (8.124)$$

The restrictive conditions are

$$\sigma \leq \sigma_{dov}, \tag{8.125}$$

which is supplemented sometimes by an energetic restrictive condition

$$P \leq P_{max}, \tag{8.126}$$

where P and P_{max} denote the real operational and maximal furnace heat power output.

The preceding dimensional mathematical model can be rewritten as a dimensionless one

$$C\frac{\partial \Theta}{\partial Fo}=\frac{1}{R}\frac{\partial}{\partial R}\left(R\Lambda\frac{\partial \Theta}{\partial R}\right)+\frac{\partial}{\partial Z}\left(\Lambda\frac{\partial \Theta}{\partial Z}\right)-Po. \tag{8.127}$$

The following holds for the boundary condition

$$\Theta_S - \Theta_n = \left(\Theta_P - \Theta_S\right)\left[Bi + St\left(\Theta_P + \Theta_S\right)\left(\Theta_P^2 + \Theta_S^2\right)\right]\frac{\Delta n}{L}, \tag{8.128}$$

and for the initial condition

$$\Theta\left(R, \Phi, Z, 0\right) = \text{const.}, \tag{8.129}$$

and for the restrictive thermal-stress condition

$$K_\alpha = \frac{\sigma}{\alpha E T_{ref}} \leq \frac{\sigma_{dov}}{\alpha E T_{ref}}. \tag{8.130}$$

The basic non-linear heat conduction Eq. (8.127) can be linearized by using the Kirchhoff transformation to the form

$$\frac{\partial U}{\partial Fo}=\frac{\Lambda}{C}\left[\frac{\partial^2 U}{\partial R^2}+\frac{1}{R}\frac{\partial U}{\partial R}+\frac{\partial^2 U}{\partial Z^2}\right]-Po, \tag{8.131}$$

In the preceding equations, the dimensionless representation of quantities is also used, namely heat conductivity Λ, specific thermal capacity C, the Biot number Bi, the Fourier number Fo, the Pomerantsev number Po, and the Stefan humber Sf.

Modeling procedure

The simulation procedure periods are given in the block diagram in Fig. 8.41.

With respect to the optimization task solution, a hybrid DSCT-type model has been chosen, realized by a universal hybrid simulation system. In building the hybrid simulation model, the last transformed basic Eq. (8.131) is converted into the difference form and so are the boundary, initial, and restrictive conditions. The procedure is introduced in [8.12] and the hybrid-system analogue-part wiring diagram is presented in Fig. 8.42, where S denotes summator, G is generator, T is setup of the initial state, Q is the inner source due to phase change,

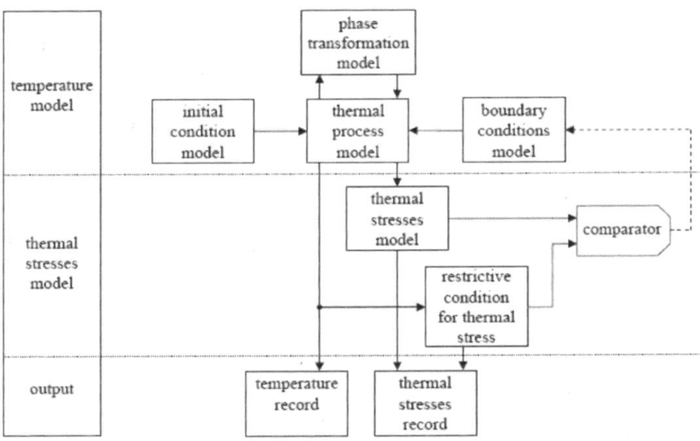

Fig. 8.41. Block simulation structure of a non-linear thermal stress task simulation on a hybrid deterministic model.

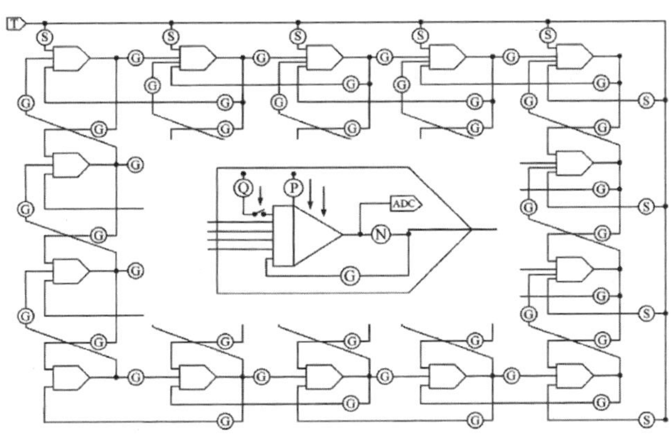

Fig. 8.42. Analogue part of the hybrid simulation model of a lid preform of a reactor vessel.

P is the digital potentiometer, ADC is analogue-to-digital converter and N is the multiplier unit. The model contains tens of integrators, adders, numerical potentiometers, and AD converters. The source non-linearity expresses the transcrystalizing heat which represents the non-linear impulse heat sink in the material. The program memory part and a non-linear internal source are realized in the numerical part. The temperature and time dependent parameters are set continuously by commands from the simulation system's numerical part. The model operates by a multiplex method in which the analogue part is always stopped after some time when the potentials found in the nodes are read and the values are stored in the numerical memory. The numerical part reads the coefficients, sets the potentiometers on a new value and starts the solution again. To consider the non-linearities, the time step must be shortened compared to the linear case. In the case considered, the model's time step corresponds to 2 minutes on the object.

Results

In Fig. 8.43, the simulation results of the initially designed three-phase heat (dashed-line curves) are compared with those of the task of optimum control (solid-line curves) of a thermal and thermally-induced stress process in a preform of a reactor vessel lid. S_{el}, S_{ep}, S_{dov} are the elastic, elastoplastic, and allowed stresses, S' are the fictive stresses. The time-temperature courses in a furnace are marked, specifically for the initially given heating (dot-and-dash line curve) and for the heating found from the optimization control task solution (short dot-and-dash line curve).

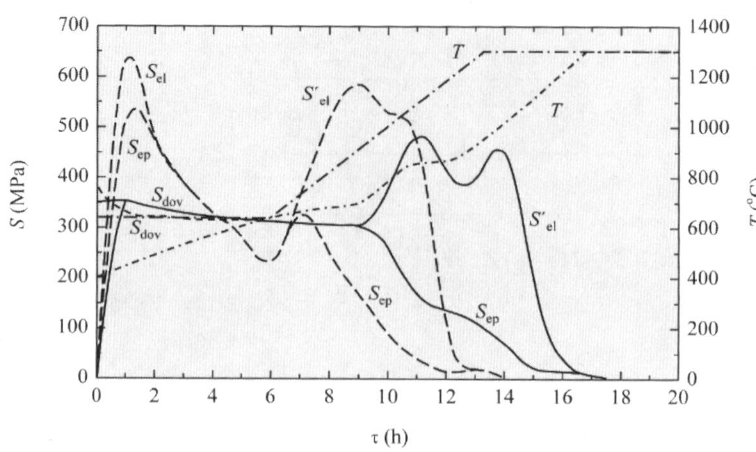

Fig. 8.43. Time course for the initial heat stress solution for the original non-optimized (dashed line) and the optimized (solid line) heating.

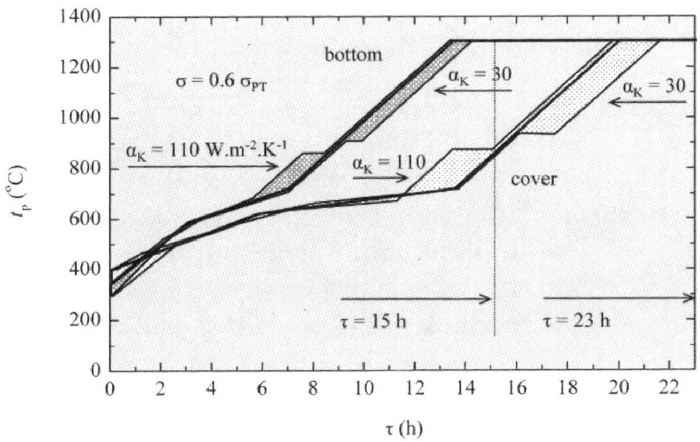

Fig. 8.44. Comparison of resulting time dependences of optimum furnace temperatures for the lid preform and the reactor vessel bottom.

In the subsequent Fig. 8.44, the resulting furnace temperature courses are compared for the case of optimal heating control: those for the lid preform with those for the reactor vessel bottom. The general optimal temperature course in the furnace is approximated by four linear sectors. In the figure, the influence of the heat transfer convective component change on the heating course is expressed by the dotted regions. The shorter heating time for the vessel bottom is due to its being less thick.

From the solution of this and other controlled-heating tasks, the advantage of a hybrid model application to solve an optimization task is clear, especially compared with the alternative solutions of many direct tasks. The disadvantage is in the fact that a more shape-complicated model needs not only tens of cells but hundreds or even thousands of them, from which complexity and high requirements follow for a hybrid simulation system.

8.4.5 Industrial medicine

In the following example, the computer model of human body heating in an environment with high temperature is created. The influence of an outer thermal barrier on the body temperature is determined. The next example is devoted to the solution of a body thermocontrol system violation during thermal overloading. The aim is to determine the maximum allowed worker's time repairing the insufficiently cooled plant ladle depending on various operation conditions.

Example 8.11 *Bio-thermomechanics. Influence of an outer thermal barrier on the temperature in a human body [4.3]*

Object of the solution The influence of an outer thermal barrier (clothing) on the transversal temperature course in a human body has to be found.

Task characterization A naked body (outer barrier thermal resistance $R_T = 0$) and normal clothing ($R_T = 0.2$ K m^2 W^{-1}) are considered. A five-layer anatomical body structure (skin, fat, muscles, inwards, and bones) and a corresponding simplified five-layer cylindrical model are assumed. An adult person (mass of 75 kg, height of 1.8 m, body surface of 1.83 m^2), extreme environmental temperature of 65°C, and the resulting heat transfer coefficient $\alpha = 15$ W m^{-2}K^{-1}) are considered. The average metabolism value in the body is q_M = 210 W m^{-2}. Further, the average thermal loss due to evaporation and heat take-away by blood is q_M = 250 W m^{-2}. The initial average internal body temperature usually approaches 36.7°C.

Applied method A numerical computer model with cylindrical coordinates and finite difference method are to be used to solve the task.

Mathematical model

The structure and functions of the layers of natural and artificial thermal barriers of a human body model are in Fig. 8.45. The physical parameters for individual anatomical elements were taken from [1.7].

For an approximate substitute mathematical model of the human body, in a multilayer cylinder shape, the following is valid

$$\frac{1}{x}\nabla\left(x\lambda_i \nabla T\right) + q_{V,i} = \rho_i c_i \frac{\partial T}{\partial \tau}. \qquad (8.132)$$

In this equation, λ_i, ρ_i, c_i are the thermophysical parameters of the corresponding layers, $q_{V,i}$ are volume densities of the heat flux, τ is time, and x is the radius. With constant thermophysical properties of each layer, Eq (8.132) can be transformed into the dimensionless form

$$\nabla^2\Theta + Po_i = \frac{\partial \Theta}{\partial Fo}, \qquad (8.133)$$

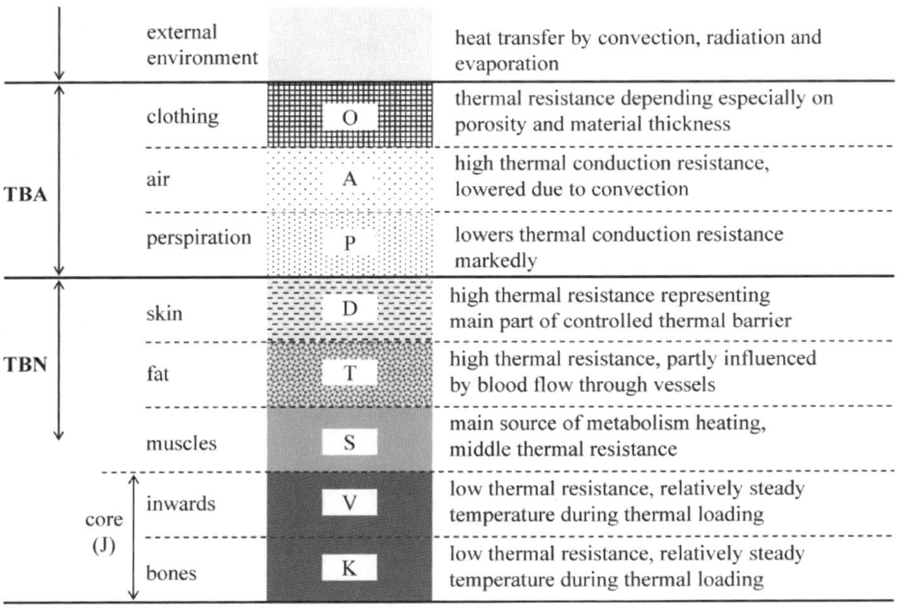

Fig. 8.45. Structure and functions of the layers of internal natural and outer artificial thermal barriers of a human body.

in which the left-side second member represents different volume densities of the heat flux (heat generation or sink) that are expressed in dimensionless form by Pomerantsev numbers

$Po_1 = Po_M$ - heat generation in the body due to metabolism (concentrated mainly in the layer of muscles)
$Po_2 = Po_C$ - heat transferred by convection
$Po_3 = Po_R$ - heat transferred by radiation (considered in the skin layer)
$Po_4 = Po_P$ - heat transferred by perspiration
$Po_5 = Po_D$ - heat transferred by breathing
$Po_6 = Po_K$ - heat transferred by blood circulation (as a heat decrease in the layer of muscles)
$Po_7 = Po_V$ - heat transferred by evaporation (considered in the skin layer)

Simulation model

For a three-layer plate model, the differential model (8.133), including the boundary conditions, can be transferred to the form that can be solved on a hybrid or analogue computer, and expressed by a differential equation set

$$\frac{1}{\Delta X_{12}}(\Theta_2-\Theta_1)+Bi(\Theta_1-\Theta_P)-Po_V+Po_R=\frac{\partial\Theta_1}{\partial Fo}\Delta X_1, \quad (8.134)$$

$$\frac{1}{\Delta X_{23}}(\Theta_3-\Theta_2)+\frac{1}{\Delta X_{12}}(\Theta_2-\Theta_1)+\Delta Po_M-Po_K=\frac{\partial\Theta_2}{\partial Fo}\Delta X_2, \quad (8.135)$$

$$\Delta Po_M-\frac{1}{\Delta X_{23}}(\Theta_3-\Theta_2)=\frac{\partial\Theta_3}{\partial Fo}. \quad (8.136)$$

In Eq. (8.134), the heat transfer by convection (Bi), by evaporation (Po_V), and by radiation (Po_R) in the outer skin layer is considered along with the heat transfer by conduction. In the layer of muscles, Eq. (8.135), the metabolic heat increment (ΔPo_M) and the heat decrease due to blood circulation (Po_K) are taken into account. In the third layer, Eq. (8.136), the heat corresponding to basal metabolism (ΔPo_M) is usually considered. The *body surface temperature* Θ_S and the *mean integral temperature in a body* Θ_{med} are considered as characteristic quantities for evaluating the effect of outer temperature on a person. The mean integral temperature can be expressed for a three-layer plate model as

$$\Theta_{med}=\frac{\Delta X_1\Theta_1+\Delta X_2\Theta_2+\Delta X_3\Theta_3}{\Delta X_1+\Delta X_2+\Delta X_3}. \quad (8.137)$$

Results

Figure 8.46 show the results of simulating the temperature course in the five-layer structure of the human body under the influence of an extreme outer environmental temperature. In case of Fig. 8.46a, the focus is the action of the inner natural thermal barrier without the outer barrier (clothing). First, the skin surface temperature increases due to the action of the thermal resistance of the internal thermal barrier. Due to subsequent heat losses by perspiration, convection, breathing, and blood circulation increase, the surface temperature starts to drop gradually, but in the body core the temperature increases. Figure 8.46b shows the strong influence of the outer artificial thermal barrier (clothing) on the character of the temperature distribution in the body. The positive influence of increasing thermal resistance of the outer thermal barrier can be seen. This causes the lowering of the temperature gradient in the subsurface layers of the inner thermal barrier and equalizing the temperature across the body. In the core of the body, the temperature changes minimally in both cases.

The above-mentioned extreme example clearly indicates the function of both thermal barriers in the human body thermocontrol

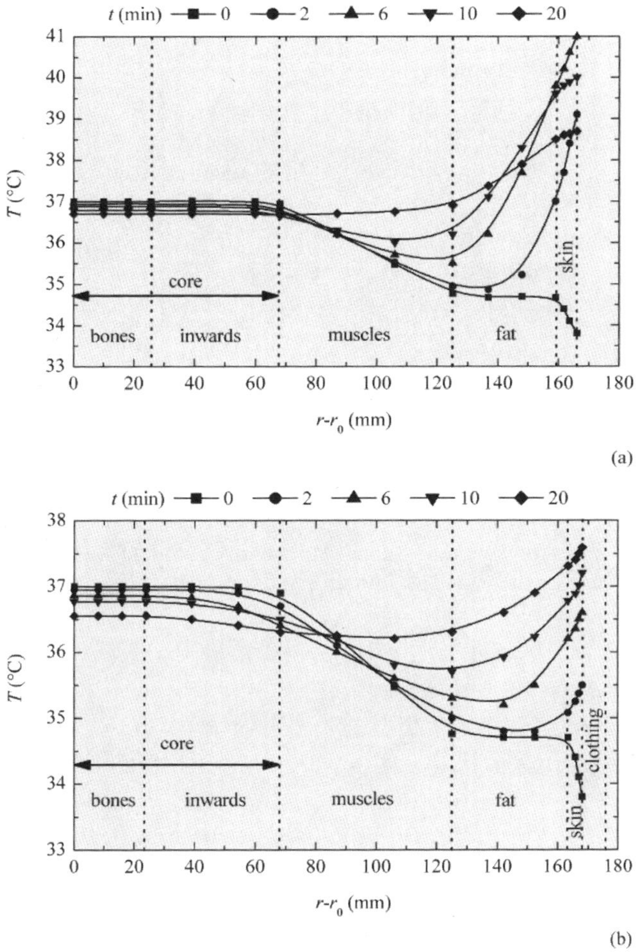

Fig. 8.46. Result of the temperature numerical simulation in a five-layer human body model for the case without clothing R_T = 0 m^2 K^{-1} W^{-1} (a) and with warm clothing R_T = 0.2 m^2 K^{-1} W^{-1} (b).

system, but it cannot be seen whether the person withstands such temperatures. For this, the mean integral temperature is decisive, and is expressed by the average blood temperature in the body. This acts on the thermocontrol center in the hypothalamus in the brain, which influences the fundamental physiological and other functions.

Figure 8.47 shows an approximation of the influence of different environmental temperatures and the length of time spent in them on a person's physiological and psychological state. Nevertheless, this cannot represent possible individual differences.

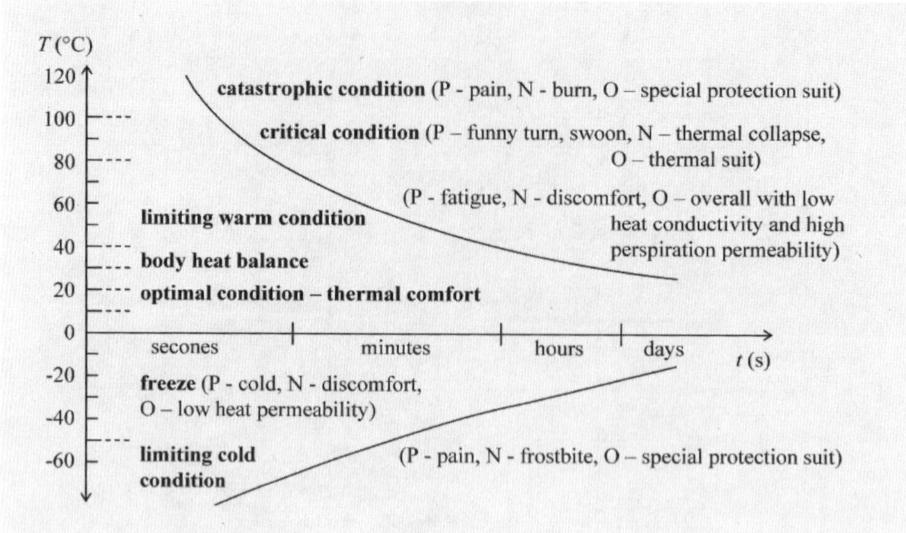

Fig. 8.47. Approximation of the influence of the environmental temperatures and the length of time spent in them on the human physiological and psychological state.

Example 8.12 *Bio-cybernetics. Thermocontrol system violation due to body overloading [4.3], [8.9]*

Object of the solution On a computer model, the thermal loading of the body of a worker who was repairing the internal surface of an insufficiently cooled big steel plant ladle has to be clarified. The worker is presumed to have sustained a heart attack due to thermal overload and is disabled permanently. The aim is to confirm or disprove the conclusion of the industrial physician that he was overloaded by heat.

Task characterization Due to the nature of the operating conditions under which the injury occurred, the task is complicated. It represents a controlled unsteady process in a heterogeneous system with internal heat sources and complicated boundary conditions. However, the task enables one to use a relatively high degree of human body geometric and anatomical simplification.

Applied method The numerical finite difference method, applied to a multilayer heterogeneous biological system composed of diverse anatomical layers.

Thermocontrol system

It follows from the aim and characteristics of the task, that the thermocontrol system in the worker's body was violated as a result of his thermal overloading. In the thermocontrol system, the blood temperature, expressed approximately by the average integral body temperature and acting directly on the thermocontrol center in the brain, is the governing function. The rise or drop of this temperature by approximately ±1.8 K results in torpidity up to collapse. This temperature can be determined approximately enough even on a simplified multilayer geometric human body model which respects the human anatomical structure. The dynamic thermal barrier, forming an integral part of the thermocontrol system, plays a key role in human body protection. The function of this thermal barrier in the human; body thermocontrol system is in Fig. 8.48 and is described in more detail in the book [4.3]. The mathematical model was described in the preceding example.

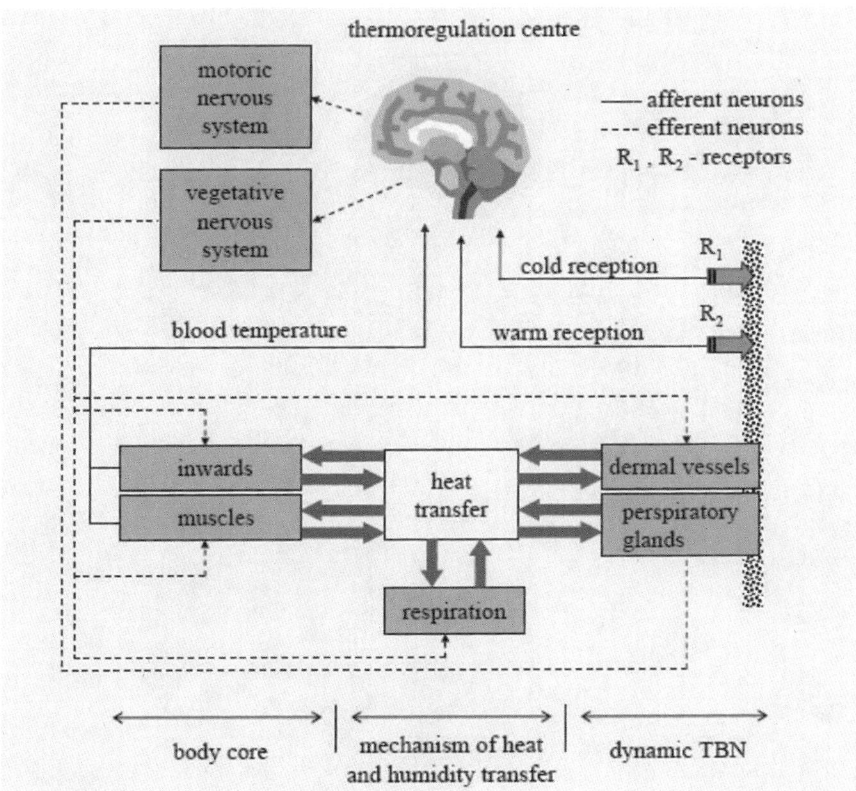

Fig. 8.48. Function of the dynamic thermal barrier in the human body thermocontrol system.

Solution procedure

The first step consists of the specification of working conditions, which is difficult when working retrospectively. While working, the worker was partly inside the ladle (less frequent case) and partly in front of it at the distance of 0.1 to 0.5 m (more frequent case). In the first case, the whole body was exposed to the radiating heat flux of the ladle lining and in the second case the body was exposed to mostly unilateral frontal radiation. A diagram with the dimensional draft of the ladle laid on side is in Fig. 8.49. There are three air temperatures measured at different places in front of the ladle. They serve first to specify the heat radiation characterization and its orientation.

In the subsequent Fig. 8.50, two basic worker's positions are displayed, representing the way he was exposed to the heat radiation inside the ladle (a) and in front of it (b). S_i, T_i, ε_i are the surface,

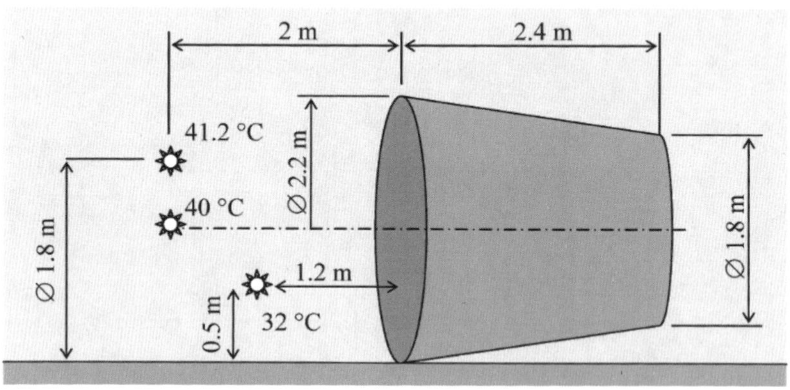

Fig. 8.49. Dimensional draft and ladle position.

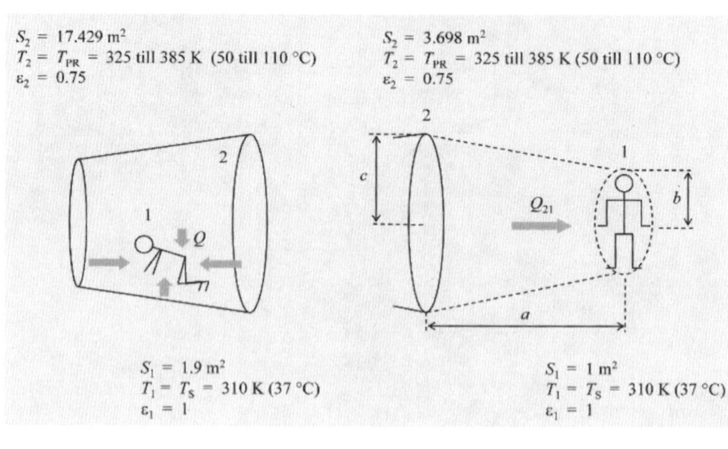

Fig. 8.50. Basic worker's positions working inside the ladle (a) and outside of it (b).

temperature, and relative surface absorptance; the index $i = 1$ is valid for the body and $i = 2$ for the ladle.

The complicated boundary conditions play important role because they involve radiation, convection, perspiration, evaporation, and blood flow, as well as the metabolism rates from the characterization of the work. The solution is realized in five alternatives of the metabolism rates, in several alternatives of the internal ladle surface temperature, and several various thermal losses due to perspiration and evaporation. Further details concerning the solution are in [4.3].

Results

Primarily, the modeling results show the time course of the average body temperature increment at presumed maximum internal surface ladle temperature (A), at various rates of metabolism (M), and different thermal losses due to perspiration and evaporation (V). For the physiologically admissible mean temperature rise, $\Delta T_S = 2.5$ K can be held. With this value, the maximum admissible worker's stay time in the thermal source proximity is limited.

In Fig. 8.51, this admissible mean temperature rise is denoted as the *thermoregulation limit,* beyond which the thermocontrol system collapses, with physiological and psychological consequences for the worker. From the analysis of the workplace conditions and the fact that the worker tried to finish his work as quickly as possible with regard

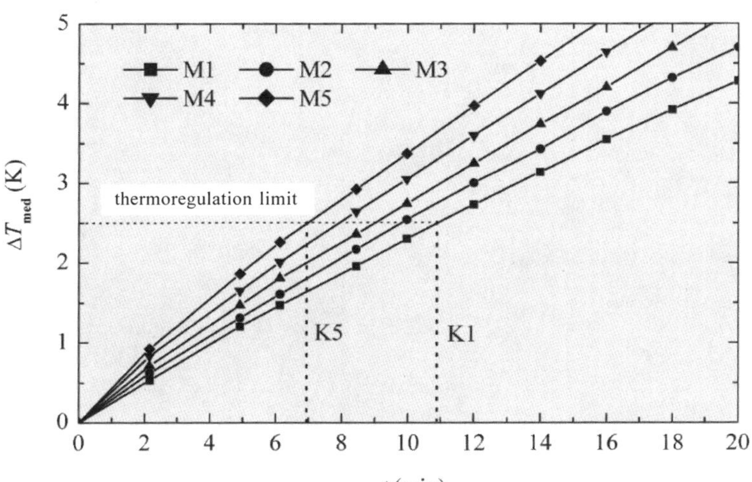

Fig. 8.51. Time course of the mean body temperature increment when considering five different metabolism rates M1 to M5, high diffusion value of humidity from the body V3, and highest ladle surface temperature A4.

to his working conditions, a high working performance followed, which corresponded to the internal rate of metabolism M4 or M5 and high values of perspiration and evaporation V3. Under these conditions, the admissible working time K5 is only 7 minutes for the supposed rate of metabolism M5. By decreasing the working performance and the metabolism rate to M1 at the same time, this admissible time is prolonged up to 11 minutes.

However in fact, the extent of perspiration and evaporation influences the average body temperature more, and the results given in Fig. 8.51 are therefore nearest to the reality. According to these results, the worker's time repairing the ladle should not have exceeded approximately 9 minutes. In fact, it lasted 15 to 20 minutes. The result proves that the thermocontrol system can break down, and the worker's after-thermal-shock reactions correspond to this finding: discomfort, dizziness and, finally, thermal collapse. The more detailed procedure and results are given in [4.3].

"When using a mathematical model, careful attention must be given to the uncertainties in the model."

Richard Phillips Feynman *(11.5.1918–15.2.1988)*

9

Stochastic Computer Models

In contrast to deterministic models, stochastic models operate over random processes and quantities which are usually arranged in time. In a discrete time case, the random processes can be interpreted as randomly selected sequences or as random functions in a continuous time case. The physical description is based on Brown's probability model describing the movement of a fluctuating particle. Its trajectory forms continuous functions only, which have no derivation anywhere. Unlike deterministic methods, stochastic models and methods are applied when their unique properties are useful for solving stochastic as well as deterministic tasks, particularly those which have a very complicated character. In biology, complex problems are solved from the molecular to ecological level. Stochastic models are also used in social sciences, operational research, in economics and finance.

Stochastic models and methods are applied widely in many scientific and engineering fields, especially those which deal with physical processes and fields.Above all, they involve solutions of various boundary tasks, for example, in the thermomechanics of radiation or heat conduction. They are also used in statistical physics, quantum chemistry and other scientific and engineering fields. A more detailed survey of stochastic models and modeling can be found on internet websites.

9.1 Characterization of Stochastic Models

The importance of stochastic models consists in their specific approach

J. Kuneš, *Similarity and Modeling in Science and Engineering*,
DOI: 10.1007/978-1-907343-78-0_9,

to solution of the tasks. Generally, *deterministic and stochastic approaches to the solution have a different concept of task complexity.* The task that appears from the deterministic point of view to be complex and can be solved with difficulties only, can be effectively solved from the stochastic point of view. Naturally, the contrary can be said too. One can say that the deterministic and stochastic models are complementary due to their specific properties.

The mathematical stochastic models of tasks can be solved in digital, analogue or hybrid ways by an appropriate computer. The mathematical stochastic operating model for an investigated process is described by the Markov continuous or discrete chains. The problem solution consists of finding certain properties of the chains. The chains are determined by probabilities of transition from a certain state into other states. With all these probabilities known, the properties of the Markov chain can be established with an unambiguous procedure. This method is called the *probability method.* In the stochastic *Monte Carlo* and *Exodus* methods, the Markov chain properties are found experimentally on the basis of a great number of actual experiments. Therefore, the stochastic models can be divided into models operating on the principle of theoretical *probability method* and experimental *Monte Carlo* or *Exodus* methods. They are used to solve tasks expressed in stochastic and deterministic ways. All three methods are described in [9.27] in more detail.

From the three stochastic methods, the probability method, whose solution accuracy is high, is suitable for solving smaller tasks. The Exodus method is faster and does not require such a big memory, but the solution accuracy is related to the solving time. The majority of vast multi-dimensional and complex-shape tasks can be solved by the Monte Carlo method. The choice of a numerical or hybrid method with their specific properties plays an important role. These specific properties can be used not only to solve direct tasks, but also, especially, indirect and optimalization ones.

9.2 Outline of General Theory

A *stochastic system* can be imagined as composed of subsystems which move randomly among a finite number of states. The subsystem can be an arbitrary *information carrier* such as, for example, a molecule, a product, a live individual, a financial capital, an attended machine, or a social group. Every carrier involves complete information about itself. The *carrier model* contains a part of this information, especially that which is essential from the point of the goal of examination. For example, it can be information on the state, extent, size, position, and properties of the carrier. The *system model* contains the necessary information on the structre of the examined system and on the conditions of carrier transitions among individual states. The system structure and the transition conditions can change randomly or depending on time, the state of the carriers, etc. The stochastic systems are described by the Markov chain properties. These are discrete random sequences fulfilling the condition that the state of the immediately following future is given by the *conditioned probability* which depends on the present state of this stochastic process. These conditioned probabilities are called *transition probabilities.*

9.2.1 Mathematical Stochastic Model

Mathematical stochastic description can be used not only for stochastic systems whose behavior is random in principle, but also for deterministic systems whose mathematical deterministic description is transformed to mathematical stochastic one.

Formation of the transition probability matrix

A stochastic system can be described by the *transition probability matrix* **P** in the form

$$\mathbf{P}=\begin{bmatrix} p_{11} & p_{12}, & \ldots, & p_{1n} \\ p_{12}, & p_{22}, & \ldots, & p_{2n} \\ \vdots & \vdots & \ddots & \vdots \\ p_{n1}, & p_{n2}, & \ldots, & p_{nn} \end{bmatrix}, \tag{9.1}$$

where n denotes the *number of possible states of the stochastic system.* The matrix **P** is of the (n,n) type. The element p_{ij} denotes the *carrier transition probability from the state i into the state j,* the element P_{ij} denotes the *carrier probability of remaining in the state i.* For the probabilities, the following must be valid

$$\sum_{j=1}^{n} p_{ij} = 1, \quad i = 1, 2, \ldots, n. \tag{9.2}$$

Occurrence probability determination

At the beginning, the carriers have a certain distribution in possible system states. The *carriers occurrence probabilities* $m_i^{(0)}$ *in individual initial states i* can be arranged into a row *vector of initial occurrence probabilities*

$$\mathbf{m}^{(0)} = \left(m_1^{(0)}, m_2^{(0)}, \ldots, m_x^{(0)}\right). \tag{9.3}$$

The information carriers walk between individual states according to given probabilities. In investigating the system, the question is how the carriers will be distributed among individual states after a certain number of steps of their random walk.

After k steps, the distribution of carriers among the states is determined by the relation

$$\mathbf{m}^{(k)} = \mathbf{m}^{(k-1)}\mathbf{P}, \tag{9.4}$$

or using the initial occurrence probabilities

$$\mathbf{m}^{(k)} = \mathbf{m}^{(0)}\mathbf{P}^{k}. \tag{9.5}$$

The elements of vector $\mathbf{m}^{(k)}$ change in each step. The sequence of vectors $\mathbf{m}^{(k)}$ is denoted as the *Markov chain,* which is determined by transition probability matrix $\mathbf{P}$. The occurrence probabilities in step k depend only on the occurrence probabilities in the previous step $k - 1$.

The stochastic system is in the *dynamic equilibrium state* when the number of carriers is constant in each state. This means that in each step, the number of carriers coming into each state equals the number of carriers leaving this state. The dynamic system equilibrium occurs after a finite number of steps if each power of the matrix $\mathbf{P}$ has positive elements only. With this condition fulfilled, a constant vector $\mathbf{m}^{(K)}$ of the matrix $\mathbf{P}$ can be found for which the following holds

$$\mathbf{m}^{(K)} = \mathbf{m}^{(K)}\mathbf{P}. \tag{9.6}$$

The number K denotes the number of steps necessary to reach the dynamic system equilibrium.

The probability for a certain carrier to be in the state j exactly after k steps if it had been in the state i at the beginning can be determined utilizing Eqs. (9.4) or (9.5). If the vector of initial occurrence probabilities of carriers $\mathbf{m}^{(0)}$ contains unity in the state i with zeros in all other states, then the vector $\mathbf{m}^{(k)}$ involves the probabilities with which the carrier is in individual states after k steps. The *vector element $m_j^{(k)}$ determines the investigated occurrence probability of the carrier being in the state j after k steps. These operations, described by Eqs. (9.4) to (9.6) and Eq. (9.10), are the basis for the Exodus method.*

The next method for determination of the carrier occurrence probability in state j, if it had been in the state i at the begining, is the matrix solution procedure. This method utilizes rearranging of the probability transition matrix **P** according to the absorption abilities of individual states of the system. In the investigated system, there can be states with the probability of remaining in the state $p = 1$. If a carrier gets into such a state, it remains here. This state is denoted as the *absorbing state.*

Let us have a system with absorbing states that is described by the transition probability matrix **P.** The absorbing Markov chain, described by this matrix **P**, is the mathematical model of this system. To be solved, the matrix must be rearranged into a canonical form so that the absorbing states should be in the first rows and columns. With r absorbing states and s non-absorbing ones, the transition probability matrix has the following form

$$\mathbf{P} = \begin{bmatrix} I & \vdots & 0 \\ \cdots & \cdots & \cdots \\ R & \vdots & Q \end{bmatrix}, \tag{9.7}$$

where **I** is a unity matrix of (r, r) type, 0 is a zero matrix of (r, s) type, **R** is a matrix of (s, r) type, and **Q** is an (s,s) type matrix.The *fundamental matrix* **N** of (s,s) type is determined by the following relation

$$\mathbf{N} = (\mathbf{I} - \mathbf{Q})^{-1}, \tag{9.8}$$

where **I** is a unity matrix of (s, s) type. *The element* n_{ij} *of the matrix* **N** *denotes the mean value of the number of carrier transitions through the non-absorbing state j, prior to ending in some absorbing state, if it had been in the non-absorbing state i initially.*

The probability for the carrier to be absorbed in the state j, if it had been in the non-absorbing state i initially, is given by the element b_{ij} *of the matrix* **B** that is determined by the relation

$$\mathbf{B} = \mathbf{NR}\,. \tag{9.9}$$

The *matrix of transition probabilities to the absorbing states* **B** of (s,r) type involves the probabilities of transitions from non-absorbing states into absorbing ones.

The occurrence probability of the carrier in the absorbing state j, if it had been in the non-absorbing state i initially, is given by the following relation

$$\mathbf{m}_A^{(K)} = \mathbf{m}_N^{(0)}\,\mathbf{B}, \tag{9.10}$$

whereas the vector of initial probabilities of occurrence of carriers in the nonabsorbing states $\mathbf{m}_N^{(0)}$ contains unity in the state i with zeros in all other states and vector $\mathbf{m}_A^{(K)}$ determines the probabilities of occurrence of the carrier in the absorbing states at the dynamic equilibrium state. *The vector element* $m_{Aj}^{(K)}$ *denotes the sought occurrence probability of the carrier being in the absorbing state j at the dynamic equilibrium state of the system, if it had been in the non-absorbing state i initially.*

These operations, described by Eqs. (9.7) to (9.9), form the basis for the probability method. In this solution procedure, the most demanding operation is to invert the matrix (9.8) of s order.

Stochastic description of physical fields

In a *stochastic description of deterministic* systems, for example, of *various physical fields,* the investigated area is first discretized in space and time, if required. The number of nodes of the solved area is denoted by n which is an analogous quantity to the number of possible states in the stochastic system. The probabilities of transition into other network nodes are determined for each node in this area,.

The computational network nodes are divided into *internal* and *boundary* ones and, further, into *absorbing* and *non-absorbing* ones. Whereas the *internal nodes* have at least one neighbouring node on all sides, some neighbouring nodes are missing for the *boundary nodes*. The *absorbing nodes* are those boundary nodes which have the probability of remaining in the node, i.e. the absorption probability, which equals unity. This means that a carrier, after having get in a absorbing node, remains there with the probability equalling unity. The *non-absorbing nodes* are other boundary nodes and all internal nodes. The absorption probability for these nodes is less than unity.

In a *stochastic description of steady physical fields,* the transition probability matrix **P** is of the (n,n) type. The number n of the computational network nodes can be expressed as $n = r + \mathrm{s}$, where r equals the number of all *absorbing nodes* and s is the number of *non-absorbing ones*. In the *stochastic description of unsteady physical fields,* the transition probability matrix **P** is of the $(2s + r,\ 2s + r)$ type. It consists of s non-absorbing nodes in the time $\tau + \Delta\tau$, s absorbing nodes in the time τ and r absorbing nodes from the time level τ or $\tau + \Delta\tau$, according to the time difference in the discretization of a differential equation of a physical field. Therefore, in unsteady physical fields, the matrix **P** consists of s non-absorbing nodes and $s + r$ absorbing ones.

The *transition probabilities for internal nodes* are derived from the corresponding differential equation of the investigated physical field and are therefore different for every equation type. They are determined by the kind of physical field, by the selected space and time discretizations, and by the physical parameters of the field. The *transition probabilities for boundary nodes* are obtained from the same parameters as for internal nodes and from the relations for boundary conditions. They are determined, in addition to the internal nodes, by the type of boundary conditions and the values of their parameters.

In stochastic modeling of physical fields, the computational network nodes have dual numbering. The *global nodes numbering* corresponds to serial number assignment to every network number. Using the global numbering, the whole stochastic system is described, i.e. the transition probability matrix and the occurrence probability vectors, etc. are assembled. To obtain the transition probability matrix from the differential equations of the physical field, a *local nodes numbering* is used, where neighbouring nodes of each node are numbered. However, in the formation of the matrix **P** from the obtained transition probabilities, one goes over from the local numbering to a global one which is used henceforth.

9.2.2 Transition Probabilities of Internal Nodes

In stochastic modeling of physical fields, the transition probabilities of internal nodes are determined from the basic equations of these fields. The determined probabilities are divided to *probabilities of transitions into neighbouring nodes* and *probabilities of remaining in the node*. If need be, an *inner source contribution in the node* can be considered, but it is not the probability. The probabilities of transitions from a node into other than neighbouring nodes are zero in physical fields. In the following text, the probabilities of transitions, probabilities of remaining in the node and inner source contributions in the node are determined for unsteady and steady two-dimensional physical fields, including inner sources and constant physical parameters.

Unsteady physical field

The Fourier equation of an unsteady two-dimensional field with inner sources and constant physical parameters has the following form

$$a\left(\frac{\partial^2 u}{\partial x^2}+\frac{\partial^2 u}{\partial y^2}\right)+\frac{q_V}{\lambda}a=\frac{\partial u}{\partial \tau}. \tag{9.11}$$

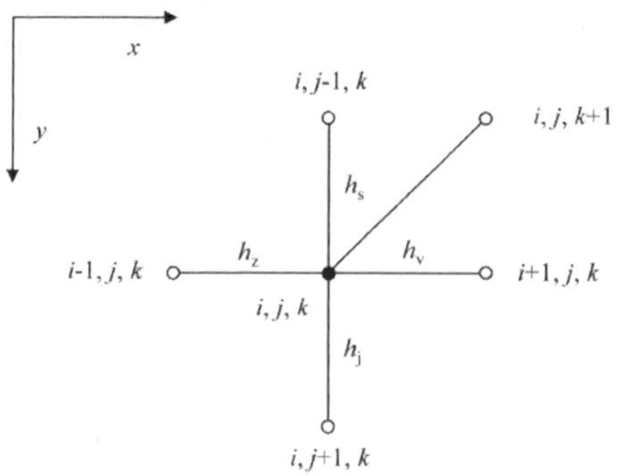

Fig. 9.1. Difference diagram of an internal node for an unsteady physical field.

After space and time discretization according to Fig. 9.1, the quantity $u(x,y,\tau)$ goes to $u_{i,j,k}$ and the internal sources $q_V(x,y,\tau)$ go to $q_{Vi,j,k}$.

Individual partial derivations are replaced by the following difference terms

$$\frac{\partial^2 u}{\partial x^2} = \frac{2}{h_z + h_v}\left(\frac{u_{i+1,j,k} - u_{i,j,k}}{h_v} - \frac{u_{i,j,k} - u_{i-1,j,k}}{h_z}\right), \tag{9.12}$$

$$\frac{\partial^2 u}{\partial y^2} = \frac{2}{h_s + h_j}\left(\frac{u_{i,j+1,k} - u_{i,j,k}}{h_j} - \frac{u_{i,j,k} - u_{i,j-1,k}}{h_s}\right), \tag{9.13}$$

$$\frac{\partial u}{\partial \tau} = \frac{1}{\Delta\tau}\left(u_{i,j,k+1} - u_{i,j,k}\right), \tag{9.14}$$

where the last equation expresses a forward time difference.

Using the relations (9.12) to (9.14), the basic Eq. (9.11) can be converted into the form

$$\begin{aligned} u_{i,j,k+1} &= \frac{2a\Delta\tau}{h_v\left(h_z + h_v\right)} u_{i+1,j,k} + \frac{2a\Delta\tau}{h_z\left(h_z + h_v\right)} u_{i-1,j,k} + \\ &+ \frac{2a\Delta\tau}{h_j\left(h_s + h_j\right)} u_{i,j+1,k} + \frac{2a\Delta\tau}{h_s\left(h_s + h_j\right)} u_{i,j-1,k} \left(1 - \left| \frac{2a\Delta\tau}{h_v\left(h_z + h_v\right)} - \right.\right. \\ &\left.\left. - \frac{2a\Delta\tau}{h_z\left(h_z + h_v\right)} - \frac{2a\Delta\tau}{h_j\left(h_s + h_j\right)} - \frac{2a\Delta\tau}{h_s\left(h_s + h_j\right)}\right) u_{i,j,k} + \frac{a\Delta\tau}{\lambda} q_{V_{i,j,k}}. \end{aligned} \tag{9.15}$$

Using this, the *probabilities of transition into neighbouring nodes in a lower time level* are stated as (9.15)

$$\left.\begin{aligned} p_v &= \frac{2a\Delta\tau}{h_v\left(h_z+h_v\right)}, \quad p_z = \frac{2a\Delta\tau}{h_z\left(h_z+h_v\right)} \\ p_j &= \frac{2a\Delta\tau}{h_j\left(h_s+h_j\right)}, \quad p_s = \frac{2a\Delta\tau}{h_s\left(h_s+h_j\right)} \end{aligned}\right\} \tag{9.16}$$

and the *probability of transition into the same node in a lower time level*, that means the *probability of remaining in the node,* is stated as

$$p_t = 1 - p_v - p_z - p_j - p_s, \tag{9.17}$$

and the *inner source contribution in the node* is

$$u_{q_V i,j,k} = \frac{a\Delta\tau}{\lambda} q_{V\,i,j,k}. \tag{9.18}$$

For all internal nodes, the carrier *probability of absorption* p_{abs} and the carrier *probability of reflection* p_{refl} can be defined in compliance with the following relations

$$\begin{aligned} p_{abs} &= p_t \neq 0, \\ p_{refl} &= p_v + p_z + p_j + p_s \neq 1. \end{aligned} \tag{9.19}$$

For the following time step, the resulting equation for unknown quantity determination in the central node of a computational cell has the form

$$\begin{aligned} u_{i,j,k+1} = p_v u_{i+1,j,k} + p_z u_{i-1,j,k} + p_j u_{i,j+1,k} + p_s u_{i,j-1,k} + \\ + p_t u_{i,j,k} + u_{qvi,j,k}. \end{aligned} \tag{9.20}$$

The probabilities p_v, p_z, p_j, p_s and p_t are incorporated into the transition probability matrix **P** of the $(2s + r, 2s + r)$ type into the row corresponding to the node $u_{i,j,k+1}$ in the time level $k + 1$. They express the probabilities of transition from the node $u_{i,j,k+1}$ into corresponding nodes in a lower time level. The probabilities of transition from this node into other nodes are zero. The inner source contribution in the node is not used before the function values in network nodes are determined.

The probabilities of transitions from non-absorbing internal nodes in the time level $k + 1$ are determined using Eq. (9.20). The probabilities of transitions for boundary nodes in the same time level can be obtained as presented in a later subsection 9.2.3. Other s absorbing nodes have the probability of absorption $p_{abs} = p_t = 1$ and probability of reflection $p_{refl} = 0$. This means that the rows in the matrix **P** corresponding to the absorbing nodes contain only one non-zero element equalling unity, which corresponds to the probability of remaining in the node. In the case of a physical field without inner sources, the additive member expressing these sources can be omitted in Eq. (9.20).

Further, the time derivation can be replaced by the following difference expression representing the backward time difference

$$\frac{\partial u}{\partial \tau} = \frac{1}{\Delta \tau}\left(u_{i,j,k} - u_{i,j,k-1}\right), \tag{9.21}$$

modified Eq. (9.15) is obtained, from which the resulting equation can be derived in the form

$$\begin{aligned} u_{i,j,k} = {} & p_v u_{i+1,j,k} + p_z u_{i-1,j,k} + p_j u_{i,j+1,k} + p_s u_{i,j-1,k} + \\ & + p_t u_{i,j,k-1} + u_{q_V i,j,k}, \end{aligned} \tag{9.22}$$

where p_v, p_z, p_j and p_s are the *probabilities of a transition into neighbouring nodes in the same time level,* p_t is the *probability of transition into the same node in a lower time level* that means the *probability of remaining in the node*, and $u_{qvi,j,k}$ is the *inner source contribution in the node.* However, the probabilities of transitions and the inner source contribution are other than those in Eqs. (9.16) to (9.18).

Steady physical field

The Poisson equation for a steady two-dimensional field with inner sources and constant physical parameters has the following form

$$\left(\frac{\partial^2 u}{\partial x^2} + \frac{\partial^2 u}{\partial y^2}\right) + \frac{q_V}{\lambda} = 0. \tag{9.23}$$

The difference diagram for an internal node is in Fig.9.2.

The probability equation can be obtained by the same procedure as for the unsteady field, i.e. by replacing the individual partial derivations with difference terms and subsequent modifications. However, Eq. (9.15) can be used directly, in which $u_{i,j,k+1} = u_{i,j,k}$ is stated. The equation is modified to the following form

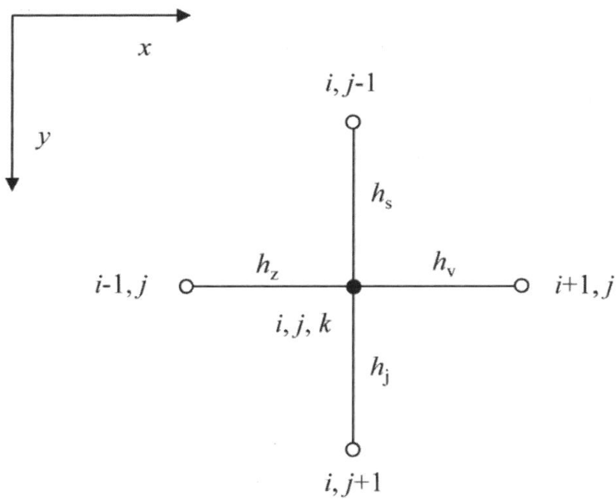

Fig. 9.2. Difference diagram of an internal node for a steady physical field.

$$u_{i,j}=\frac{h_zh_jh_s}{(h_z+h_v)(h_jh_s+h_vh_z)}u_{i+1,j}+\frac{h_vh_jh_s}{(h_z+h_v)(h_jh_s+h_vh_z)}u_{i-1,j}+$$
$$+\frac{h_vh_zh_s}{(h_s+h_j)(h_jh_s+h_vh_z)}u_{i,j+1}+\frac{h_vh_zh_j}{(h_s+h_j)(h_jh_s+h_vh_z)}u_{i,j-1}+$$
$$+\frac{h_vh_zh_jh_s}{2\lambda(h_jh_s+h_vh_z)}q_{V\,i,j}. \tag{9.24}$$

Using this, the *probabilities of transition into neighbouring nodes* are stated as

$$p_v=\frac{h_zh_jh_s}{(h_z+h_v)(h_jh_s+h_vh_z)}\quad p_z=\frac{h_vh_jh_s}{(h_z+h_v)(h_jh_s+h_vh_z)},$$
$$p_j=\frac{h_vh_zh_s}{(h_s+h_j)(h_jh_s+h_vh_z)}\quad p_s=\frac{h_vh_zh_j}{(h_s+h_j)(h_jh_s+h_vh_z)}, \tag{9.25}$$

and the *inner source contribution in the node* is

$$u_{q_V i,j}=\frac{h_vh_zh_jh_s}{2\lambda(h_jh_s+h_vh_z)}q_{V\,i,j}. \tag{9.26}$$

For the probabilities of transition into neighbouring nodes, the following holds

$$p_v + p_z + p_j + p_s = 1, \tag{9.27}$$

from which the *probability of remaining in the node* follows

$$p_t = 0. \tag{9.28}$$

For all internal nodes, the carrier *probability of absorption* P_{abs} and the carrier *probability of reflection* P_{refl} can also be expressed in compliance with the following relations

$$\begin{aligned} p_{abs} &= p_t = 0, \\ p_{refl} &= p_v + p_z + p_j + p_s = 1. \end{aligned} \tag{9.29}$$

The resulting equation for determination of the unknown quantity in the central node of a computational cell has the following form

$$u_{i,j} = p_v u_{i+1,j} + p_z u_{i-1,j} + p_j u_{i,j+1} + p_s u_{i,j-1} + u_{q_V i,j}. \tag{9.30}$$

The transition probability matrix **P** of the $(r + s,\ r + s)$ type into the row of the corresponding node u_{ij}, included the probabilities of transitions $p_{v,}$ $p_{z,}$ p_j and p_s from this node into neighbouring ones. The probabilities of transition into other nodes are zero. In the node, the inner source contribution is not used before function values in the network nodes are determined.

The probabilities of transitions from internal network nodes are determined using Eq. (9.30). The probabilities of transitions for boundary network nodes are obtained as shown in the following subsection 9.2.3. In the case of a physical field without inner sources, the additive member expressing these sources can be omitted in Eq. (9.30).

9.2.3 Transition Probabilities of Boundary Nodes

The transition probabilities for boundary nodes, i.e. those not having some of the neighbouring nodes, are solved by considering the boundary conditions of the investigated process. For the sake of simplicity, the boundary of a discretized area can be assumed to be as in Fig. 9.3.

In the case of an unsteady task, the general Eq. (9.20) holds for internal nodes, as does Eq. (9.30) for a steady task. Certain relations for the probabilities of transitions into neighbouring nodes p_v, p_z, p_j, p_s, for the probability of remaining in the node p_t, and for the inner source contribution in the node $uqv_{i,j,k}$ are valid for these equations. In the case of the boundary nodes, these probabilities are replaced by

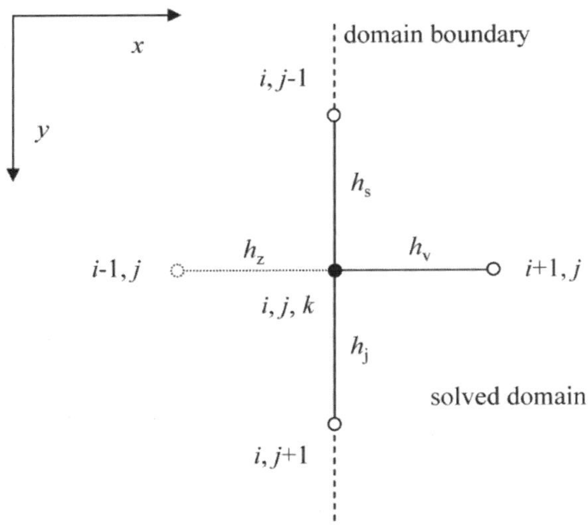

Fig. 9.3. Difference diagram of a boundary node.

the probabilities $p_v^*, p_z^*, p_j^*, p_s^*$, and p_t^*, which depend on the original transition probabilities and on the boundary conditions. For the boundary node, the *probability of absorption* p_{abs}^* and the *probability of reflection* p_{refl}^* are determined and the boundary condition can be written in the general form as

$$u_{i,j,k} = p_{refl}^* u_{i+1,j,k} + p_{abs}^* f_{i,j,k}, \tag{9.31}$$

where $f_{i,j,k}$ is the prescribed value of $f(x, y, \tau)$ for the sought unknown quantity $u(x, y, \tau)$ on the area boundary. The following holds for the probabilities,

$$p_v + p_z + p_j + p_s + p_t = 1, \tag{9.32}$$

$$p_v^* + p_z^* + p_j^* + p_s^* + p_t^* = 1, \tag{9.33}$$

$$p_{abs}^* + p_{refl}^* = 1. \tag{9.34}$$

1st-kind boundary condition

This condition is expressed by the known function value $f(x, y, \tau)$ of the quantity $u(x, y, \tau)$ on the area boundary

$$u(x, y, \tau) = f(x, y, \tau), \tag{9.35}$$

and it can be written after discretization as

$$u_{i,j,k} = f_{i,j,k}. \tag{9.36}$$

By comparing this equation with the general Eq. (9.31) for the boundary condition, the absorption and reflection probabilities can be determined as

$$p^*_{abs} = 1, \quad p^*_{refl} = 0. \tag{9.37}$$

Because the quantity value is given by the prescribed value and does not depend on values in neighbouring network nodes, the *probabilities of transition into neighbouring nodes* are

$$p^*_v = p^*_z = p^*_j = p^*_s = 0. \tag{9.38}$$

Therefore, it follows from Eq. (9.33) that the *probability of remaining in the node* is

$$p^*_t = 1. \tag{9.39}$$

If the carrier arrives into this node, it is absorbed with the probability $p^*_{abs} = 1$. Therefore, *boundary nodes with the 1st-kind boundary condition are absorbing nodes.*

2nd-kind boundary condition

In this condition, the surface flux $q_n(x, y, \tau)$ is prescribed as

$$-\lambda \frac{\partial u(x, y, \tau)}{\partial n} = q_n(x, y, \tau), \tag{9.40}$$

where λ is the conductivity of the inner environment. After discretization, it can be written as

$$-\lambda \frac{u_{i,j,k} - u_{i+1,j,k}}{h_v} = q_{n\,i,j,k}, \tag{9.41}$$

or

$$u_{i,j,k} = u_{i+1,j,k} - \frac{h_v}{\lambda} q_{n\,i,j,k}, \tag{9.42}$$

where h_v is the distance of a boundary node from its internal neighbour. Because the flux is set in the direction perpendicular to the area boundary, the probabilities of a transition into the nodes $(i, j - 1, k)$ and $(i, j + 1, k)$ remain unaffected by the boundary condition. Because the node $(i - 1, j, k)$ does not exist, one can write

$$p_j^* = p_j, \quad p_s^* = p_s, \quad p_z^* = 0. \tag{9.43}$$

For $q_{ni,j,k} = 0$, from comparison of Eq. (9.42) with general Eq. (9.31), the absorption and reflection probabilities follow directly

$$p_{abs}^* = 0, \quad p_{ref}^* = 1. \tag{9.44}$$

This means that the whole original probability p_z is added to the probability p_v, and one obtains

$$p_t^* = p_t, \tag{9.45}$$

$$p_v^* = p_v + p_z. \tag{9.46}$$

The following can be written for $q_{ni,j,k} \neq 0$ in a general case

$$q_{n\,i,j,k} = -\lambda \frac{f_{i,j,k} - u_{i,j,k}}{h_z}, \tag{9.47}$$

where the prescribed flux $q_{n\,i,j,k}$ is replaced by the outer environment conductivity λ and by the outer environment functio $f_{i,j,k}$ in the node distant by h_z from the boundary node. Therefore, the two prescribed variables λ and h_z can be chosen. With $\lambda = h_z = 1$ chosen, the equation changes to the following form

$$q_{n\,i,j,k} = -\left(f_{i,j,k} - u_{i,j,k}\right). \tag{9.48}$$

By inserting this into Eq. (9.42), one obtains

$$u_{i,j,k} = \frac{\lambda}{\lambda + h_v} u_{i+1,j,k} + \frac{h_v}{\lambda + h_v} f_{i,j,k}. \tag{9.49}$$

By comparing this equation with the general equation (9.31), the absorption and reflection probabilities follow directly for the boundary node

$$p_{abs}^* = \frac{h_v}{\lambda + h_v}, \quad p_{refl}^* = \frac{\lambda}{\lambda + h_v}. \tag{9.50}$$

The original unaffected transition probability p_z is divided into the absorption–reflection probability ratio and added to the probability of remaining in the node and to that of transition into the node $(i + 1, j, k)$

$$p_t^* = p_t + p_{abs}^* p_z, \tag{9.51}$$

$$p_v^* = p_v + p_{refl}^* p_z. \tag{9.52}$$

It can be said, therefore, that *boundary nodes with a 2nd-kind boundary condition are non-absorbing ones,* because $p^*_{abs} \neq 1$.

3rd-kind boundary condition

In the case of the 3rd-kind boundary condition, the transfer coefficient α, e.g. for heat, mass, etc., and also the value of the quantity $f(x, y, z)$ are given outside of the zone by the relation

$$-\lambda \frac{\partial u(x,y,\tau)}{\partial n} = \alpha\big(u(x,y,\tau) - f(x,y,z)\big), \tag{9.53}$$

which can be written after discretization

$$-\lambda \frac{u_{i,j,k} - u_{i+1,j,k}}{h_v} = \alpha\left(u_{i,j,k} - f_{i,j,k}\right), \tag{9.54}$$

or

$$u_{i,j,k} = \frac{\lambda}{\lambda + \alpha h_v} u_{i+1,j,k} + \frac{\alpha h_v}{\lambda + \alpha h_v} f_{i,j,k}. \tag{9.55}$$

With these, the absorption and reflection probabilities are determined for the boundary node

$$p^*_{abs} = \frac{\alpha h_v}{\lambda + \alpha h_v}, \quad p^*_{refl} = \frac{\lambda}{\lambda + \alpha h_v}, \tag{9.56}$$

and the resulting general equation to determine the value in the boundary node is derived, this equation being expressed by the values $u_{i+1,j,k}$ in an internal node and the boundary condition $f_{i,j,k}$ in the following form

$$u_{i,j,k} = p^*_{refl} u_{i+1,j,k} + p^*_{abs} f_{i,j,k}. \tag{9.57}$$

With the boundary condition determining the transfer perpendicular to the zone boundary, the probabilities of transition into the nodes $(i, j-1, k)$ and $(i, j+1, k)$ remain unaffected by the boundary condition and, because the node $(i-1, j, k)$ does not exist, the following holds

$$p^*_j = p_j, \quad p^*_s = p_s, \quad p^*_z = 0. \tag{9.58}$$

The original unaffected transition probability p_z is divided in the absorption–reflection probabilities ratio and is added to the influenced probabilities for remaining in the node and transition into the node $(i+1, j, k)$. The following can then be written

$$p^*_t = p_t + p^*_{abs} p_z, \tag{9.59}$$

$$p_v^* = p_v + p_{refl}^* p_z. \tag{9.60}$$

Boundary nodes with the 3rd-kind boundary condition are non-absorbing nodes because the absorption probability $p_{abs}^* \neq 1$.

9.3 Numerical and Hybrid Stochastic Models

It follows from the characterization of the stochastic models that they can be divided according to how the Markov chain properties are determined. The models for which these properties can be determined unambiguously with known probabilities are usually called *models based on the probability method principle.* The models for which these properties are determined experimentally on the basis of a large number of random walk paths are called *models based on the Monte Carlo method principle* or *on the Exodus method principle.* The corresponding properties of the Markov chains in these models are estimated statistically. The models working on the Monte Carlo method principle are either numerical or hybrid; in the case of the Exodus method, they are numerical. Sometimes the Exodus method is considered in the literature as a variant of the Monte Carlo method. The main difference between the methods is that in the Monte Carlo method random walk paths are modeled sequentially, whereas in the Exodus method they are modeled simultaneously.

According to the principle of the methods, the probability method solves tasks in all computational nodes together at one time while the Exodus and Monte Carlo methods solve tasks in selected computational nodes indepedently. This specific property can be utilized advantageously. As already mentioned, the stochastic models serve to solve not only stochastic but also deterministic processes and fields. Primarily, they are used because of their unique advantages especially in complex tasks, where they can change the way of looking at the

Tab. 9.1 Distribution and characterization of the stochastic models and methods

Stochastic models and methods	theoretical	experimental		
	numerical		hybrid	
	probability	Exodus	Monte Carlo	
Solution principle	calculation of all possible random walk paths	simultaneous generation of random walk paths	sequential generation of random walk paths	
Investigated system	discrete or discretized		continuous	
Number of nodes	10^3	10^4	10^5	$\sim 10^9$

complexity of the task to be solved. Usually, in direct tasks solution, the stochastic methods do not have any important advantages as compared to deterministic methods. Their specific properties are often used to solve indirect and, in particular, optimization tasks. The distribution and characterization of the stochastic models and methods are given in Table 9.1.

When applying the stochastic methods, the probabilities of transitions between individual states or nodes of an investigated system must be known in advance. These are given, known, presumed or experimentally found values for an object or a physical model of it. The probabilities of transitions between the nodes of the investigated physical field depend on the space and time discretization, the choice of the difference diagram, the character of the physical process, physical parameters, boundary and initial conditions.

A stochastic model operating on the Monte Carlo method principle can be a highly effective method everywhere, especially in the cases where the deterministic methods become too complicated or impose too great computational demands. Models based on the Monte Carlo method are also highly effective for simulation of process-and-space-complex stochastic systems. For example, this could include randomly changing heat transfer in the furnace space with sprayed particles having diverse mass and ability to radiate. It could be about the stochastic nature of the heat transfer between a brake disk surface and a brake pad or, similarly, between a workpiece and a tool in machining. In these examples, the random quantities are, for example, internal heat sources, caused by surface roughness or represented by randomly sprayed particles in the furnace space.

9.3.1 Models Based on the Probability Method Principle

The *probability method* is based on problem description by discrete Markov chains. This method is suitable for small tasks with thousands of states or nodes in the order. The solution accuracy is high. The solution time depends on the dimensionality and geometric complexity, the number of nodes, and does not depend on the required accuracy of the solution. All possible random walk paths of the information carrier are solved in an unambiguous way from the knowledge of the probabilities of transitions between the states of the investigated system, or between the nodes of the investigated zone.

During the solution procedure, the whole matrix **B** of transition probabilities from the non-absorbing states to the absorbing ones is evaluated and the task solution is determined in all nodes of the computational network at once. The probability method also evaluates the fundamental matrix **N** and, therefore, the tasks with distributed inner sources can be solved.

With the probability method, tasks are solved in several steps involving task formulation, solution zone discretization, formation of the transition probability matrix and the inner source contribution vector, solution of the fundamental matrix and the matrix of transition probabilities to the absorbing nodes, the formulation and solution of the set of equations, and task evaluation.

Modeling procedure

The task can be defined by the differential equation (9.11) for an unsteady field, or by the differential equation (9.23) for a steady field. The equation is completed with boundary conditions or, if necessary, with the initial conditions.

The solution zone is approximated using the discrete network of nodes that forms the computational network. Space discretization is performed for steady tasks, while space and time discretization for unsteady tasks. Using the procedure already mentioned in the subsection 9.2.2, the differential equation can be converted to the difference form and the transition probabilities of internal nodes are gained. Discretization of boundary conditions, mentioned in the subsection 9.2.3, gives the transition probabilities of boundary nodes. The transition probabilities obtained are assembled to the transition probability matrix **P,** see the subsection 9.2.1.

In the case of the physical field with inner sources, the column vector of contributions of inner sources $\mathbf{u}_{\mathbf{qv}}$ of the $(s,1)$ type is composed, where s is the number of non-absorbing nodes. The values $u_{qV\,i}$ for $i = 1, 2,\ldots,s$ are obtained during the determination of transition probabilities of internal nodes. The contributions of inner sources are given in Eq. (9.18) for an unsteady task and in Eq. (9.26) for a steady task.

The transition probability matrix **P** is rearranged into a canonical form, see Eq. (9.7), whereas s non-absorbing nodes and r absorbing nodes are supposed for a steady physical field, or s non-absorbing and $r + s$ absorbing ones for an unsteady physical field. Then, the fundamental matrix **N** is determined using Eq. (9.8). Using Eq. (9.9), the matrix of the transition probabilities to the absorbing nodes **B** is obtained.

During the determination of matrix **N,** the inversion of the matrix of (s,s) type is executed. For the values of s over 10^2 in the order, the matrix inversion is a very time-consuming process and it is suitable to use another stochastic method to solve the task.

Formulation of the set of equations for the calculation of the unknown fuction values u in the solved region depends on the character of the task, whether it is direct or indirect, steady or unsteady, etc. The *solution of the direct steady task* consists of determination of the function values u in the non-absorbing nodes, i.e. in the internal nodes of the zone and boundary nodes with the boundary condition of the 2nd and 3rd-kind. The knowledge of the function values u in the absorbing nodes, i.e. in the boundary nodes with the boundary condition of the 1st-kind, is supposed.

The required function values u in the non-absorbing nodes are aranged into the column vector $\mathbf{u}$ of $(s, 1)$ type and the prescribed function values u in the absorbing nodes are arranged into the column vector $\mathbf{f}$ of $(r, 1)$ type.

Function values in the non-absorbing nodes are determined by the matrix equation in the following form

$$\mathbf{u} = \mathbf{Bf} + \mathbf{Nu}_{\mathrm{qV}}, \tag{9.61}$$

where the matrix of transition probabilities to the absorbing nodes $\mathbf{B}$ is of the (s, r) type and fundamental matrix $\mathbf{N}$ is of the (s, s) type.

The *solution of the direct unsteady task* consists of determination of the function values u in the non-absorbing nodes for all discrete times in the investigated time interval. The required function values u in the non-absorbing nodes are arranged into the column vector $\mathbf{u}^{(k)}$ at time $\tau + \Delta\tau$, that is of $(s, 1)$ type. The prescribed function values u in the absorbing nodes are arranged into the column vector $\mathbf{f}^{(k)}$ of $(r + s, 1)$ type, whereas using the forward time difference, see Eq. (9.14), it contains s already known function values u at time τ in the internal nodes and boundary nodes with the boundary conditions of the 2nd and 3rd-kind and r known function values u at time τ in the boundary nodes with the boundary condition of the 1st-kind.

Function values in the non-absorbing nodes at time $\tau + \Delta\tau$ are determined by the matrix equation in the following form

$$\mathbf{u}^{(k)} = \mathbf{Bf}^{(k)} + \mathbf{Nu}_{\mathrm{qV}}^{(k)}. \tag{9.62}$$

The matrix of transition probabilities to the absorbing nodes $\mathbf{B}$ is of the $(s, r + s)$ type and fundamental matrix $\mathbf{N}$ is of the (s, s) type.

At time $\tau = 0$, the initial condition must be prescribed in all internal nodes and boundary nodes with the boundary conditions of the 2nd and 3rd-kind. These values are used with the boundary condition of the

1st-kind for calculation of function values in the following time level. This procedure is similar for all time levels, the calculated function values in certain time level are used as the initial condition for the determination of the function values in the following time level. The unsteady task is solved as a sequence of Eq. (9.62) for individual time levels in the investigated time interval.

The *stochastic solution of direct steady tasks* with boundary conditions leads to the matrix equation (9.61) and for the *direct unsteady tasks* the matrix equation (9.62) is obtained. Solution of these equations is not time-consuming and the function values u in the non-absorbing nodes are determined. With the composition of vectors **u** and **f**, the function values u in all nodes of the solved region are determined.

9.3.2 Models Based on the Exodus Method Principle

The *Exodus method* principle is based on problem description by discrete Markov chains. In the literature, it is often identified with the Monte Carlo method because its principle is similar, but it can be considered a modification of it. The simultaneous random walking of a large number of information carriers is modeled using the knowledge of the probabilities of transitions between the states of the investigated system, or between the nodes of the investigated zone, otherwise a large number of random walk paths are generated simultaneously. The solution time depends on the dimensionality and geometric complexity, the number of nodes, and also on the required accuracy of the solution. This method utilizes the multiplication of matrices and vectors from the mathematical point of view and, therefore, is suitable for larger tasks with tens of thousands of states or nodes in the order.

In many practical tasks, the solution is required for a small number of individual points only, without solving the task at other points or times. Therefore, the advantage of the Exodus method as compared to the probability method is in the fact that the selected rows of the matrix **B,** corresponding to the nodes where the solution is sought, can be determined independently. The rows of the fundamental matrix **N** can also be evaluated using the Exodus method. therefore this method is suitable solving tasks with distributed inner sources.

With the Exodus method, tasks are solved in several steps involving task formulation, solution zone discretization, formation of the transition probability matrix and the inner source contribution vector. Then, for the selected node in the solution zone where the function value is sought, the corresponding rows in the fundamental matrix and in the matrix of transition probabilities to the absorbing nodes are determined. The last step is the formulation and solution of the equation and the function value in the selected node is obtained from the result.

Modeling procedure

Task formulation, solution zone discretization, formation of the transition probability matrix **P** and the inner source contribution vector $\mathbf{u}_{qv}$ are performed in the same way as in the probability method (subsection 9.3.1).

The physical field is described by the transition probability matrix **P**, whereas *s* non-absorbing and *r* absorbing nodes are meant for a steady task, and *s* non-absorbing and *r* + *s* absorbing nodes for an unsteady task. The random walking process starts in the node *i*, where the solution is sought. It is made using the initial condition in the form of the vector of initial carriers occurence probabilities $\mathbf{m}^{(0)}$ that contains unity in the state *i* ($m_i^{(0)} = 1$) with zeros in all other states. This condition means that all carriers are initially in the node *i* with the probability of 1. Equation (9.5) is solved for the kth-step of the random walking process and the vector of occurence probabilities $\mathbf{m}^{(k)}$ is obtained.

The carriers occurence probabilities change in every step of the process with some random walk paths ending in the absorbing nodes. After completing a very large number of steps, the estimations of the probabilities b_{ij} of the carrier to be absorbed in the node *j* after *k* steps, if it had been initially in the non-absorbing state *i*, are calculated.

Equation (9.10) is valid in the dynamic equilibrium state. If the vector of carrier initial *occurence probabilities in the non-absorbing nodes* $\mathbf{m}_N^{(0)}$ contains unity in the node *i*, then the following equation is obtained

$$\mathbf{b}_i = \mathbf{m}_A^{(K)}. \tag{9.63}$$

From this equation it follows that the ith row $\mathbf{b}_i$ of the matrix **B** equals the carrier *occurence probabilities in absorbing nodes* $\mathbf{m}_A^{(K)}$ *in the dynamic equilibrium state.*

The vectors $\mathbf{m}_A^{(k)}$ represents the estimate of the row of the matrix **B** at the kth-step of random walking process with certain accuracy and the following relation can be written

$$\mathbf{b}_i = \left(\mathbf{m}_A^{(k)}\mathbf{c}\right)^{-1} \mathbf{m}_A^{(k)}, \tag{9.64}$$

where $\mathbf{m}_A^{(k)}$ is a row vector of (1, *r*) type containing the carrier occurence probabilities in absorbing states at the kth-step of the random walking process, $\mathbf{c}$ is the unity column vector of $(r, 1)$ type with all elements equal to 1. The product in the parentheses expresses the probability that the carriers have been absorbed at the kth-step. For $k \to K$, $\mathbf{m}_A^{(k)}\ \mathbf{c} \to 1$ is valid and then Eq. (9.64) transforms to Eq. (9.63). Using Eq. (9.64), the row of matrix **B,** corresponding to the node where the solution is sought, is determined.

In the Exodus method a certain error is allowed, which is due to the fact that the random walking process is finished before all carriers are absorbed. The random walking process is finished after a certain number of steps or when the solution accuracy improvement is less than the acceptable deviation. This provides the condition that the carriers occurence probability after the termination of the random walking process resembles to that of the dynamic equilibrium state.

The element n_{ij} of the fundamental matrix denotes the mean value of the number of carrier transitions through the non-absorbing node j, prior to ending in some absorbing node, if it had been initally in the non-absorbing node i. The row n_i of the fundamental matrix can be determined using the following relation

$$\mathbf{n}_i = \sum_{l=1}^{k} \mathbf{m}_N^{(l)}, \tag{9.65}$$

whereas the vector of initial occurence probabilities of carriers in the non-absorbing nodes $\mathbf{m}_N^0$ contains unity in the position i with zeros in all other nodes. The sum of the vectors of occurence probabilities in the non-absorbing nodes is performed for all steps k of the random walking process.

Formulation of the equation for the calculation of the unknown function value u in the selected node of the computational zone depends on the character of the task, whether it is direct or indirect, steady or unsteady, etc. The following relation is valid for the *solution of direct steady task*

$$u_i = \mathbf{b}_i\mathbf{f} + \mathbf{n}_i\mathbf{u}_{\mathrm{qV}}, \tag{9.66}$$

where u_i is the function value sought in the node i, $\mathbf{b}_i$ is the row of the matrix of transition probabilities to the absorbing nodes of $(1\ ,r)$ type, column vector $\mathbf{f}$ of $(r, 1)$ type contains prescribed function values u in the absorbing nodes, $\mathbf{n}_i$ is the row of the fundamental matrix of $(1, s)$ type and $\mathbf{u}_{\mathbf{qv}}$ is the column vector of $(s, 1)$ type of contributions of inner sources.

The *solution of the direct unsteady task* consists of the determination of function value u in the selected non-absorbing node for all discrete times in the investigated time interval. The following equation holds

$$u_i^{(k)} = \mathbf{b}_i \mathbf{f}^{(k)} + \mathbf{n}_i \mathbf{u}_{\mathbf{qV}}^{(k)}, \tag{9.67}$$

where $u_i^{(k)}$ is the function value in the non-absorbing node i at the time $\tau + \Delta\tau$, the row of the matrix of transition probabilities to the absorbing nodes $\mathbf{b}_i$ is of $(1, s + r)$ type, $\mathbf{f}^{(k)}$ is the column vector of $(s + r, 1)$ type that contains the prescribed function values u in the r absorbing boundary nodes and s internal nodes in the preceding time level that form initial condition, $\mathbf{n}_i$ is the row of the fundamental matrix of $(1, s)$ type and $\mathbf{u}_{\mathbf{qV}}^{(k)}$ is the column vector of $(s, 1)$ type of contributions of inner sources in the time $\tau + \Delta\tau$.

The *stochastic solution of the direct steady task* with boundary conditions using the Exodus method leads to Eq. (9.66), while Eq. (9.67) is obtained for the *direct unsteady task.* Solution of these equations is not time-consuming and the function value u in the non-absorbing node is determined.

The form of the vector of initial occurence probabilities $\mathbf{m}^{(0)}$ can be utilized with advantage for the *solution of direct tasks.* The function value in the selected node, the difference of the function values between two nodes, and the mean function value in the selected region or other complicated characteristics can be determined using one calculation procedure. When the function value u_i in the node i is sought, the vector of initial occurence probabilities $\mathbf{m}^{(0)}$ contains unity in the position i with zeros in all other positions. Determining the difference of the function values between the nodes i_1 and i_2, the vector of initial occurence probabilities $\mathbf{m}^{(0)}$ contains value of +1 in the node i_1 and the value of −1 in the node i_2. If the vector $\mathbf{m}^{(0)}$ contains the value of $1/n$ in the positions of individual n nodes, the mean function value u in the region comprising these n nodes is determined.

9.3.3 Models Based on the Monte Carlo Method Principle

The *Monte Carlo method* uses the description of the solved problem by discrete or continuous Markov chains. The sequential random walking of a large number of information carriers is modeled on the basis of the knowledge of the probabilities of transitions between the states of the investigated system, or between the nodes of the investigated zone, otherwise a large number of random walk paths are generated sequentially. The solution time does not depend on the dimensionality, geometric complexity and number of nodes it depends only on the required accuracy of the solution.

The Monte Carlo methods can be divided into *numerical Monte Carlo methods with fixed of variable steps* and *the continuous hybrid Monte Carlo* method. The numerical Monte Carlo methods are based on the generation of random numbers. These methods are suitable for large-scale multidimensional and geometry-complex tasks with hundreds of thousands of states or nodes in the order. The hybrid method is suitable for modeling of continuous systems without their discretization. The analogue components applied in the hybrid method enable distinguish thousands of states in one direction which correspond to the tasks with 10^9 states in the order.

The advantage of the numerical Monte Carlo methods as compared to the probability method is the fact that the solution can be obtained in a small number of individual points only, without the need of knowledge of the solution in another points or times. It is based on the solution procedure that exploits the independent determination of the selected rows of the matrix **B** which correspond to the nodes where the solution is sought. Using the numerical Monte Carlo method with a fixed step, the rows of the fundamental matrix **N** can also be evaluated. Therefore, this method is suitable for solving tasks with distributed inner sources.

With the numerical Monte Carlo methods, tasks are solved in several steps involving task formulation, solution zone discretization and formation of the transition probability matrix. Then, for the selected node in the solution zone where the function value is sought, the corresponding row in the matrix of transition probabilities to the absorbing nodes is determined. The vector of inner source contributions and the row of the fundamental matrix which corresponds to the node where the solution is sought, are determined for the numerical Monte Carlo method with the fixed step. The last step is the formulation and the solution of the equation from which the function value in the selected node is obtained.

Modeling procedure for fixed step numerical method

Task formulation, solution zone discretization, formation of the transition probability matrix **P** and the inner source contribution vector $\mathbf{u}_{qv}$ are performed in the same way as in the probability method (subsection 9.3.1).

The random walking process starts in the non-absorbing node i, where the solution is sought. The following relation holds for the sum of transition probabilities of the carrier from the node i to the other nodes and the probability of remaining in the node i

$$\sum_{j=1}^{n} p_{ij} = 1, \quad i = 1, 2, ..., n. \tag{9.68}$$

These transition probabilities are ordered one after another, so that the whole interval ⟨0; 1⟩ is filled in. The random number is generated at interval ⟨0; 1⟩. The next step of the carrier is determined using the value of the generated random number. This procedure provides the condition that the required stochastic path properties are fulfilled.

The random path of the carrier comes through the non-absorbing nodes and ends in some absorbing node j at the zone boundary. The absorbing node in which the random path has ended is recorded, and one proceeds to the next random walk path. The measuring error is estimated continuously; the generation of walking paths is completed with the estimation less than the given limit. As a result, the number of walking paths that ended in a certain absorbing node are obtained for all absorbing nodes. The relative frequencies of reaching the individual absorbing nodes are determined by dividing these numbers by the total number of generated random walk paths. These relative frequencies are used for estimates of the probability transitions to absorbing nodes, i.e. the elements of the row of matrix **B** corresponding to the node i.

The elements n_{ij} of fundamental matrix **N** denote the mean value of the number of carrier transitions through the non-absorbing node j, prior to ending in some absorbing node, if it had been initially in the non-absorbing node i. Therefore, the carrier transitions through non-absorbing states are recorded during generation of random walk paths, These values in each non-absorbing node are then averaged for all steps of all random walk paths and determine the elements n_{ij}.

The simulation speed is limited by random number generators. That is to say, for a solution in one node of the zone at least 10^7 random numbers are necessary, with strictly kept stochastic properties.

Formulation of the equation for calculating the unknown function value u in the selected node of the computational zone depends on the character of the task, whether it is direct or indirect, steady or unsteady, etc. Formulation of the equation for *solving the direct task* is the same as in the Exodus method. Equation (9.66) is obtained for the direct steady task, while Eq. (9.67) holds for the direct unsteady task. Solution of these equations is not time-consuming and the function value u in the non-absorbing node is determined.

With *one alternative of the direct steady task* solved, another equation for the determination of function value u can be derived. It is assumed that the kth random walk path begins in the node i in which the solution is sought, and ends in some absorbing node j at the zone boundary. The following relation can be written for the function value sought

$$u_i = \lim_{K\to\infty} \frac{1}{K} \sum_{k=1}^{K} \left(f_j^{(k)} + \frac{1}{L^{(k)}} \sum_{l=1}^{L^{(k)}} u_{q_V,l} \right), \tag{9.69}$$

where $\boldsymbol{K}$ is the total number of random walk paths generated. The left member in the parentheses denotes the contribution of prescribed function value in the absorbing node j, where the kth path ends. The right member is the contribution of the inner sources during the kth walking path. This path has $\boldsymbol{L}^{(k)}$ steps through the non-absorbing nodes l, where the contribution of the inner source is involved. The quantity $u_i^{(k)}$ for $k = 1, 2, ..., K$, evaluated for each k continuously can be held as a random quantity with the normal distribution. The interval in which the correct result can be expected with the given probability is calculated. If k for a finite K is narrower than the given limit, the solution is completed. With the accuracy estimated, the results become more valuable, which is an important advantage of the method.

Modeling procedure for variable step numerical method

The numerical Monte Carlo method with the variable step introduces the faster modeling procedure compared to the fixed step method. The disadvantage of this metho is that only certain types of tasks can be solved.

Task formulation and solution zone discretization are the same as in the case of the numerical Monte Carlo method with the fixed step. According to Eqs. (9.25) and (9.27), the following relation is valid for a steady physical field with a regular square network of nodes and without the inner sources,

$$p_v = p_z = p_j = p_s = 0.25. \tag{9.70}$$

Using this presumption, the transition probabilities to all neighouring nodes are identical and the numerical Monte Carlo method with the variable step can be employed.

The inscribed-spheres technique is one way to accelerate the simulation because it represents the shape of the zone better for three-dimensional zones. Similarly, the inscribed-circles technique is useful for two-dimensional zones. Variably-long step random walking paths are generated [2.6]. The procedure using the variable step method is presented in Fig. 9.4 for the case of a two-dimensional field. The inscribed circle of the zone G is created from the node 1 with coordinates (X_P, Y_P), in which the solution is sought. It is a

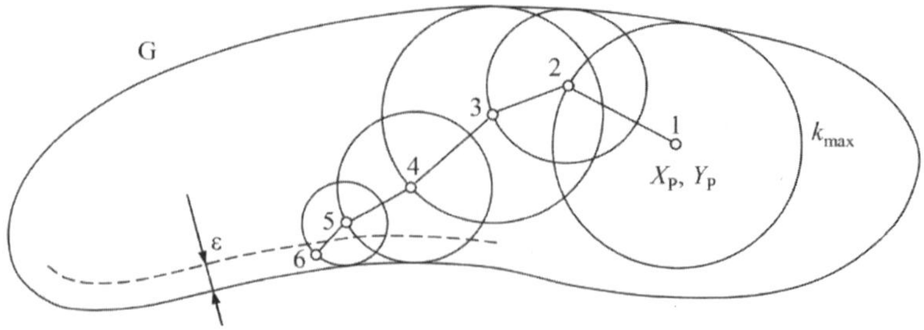

Fig. 9.4. Diagram of random walk paths for the numerical Monte Carlo method using inscribed circles.

largest circle contained in the zone that touches the zone boundary. The random number is generated at interval $\langle 0; 1\rangle$ and according to this the corresponding plane angle is chosen in the interval $\langle 0; 360^{\circ}\rangle$ is chosen. This angle on the inscribed circle defines the point 2 as the next step of walking path. From this point, another circle is inscribed again and the procedure is repeated until the carrier approaches the boundary nearer than the given value ε. The nearest boundary node is assumed to be the end of the random path. The absorbing node in which the random path has ended is recorded, and one proceeds to the next random walk path. In general, this numerical method leads to complex algorithms to inscribe the circles or spheres.

The row of the matrix of probability transitions to absorbing nodes **B**, which corresponds to node *i*, is determined using the same technique as in the case of the numerical fixed step method.

Formulation of the equation for the calculation of the unknown function value u in the selected node of the computational zone for direct steady task without inner sources is the same as in the numerical fixed step method. From this equation, the function value u in the non-absorbing node is determined.

Modeling procedure for the continuous hybrid method

The continuous hybrid Monte Carlo method is derived from the concept of Markov chains. The hybrid Monte Carlo method is used to simulate continuous systems, especially multidimensional and complex ones, in terms of shape. For example, the mathematical model for a linear unsteady three-dimensional temperature field with internal heat sources and a movable boundary with a 1st-kind boundary condition is a set of general differential stochastic equations. After transforming it into a dimensionless form, described in [2.6], one obtains

$$\frac{\mathrm{d}X_i(T_M)}{\mathrm{d}T_M} = -W_i\left[X(T_M)\right]PeFo + No_i(2Fo)^{\frac{1}{2}}, \tag{9.71}$$

$$\mathrm{d}T = -\mathrm{d}T_M, \quad i = 1,2,3, \tag{9.72}$$

where T_M is the model relative time and No_i is the dimensionless noise for which the following holds

$$No_i = N_i T_s^{\frac{1}{2}} S^{-\frac{1}{2}}, \tag{9.73}$$

where N_i are mutually non-correlated random processes with the zero mean value which approximate the white noise, T_s is relative time, and S is the spectral power density of the processes.

The initial condition of the set consists of the coordinates (X_{p1}, X_{p2}, X_{p3}, $X_{p4}=T_p$) of the point at which the solution is sought

$$X_i(0) = X_{pi}, \quad i = 1,2,3,4. \tag{9.74}$$

For the case of the *steady-state field solution* $T_p \rightarrow \infty$. Solving the set (9.71) gives the trajectory of the carrier random walk in the defined zone G. Figure 9.5 shows the possible course of three trajectories for a two-dimensional zone. The random walk path ends with the first reaching of the zone boundary. For the kth chain, the following expression is derived

$$\Phi_k = U_g\left[X(T_g)\right] + PoFo\int_0^{T_g} \mathrm{d}T_M, \tag{9.75}$$

where U_g $[X(T_g)]$ is the 1st-kind boundary condition value at the boundary point at which the realization has ended at time T_g. At the place from which the random walk path processes had begun, the potential function value U (X_p) is given by the mean value

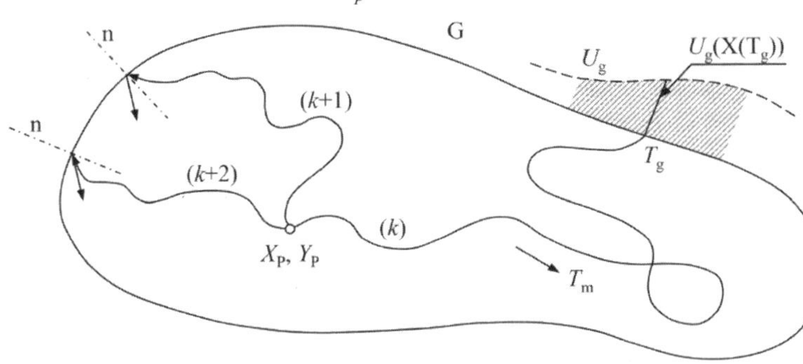

Fig. 9.5. Diagram of random walk paths for the continuous hybrid Monte Carlo method.

$$U\left(X_p\right)=U\left(X_{p1,}X_{p2,}X_{p3,}T_p\right)=\lim_{K\to\infty}\frac{1}{K}\sum_{k=1}^{K}\Phi_k. \tag{9.76}$$

The number of realizations K is finite but is very large, however, with respect to the accuracy of the solution.

For the case of the *unsteady field solution* time T_p is finite. The random walk process in the zone G is described by Eq. (9.71), with the boundary condition. The carrier movement is described in the time from the given value to zero. The process ends either at the zone boundary, as for the steady-state field, or it stops after passing the time T_p. In Eq. (9.75), the initial condition for the place the process has ended in the zone is taken as the boundary condition.

The *structural diagram of a hybrid stochastic model* for a three-dimensional unsteady field is shown in Fig. 9.6. The source 1 generates the random process and from it the independent random processes N_1 to N_3 are formed using the divider 2. In the analogue hybrid-system part, circuits 3 solve the set of stochastic Eqs. (9.71) with the boundary conditions X_{p1}, X_{p2}, X_{p3}, T_p, for circuit 4 the integral in Eq. (9.75) is calculated. Circuit 5 tests whether the random walking carrier has reached the boundary. When the boundary is reached, the solution in circuits 3 and 4 is stopped by a control signal. The boundary point coordinates at which the process ended are led by analogue-digital converters into a digital sub-system; the corresponding boundary condition value U_g $[X(T_g)]$ is determined by program 6 and an intermediate result Φ_k is calculated using Eq. (9.75). The average value (9.76) from the intermediate results Φ_k of repeated experiments,

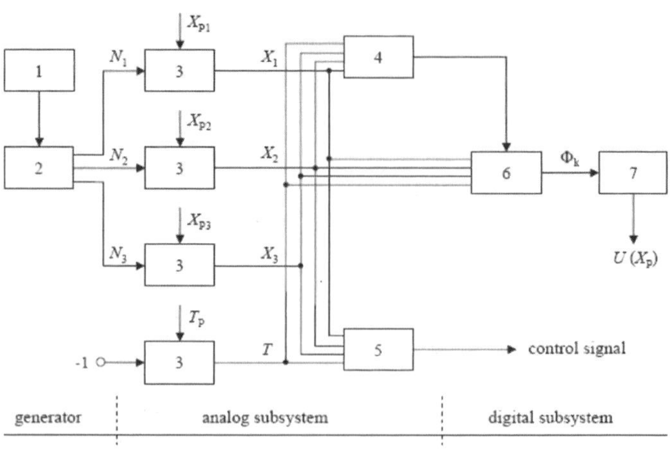

Fig. 9.6. Structural diagram of hybrid simulation of a three-dimensional unsteady-state field by the Monte Carlo method.

as well as the solution error estimation, are calculated by program 7. The generator of pseudorandom sequences with a shift register can be used as a random process source. The generator and calculating network parameters and modeling scales must be chosen so that the spectral power density may be held as a constant. With the systematic errors eliminated, the simulation accuracy depends only on the number of the realized processes. The accuracy estimation can be evaluated similarly as in the case of numerical simulation, but in parallel with the actual solution. A detailed analysis of the conditions used to reach the required accuracy is given in [9.7].

In addition to the already mentioned possibility of solving tasks which are complex in space and time, the advantages of the Monte Carlo methods enable one to determine relatively easily the integral characteristics and, for example, the average volume temperature. They also enable one to determine directly the investigated function gradient and to solve tasks with various boundary conditions and indirect tasks. The method is suitable for solving some optimization tasks in which the whole multidimensional field need not be solved, but the solution must be sought only in some exposed points.

9.4. Stochastic Computers and Models

Stochastic computers and models are based on the idea of applying a random process for information coding. The units executing basic calculating operations are completely unlike those in analogue or digital computers. They are simpler in many cases. To cooperate with digital and analogue devices, and for information input and output, stochastic computers must be equipped with converters of digital and analogue quantities to stochastically coded ones and vice versa.

9.4.1 Stochastic Coding and Operations

In stochastic coding, the random, mostly binary, process is the carrier of information. In it, the information is coded as the occurring probability of the logic 1. For example, the processed quantity X is transformed into the interval $\langle 0,1 \rangle$. The probability $p = 0$ is assigned in a carrier binary signal X_s to the value of 0 and the probability $P = 1$ is assigned to the value of 1. The coded quantity is then assigned to the probability by the following relation

$$p(X_s = 1) = p_X = X. \tag{9.77}$$

This coding is called unipolar, but bipolar coding is also used, too.

Figure 9.7 shows the course of a time-variable quantity. The digital or analogue signals are converted into stochastic form. The coded signal X is compared with the reference random signal R. With the coded signal value greater than the reference signal value, the new signal X_s will have the level 1. The probabilities of 1, occurring in the carrier signal, depend on the value of X. Otherwise, it can be said that the time average output signal value depends on that of the input signal. This signal can be time variable, deterministic or stochastic. The reference signal with the uniform probability density distribution can be continuous or discrete. The output signal value can change at any time or at certain times given by the cycle frequency. Depending on which case it is, the terms *asynchronous* or *synchronous binary signal* are used. The stochastic independence of coded and reference signals and the precisely kept condition of uniform probability density distribution in the whole range $\langle 0;1 \rangle$ are the necessary conditions for accurate and precise conversion.

The *digital-to-stochastic converter DSC* (a) and its schematic symbol (c) are presented in Fig. 9.8. The digital value X_D is compared by a comparator with random numbers from a reference source. Similarly, an *analogue-to-stochastic converter ASC* can be arranged. Figure 9.8b shows another principle with a feedback. This arrangement has a linear characteristic even if the reference signal does not have a uniform distribution. In the comparator output, there is a random process $\boldsymbol{X}_s$ with statistically coded information. This signal is converted back to an analogue one by means of a stochastic-to-analogue converter SAC. If it differs from the input analogue signal X_A, the reference voltage is changed in the comparator until a minimal deviation is reached. The

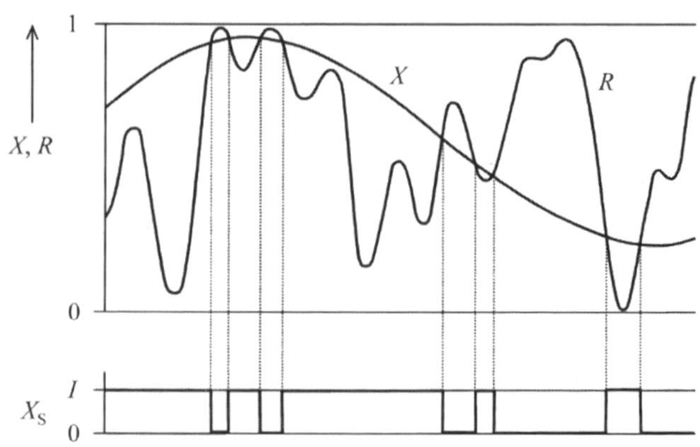

Fig. 9.7. Stochastic coding principle for a time variable quantity.

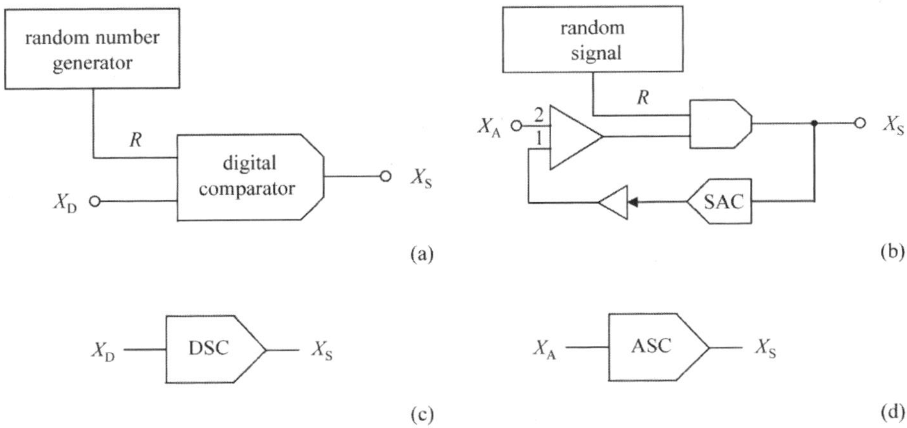

Fig. 9.8. Circuit diagram of a digital-to-stochastic (a) and analogue-to-stochastic (b) converter and their symbols (c, d).

schematic symbol of this analogue-to-stochastic converter is shown in Fig. 9.8d.

The *stochastic-to-analogue conversion* is executed simply by analogue integration. Similarly, the *stochastic-to-digital conversion* can be performed by an integrator with a memory. The stochastic operations and devices, such as multiplier, quadrator, adder, subtracter, and integrator, are described in [2.6].

9.4.2 Properties of Stochastic Computers and Models

The results of stochastic operations are obtained as estimates of random process mean values. This has several important consequences. The speed of the solution is given by accuracy requirements. The longer the solution is in progress, the more precise are the results. With time-variable information being processed, a compromise between the admissible speed of changes and the accuracy of the solution must be found. The advantage of the stochastic model is its resistance against disturbances. In addition, an erroneous pulse does not cause any error in the results if there are many transmitted pulses.

Stochastic computers and models can be used to solve sets of linear equations, ordinary and partial differential equations, to filter signals, and to form adaptive automatic machines, among other things. Figure 9.9 compares the application of stochastic (S), digital (D) and hybrid (H) computers to solve sets of differential equations. The hybrid computers are fast but the number of amplifiers is limited. Considering the simplicity of operational units, the stochastic computers can also be used for complicated tasks. Thanks to the parallel operation, the time

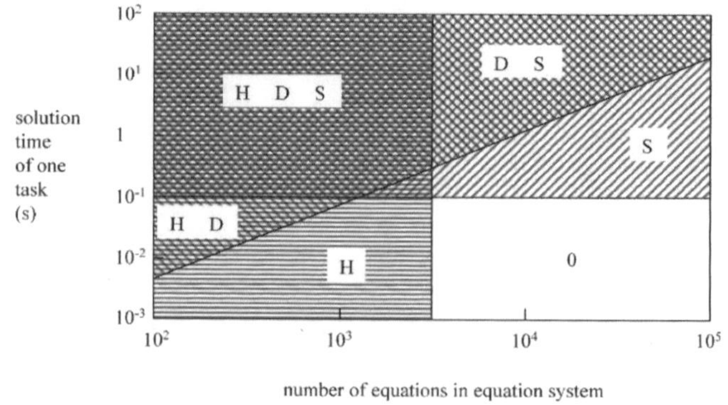

Fig. 9.9. Comparison of the application of stochastic (S), digital (D) and hybrid (H) computers to solve sets of differential equations.

to solve one task with a large number of equations is shorter than in the case of a digital computer.

The operational elements of the stochastic models and computers are simple and can be manufactured with the same technology as those of digital computers. They can be cheap, reliable and have a high degree of integration. From the operational elements, simultaneous models can be formed. They work in parallel and their application is suitable for solving complex tasks swiftly with no accuracy requirements. With their parallel structure and ability to solve complicated tasks, the stochastic models can also be included among cybernetic ones.

9.5 Applications

All the presented examples show utilization of the computer stochastic models in various areas of thermomechanics. Different stochastic methods are presented in these examples.

9.5.1 Thermomechanics

In the following example, the heat transfer coefficient is determined for a casting-mould system on the basis of experimentally obtained temperatures during solidification. The temperature field during the cooling of the casting of a water turbine blade in a sand mould is simulated in the next example. Controlled heating of a steel plate in a furnace is solved in the last example. The temperatures and thermal stresses are solved in the dimensionless form.

Example 9.1 *Thermomechanics. Indirect heat transfer task modeling*

Object of the solution	In a steel-casting-sand-mould system, the local heat transfer coefficient has to be determined on the basis of experimentally obtained temperature courses in a sand mould and in a location just below the surface of a solidification and cooling-down casting.
Task characterization	The solution of indirect unsteady one-dimensional task for heat transfer between the steel cylinder casting and the sand mould.
Applied method	Considering task characterization, the probability method, distinguished by simplicity and relatively high accuracy for simple tasks, has been chosen.

Casting–mould system

In the first phase, the task involves temperature–time dependence measurement in at least four places of the *casting–sand–mould system*. In Fig. 9.10, these places are denoted as P_1 to P_4. In place P_1, the temperature is measured just under the casting surface by a thermocouple connected to the molten material and the casting, and the measured temperature represents that of the metal surface. In the mould, the measurement places P_2 and P_3 are located so that place P_2 is most important and must be as close as possible to the surface, with the precisely defined distance from it. The temperature gradient near the surface is most important for the accuracy of temperature measurements and indirect task solution. The temperature gradient depends on precise temperature determination in places P_1 and P_2 with respect to the fact that the thermal conductivity of the casting differs by two orders from that of the mould. In the mould, other measurement

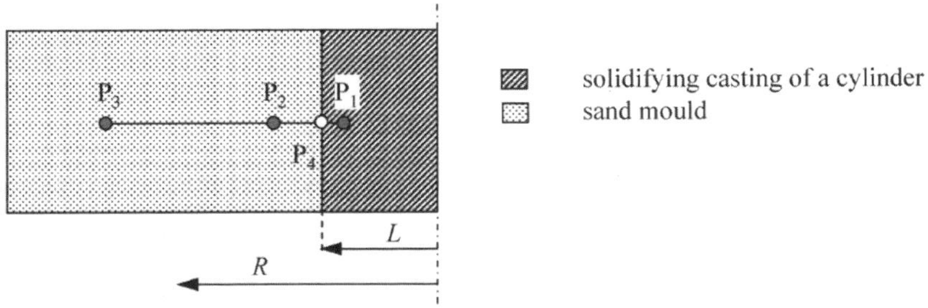

Fig. 9.10. Diagram of a temperature measuring method in the casting–mould system and heat transfer determination in the system.

places have less weight the more distant they are from the surface, and serve only to make the temperature course extrapolation on the inner mould surface more precise. The temperature in place P_4 is determined by extrapolation the measured temperature course in the mould to the mould surface. Quantity L represents characteristic length (thickness), while R is the cylinder radius.

Surface heat flux determination

Figure 9.11 outlines the way to determine the surface heat flux in the normal direction from the temperature difference ΔT between the surface temperature $T_S(\tau)$ and the temperature $T_x(\tau)$ at a certain depth x. This depth depends strongly on the kind of material and its thermal conductivity λ. The greater the material's thermal conductivity, the smaller must the depth be and more precisely it must be determined, which is often very difficult. On the contrary, with a material with lower thermal conductivity the requirement is not so rigorous with respect to the depth Δx and its determination. For example, in solving the heat flux in the casting–sand–mould system, the thermal conductivities ratio has two orders. In the casting which is more conductive material, the temperature sensor must be placed closely under the surface. In the sand mould, the highest-influenced sensor can be placed deeper by one or two orders.

The surface heat flux can be determined from the relation

$$q_A \approx \frac{\lambda\left(T_S(\tau) - T_x(\tau)\right)}{\Delta x}. \tag{9.78}$$

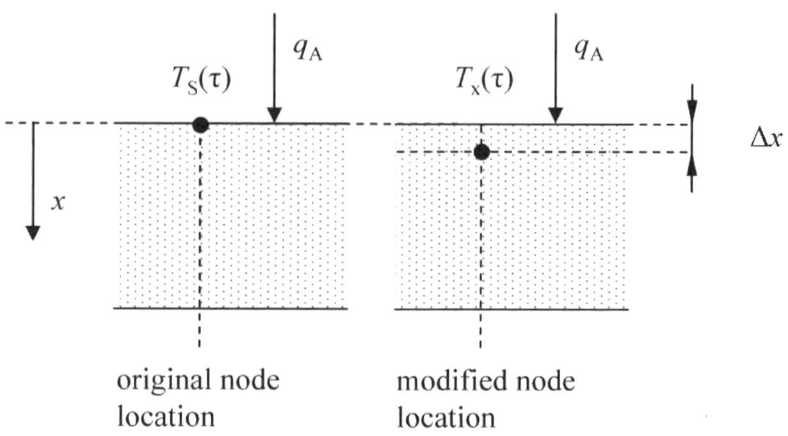

Fig. 9.11 Diagram of sub-surface heat flux determination.

Basic equation

The *basic equation for unsteady heat conduction* in a cylindrical steel casting with internal heat sources has the following dimensionless form

$$\frac{\partial^2 \Theta}{\partial R} + \frac{1}{R}\frac{\partial \Theta}{\partial R} + Po = \frac{\partial \Theta}{\partial Fo}, \tag{9.79}$$

where the dimensionless quantities are: Θ – temperature, R – radius, Po – Pomerantsev number (dimensionless heat source), and Fo – Fourier number (dimensionless time).

To solve the indirect task of finding the boundary condition from the experimentally found temperatures in the casting–mould system, the difference diagram is chosen (Fig. 9.12) and the constant geometric step H and the time step ΔFo are chosen for it. Using the heat balance method, the probability of transitions p_1 to p_3 can be determined, and also the contribution g_{ij} in the case of inner sources for which the following holds

$$\left.\begin{aligned} p_1 &= \frac{n^2 \Delta Fo\, a}{2a_{ref} + 2n^2 \Delta Fo\, a}\left(1 + \frac{1}{2nR_i}\right) \\ p_2 &= \frac{n^2 \Delta Fo\, a}{2a_{ref} + 2n^2 \Delta Fo\, a}\left(1 - \frac{1}{2nR_i}\right) \\ p_3 &= \frac{a_{ref} - n^2 \Delta Fo\, a}{a_{ref} + n^2 \Delta Fo\, a}, \quad g_{i,j} = \frac{Po\, \Delta Fo\, a}{a_{ref} + n^2 \Delta Fo\, a} \end{aligned}\right\} \tag{9.80}$$

where $\Delta Fo = a_{ref}\, \Delta\tau L^{-2}$ is the time step (the Fourier number), $Po = q_V L^2 (\lambda T_{ref})^{-1}$ is the Pomerantsev number, $R_i = r_i L^{-1}$ is the

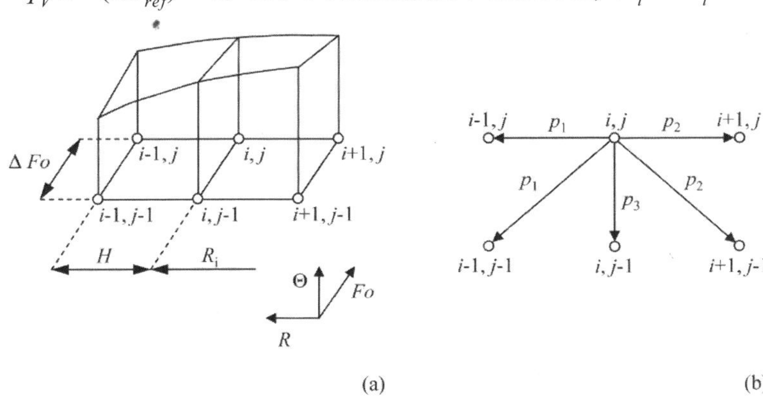

Fig. 9.12. Diagram of difference network arrangement for the probability method (a) and the probability of transition from the internal node (i, j) to a new time level (b).

dimensionless radius, L is the reference length, r_i is the radius of ith cell node, $H = hL^{-1}$ is the dimensionless geometric step of the cell, a, a_{ref} are the real and reference thermal diffusivity, and $n = Lh^{-1}$ is the number of geometric steps related to reference length L.

The subsequent matrix solution procedure for the indirect heat transfer task with unsteady conduction is presented in [9.8] and [9.11].

Results

In Fig. 9.13a the dependences of measured temperatures Θ are expressed in dimensionless form in three places P_1 to P_3, and the inner mould surface temperature in place P_4 found by extrapolating the measurement results. The simulation results of the indirect task of unsteady heat transfer in the 3rd-kind boundary condition (the Biot number) are in case (b) dependent on time.

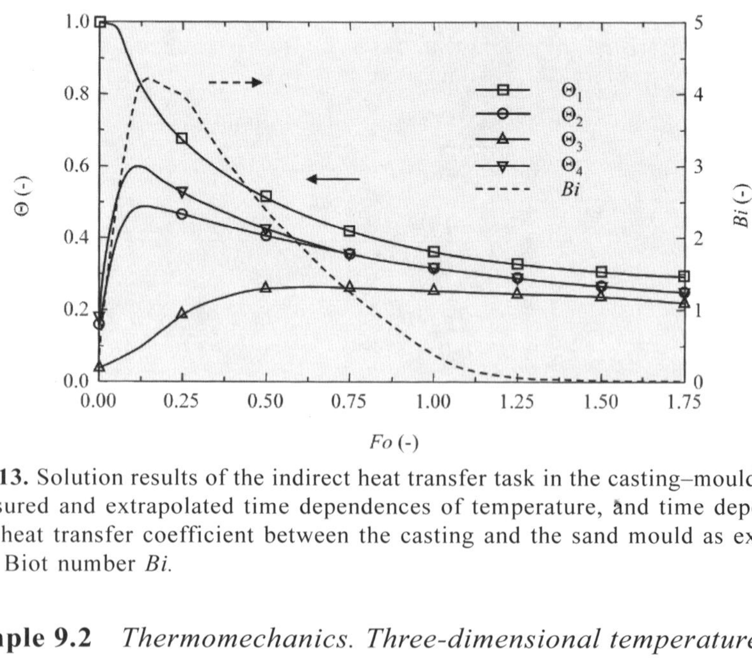

Fig. 9.13. Solution results of the indirect heat transfer task in the casting–mould system – measured and extrapolated time dependences of temperature, and time dependence of the heat transfer coefficient between the casting and the sand mould as expressed by the Biot number *Bi*.

Example 9.2 *Thermomechanics. Three-dimensional temperature field in a casting of complicated shape*

Object of the solution To compare the results of the stochastic temperature-field simulation of a several-ton weighting blade casting for a Kaplan water turbine during solidification and cooling down in a sand mould with the direct temperature measurement results for the casting surface.

Task characterization A three-dimensional unsteady task in a heterogeneous system consisting of shape-complex casting and a sand mould.

Applied method The hybrid stochastic Monte Carlo method.

Solution procedure

The blade shape is displayed in Fig. 9.14a. In the Z-axis it is limited by balls with radii R_1 and R_2, by a general cylinder with the base given by the functions $f_1(X)$and $f_2(X)$ and, finally, by a rotationally symmetrical area with a surface line given by the function $f_3(Z)$. To solve this, the hybrid Monte Carlo method has been used, arranged according to the diagram in Fig. 9.6, in which the simulation procedure is presented. The complex shape casting is expressed by continuously working function converters, quadrators, and comparators.

Results

In Fig. 9.14b the results obtained by the hybrid Monte Carlo method are compared with the temperature values found by direct measurement

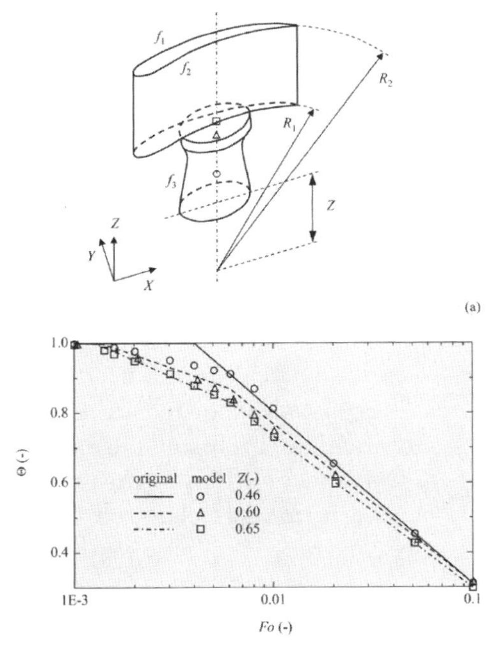

Fig. 9.14. The shape of a Kaplan water turbine blade (a) and comparison of experimentally obtained temperature–time courses during cooling of the blade, with the temperatures determined on the stochastic model by the Monte Carlo method (b).

on the object. The dimensionless temperatures Θ are expressed by the Fourier number *Fo* in time dependence. The comparison has been made for three different places in the *Z*-axis on the casting surface during its solidification and cooling down. In the initial part of solidification and cooling down, especially in the cylindrical pivot part (f_3) greater deviations are primarily due to not considering the influence of non-linear thermally physical properties of the steel in the stochastic model.

Summary

The solution shows how the stochastic Monte Carlo method is suitable for solving such a complicated space task. Except for the initial solidification phase, with theinfluence of not considering the non-linearity in the system during the MC method solution, the deviation from the measured surface temperatures in three places (riser, transition into the leaf, and blade leaf) is approximately within the limits of 3 to 4%.

Example 9.3 *Thermomechanics. Controlled material heating*

Object of the solution To solve bilateral controlled heating of a steel plate in a furnace so that the stress in the plate does not exceed a prescribed limit and that the difference between real and prescribed values at the end of heating should be less than the given value. The solution has to be executed in a dimensionless form so that the results may be applicable to physically similar tasks. The example has to explain the application of the Exodus method procedure.

Task characterization Unsteady controlled heat conduction with a symmetrical boundary condition expressed by a constant heat transfer coefficient and constant physical material parameters. To ensure the given conditions, the temperature velocity in the furnace is changed linearly.

Applied method The numerical stochastic Exodus method.

Solution procedure

First of all, the dimensionless quantities and the dimensionless basic equation are determined for the heat conduction through the plate thickness.

The *dimensionless quantities* are: Θ – temperature, S – thermoplastic stress, L – characteristic length dimension which equals half of the plate thickness, Z – z-axis coordinate, Fo – Fourier number, Bi – Biot number, Pd – Predvoditel number.

The *dimensional quantities* are: T – local temperature (K), T_{ref} – reference temperature (K), E – Youngs' modulus (Pa), σ – stress (Pa), ν – Poisson number (-), α_L – linear thermal expansion coefficient (K^{-1}).

The *basic equation* for one-dimensional heat conduction in a symmetrical non-limited plate has the following form

$$\frac{\partial^2\Theta}{\partial Z^2} = \frac{\partial\Theta}{\partial Fo}, \tag{9.81}$$

and together with the 3rd-kind boundary condition, expressed by the Biot number Bi, this is a mathematical unsteady heat conduction model. For dimensionless expression of the thermal stress, the following is valid

$$S = \frac{\sigma(1-\nu)}{E\alpha_L T_{ref}}. \tag{9.82}$$

In solving the basic Eq. (9.81) with the Exodus method, the one-dimensional zone $\langle 0;L\rangle$ is divided into m equal parts h. For the internal network nodes (Fig. 9.15), the probabilities of transition into a higher time level can be deduced from elementary heat balances

$$p_1 = m^2\Delta Fo, \quad p_2 = 1 - 2m^2\Delta Fo. \tag{9.83}$$

Then, depending on the boundary condition, the transition to a place on the plate surface in subsequent time occurs with the probability

$$p_1 = 1 - 2m^2\Delta Fo - 2mBi\Delta Fo, \tag{9.84}$$

or reflection or absorption occurs

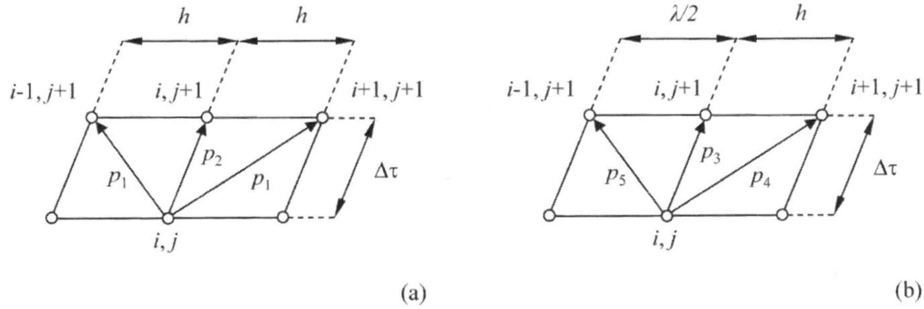

Fig. 9.15. Diagram of transition probability determination for internal zone node (a) and surface zone node (b).

$$p_{refl} = p_4 = 2m^2\Delta Fo, \quad p_{abs} = p_5 = 2mBi\Delta Fo. \tag{9.85}$$

The matrix **P** is composed from these probabilities.

To determine the temperature in the required place, the value of 1 is inserted into the corresponding element of the initial occurrence probability vector $\mathbf{b}^{(0)}$, with zeros in other elements.

The thermoplastic stress in the plate is

$$\sigma_{xx} = \sigma_{yy} = \frac{\alpha_L E}{1-\nu}\left(-T + \frac{1}{2L}\int_{-L}^{L} T\,\mathrm{d}z + \frac{3z}{2L^3}\int_{-L}^{L} T\,\mathrm{d}z\right). \tag{9.86}$$

The maximum stress is on the surface of the plate, which is heated symmetrically, so that the stress can be written after modification and conversion into a dimensionless form

$$S = \int_0^1 \Theta \mathrm{d}Z - \Theta_s, \tag{9.87}$$

where Θ_s is the temperature on the surface in the place 1. To check the accuracy of the Exodus method, the stresses on the plate surface are solved dependent on the Fourier number and compared with the accurate values specifically for the case of the step change of the external environment temperature in the time $\tau = 0$ $(Pd = \infty)$ and for various boundary condition values (Biot numbers Bi), The comparison results and other details concerning the solved case are presented in [9.11].

Results

Figure 9.16 shows the example of results of the controlled plate heating simulation using the Exodus method. Here, the 3rd-kind boundary

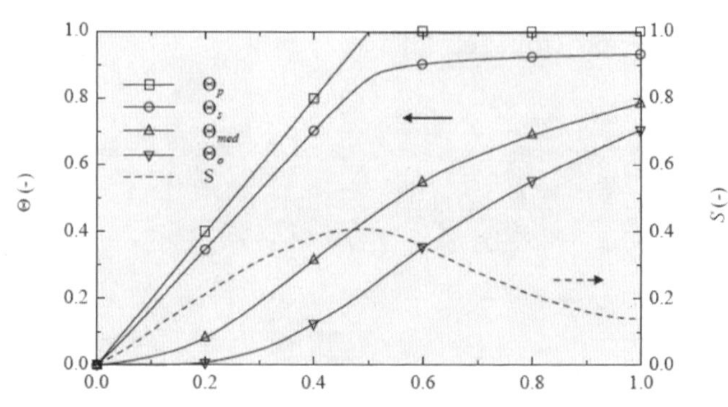

Fig. 9.16. Results of temperature and thermal stress simulation of a heated plate with the 3rd-kind boundary condition.

condition is given as expressed by the Biot number $Bi = 10$ and by the initial linear temperature rise of the external environment, which is described by the Predvoditel number for the temperature change rate $Pd = 2$ and subsequent restriction $\Theta_p = 1$. In the figure, the courses express, in dimensionless form, the time dependences of the linearly changing environment temperature Θ_p, surface temperature Θ_s, average integral plate temperature Θ_{med}, plate axis temperature Θ_o, and thermal stress S on the plate surface.

Above all, the example shows how the Exodus method can be used to solve a direct process control task with a restrictive condition.

"We have at our disposal three important tools: the observation of Nature, analytic thought and experiment. Experiment gathers facts, analytic thought combines them: experiment verifies the results of the combination. Perseverance in the observation of Nature, profundity in thought and accuracy in experimentation are essential."

Denis Diderot *(5.10.1713–31.7.1784)*

10

Cybernetic Models

It is implied from the concept of cybernetics that cybernetic models are those of complex controlled dynamic systems above all. They represent the highest abstraction degree in modeling, where the geometric or other physical similarity disappears and external behavior similarity appears. Besides the functional model, operating on the *black-box* principle, they are the *simultaneous models* and those on the principle of a *neural network*. All these models are extraordinarily important, above all in the solving of present and future global problems of mankind. Practically, they concern all scientific or engineering disciplines. For example, in medicine and biology they are models of malignant tumours, human genomes, and models in computer-oriented diagnostics; in chemistry the models of chemical and nuclear reactions; in geophysics models of seismic activity, shifts of tectonic plates and movements of tsunami waves; in meteorology the weather and terrestrial climate models; in astronomy the models of planetary and galactic orbits. The models are very important in engineering to control complex machine systems and technological processes with optimization thereof. They are widely applied in economics, finance, commerce and transport as well. Advantageously, cybernetic models, especially the simultaneous ones, enable the solution of broad systems and save time because multiple operations are solved simultaneously. In comparison to super-computer modeling, the parallel solution using a simultaneous model is usually more economical, for example, because of using multiple chip cicuits with a higher degree of elements itegration.

J. Kuneš, *Similarity and Modeling in Science and Engineering*,
DOI: 10.1007/978-1-907343-78-0_10,

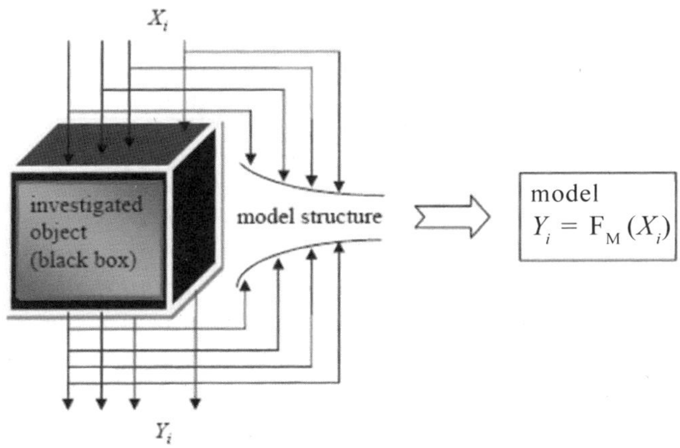

Fig. 10.1. Functional model composition and assignment diagram of an investigated object.

10.1 Black-box-type Models

Figure 10.1 shows schematically the procedure of creating a functional black-box-type model. Here, the *black box* represents an investigated system about the internal structure and behavior of which one knows nothing or only a little, which need not always be substantial. In the latter case, one speaks often about a *grey box*.

With the black-box-type models, we are interested mainly in the output signal reaction (response) to an input signal. These cases are innumerable in various fields of human activity. On the above mentioned figure, one proceeds from observing the input quantities X_i and the output ones Y_j. The functional dependence

$$Y_i = F_M(X_i) \tag{10.1}$$

is the model expression of the behavior of quantity X_i on an object. F_M is the *model operator*. Its determination is one of the most complicated tasks in using these models. The *verification* problem is connected with it and represents the qualitative evaluation of the model correctness. In applying it one verifies whether the model represents the system acceptably for both structure and behavior. On the contrary, the *credibility (validity)* expresses the consent degree of the quantitative behavior of a model and an investigated system. Among other important characteristics of the functional black-box-type models are the *validity (valence)*, the *variability* and the *adaptability* of a model, the characterization of which has already been presented in Chapter 2.

10.2 Simultaneous Models

The *simultaneous models,* otherwise named *parallel models* or *homogenous structures* or *parallel processor fields,* result logically from two independent development lines of the simulation technique, namely the development of numerical computers and that of analogue network models. One of the ways to extend the possibilities of numerical computers is to create structures composed of numerical processors operating in parallel, exchanging mutually the information, and controlled by a superior computer. The interconnection of many. numerical computers into higher complexes is a contemporary reality which also enables the solution of very complicated new tasks. These models can be said to represent the apex of the evolutionary development of serially operating computers and the approximation thereof to reality.

10.2.1 Structure and Properties

In simulation models, the main role is being played by microprocessors, which enable one to perform various optional operations. The microprocessors process the information autonomously. The inner structure of an operational element can be be programmed or selected from at least several possibilities. Greater possibilities can be offered by a numerical microprocessor which can be reprogrammed simply by one order to solve one or more operations. First of all, the compactness and reliability of a model are supported by using integrated circuits. The microprocessor simultaneous model is universal and applicable for various types of tasks. Usually, the internal element structure must perform arithmetic and logic operations, integration and derivation, operations over matrices, vectors and tensors, and stochastic operations, etc. One-dimensional, two-dimensional and three-dimensional models can be built, the individual cells of which consist of an operational element which is surrounded by a field of interconnecting ones. In a model, the values of boundary and initial conditions can be set. The actual solution proceeds in parallel in all operational elements.

In a simultaneous model, the operational elements are interconnected mutually with connecting elements. The inner structure of individual elements and their mutual interconnection can be changed in compliance with a previously given program or on the basis of the processed information. The program change of the inner element structure enables the solution of various tasks with diverse parameters with a single model. On the base of the processed information, the inner element

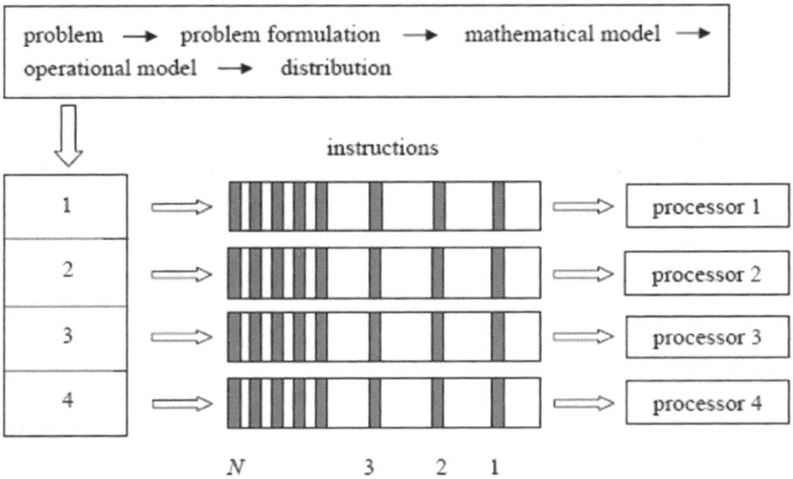

Fig. 10.2. Simplest diagram for parallel numerical modeling.

structure change enables the solution of the tasks in dependence on internediate results of the solution. This enables, for example, the solution of changing-phase tasks, material microstructure change tasks, transcrystallization tasks, tasks with turbulence origin, with interlayer development, with moving sources, etc.

Figure 10.2 show the simplest diagram of the parallel modeling principle. After formulation, establishing an operational and mathematical model, the problem to be solved is divided into several parts which can be solved in parallel. Further, every part is transformed into a series of instructions. The instruction from each part is being performed in various processors simultaneously. The solution proceeds either in one multiple processor computer or in an arbitrary number of computers which are interconnected in a network or, possibly, in a combination of both ways.

Hence, the properties of the solution in simultaneous models are expressed usually by characteristics such as the possibility to divide the problem into parts which can be solved simultaneously. In addition to this, the multiple program instructions can be performed in an arbitrary time moment span. The possibility of utilizing the rest of the time in multiple calculations for solving in an individual computer is also important.

10.3 Neuron-web-type Models

The *neuron-web model* is a simultaneous stochastic model with many stochastic operational elements and with the possibility to establish many random connections even between elements distant in space. The neuron-web models enable the modeling of some properties of biological neural networks but they are used mainly to model other extensive engineering and non-engineering systems. The artificial neural network aplication is treated in medicine in survey, by solution of various problems in modeling and diagnostics of cardiovascular systems, the use in telemedicine, and pathology to recognize the appearances, etc.

10.3.1 Neural Networks

The *biological neural networks* proceed from the human brain structure. However, in modeling, the concept of neural networks means *artificial neural networks.* The brain is a complicated, non-linear and parallel information processing system consisting of small structural units called *neurons.* Its great advantage is the ability to create its own rules, usually called *experience.* The development of neurons enables the nervous system to be adapted to surroundings and subsequently to fulfil taught tasks, with great speed and accuracy.

The *biological neuron* consists of a *nucleus* (cellular body), *dendrites* and an *axon.* The nucleus processes the information, the dendrites receive the information and the axon passes over the information to other neurons. The biological neuron is presented as a diagram in Fig. 10.3.

The neuron has a lot of possible information inputs and mostly one output, which is forked at the end and establishes *connections* (*synapses*) to other neurons. One neuron can accept up to 10^4 synaptic contacts and deliver the output information to hundreds of other neurons. The axons can be very long and enable one neuron to establish a connection to arbitrary distant system neuron. In the brain, the information processing is organized in several scales from molecules,

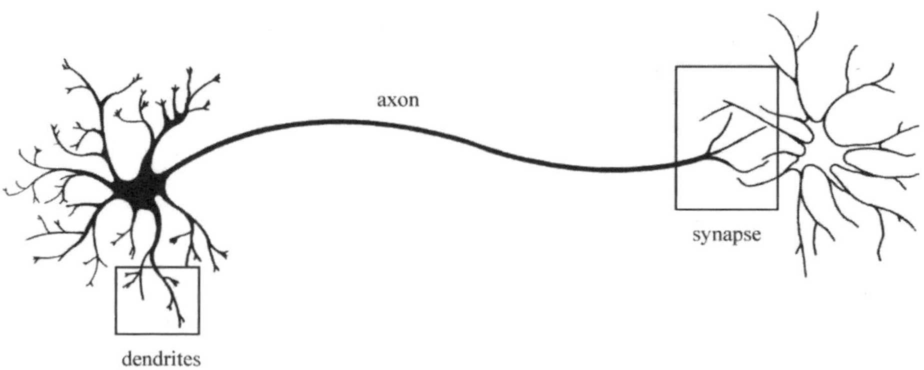

Fig. 10.3 Diagram of a biological neuron.

over individual neuron parts, neurons, local neural circuits up to the central system of nerves. The human brain contains 10^{10} to 10^{11} neurons approximately. The *neuron body* (*soma*) can be 4 to 100 μm large depending on the neuron function. Usually, the dendrites are 1 to 3 mm long and the axons can be longer than 1 m. The information is transferred between neurons by means of chemical couplings which can be exciting or damping.

The artificial neuron network (hereafter the *neuron network* only) is a system of simple processing elements, the neurons, which are connected into a network by means of a set of synaptic couplings. These enable to connect two neurons so that the information can be passed over. The significance of individual inputs into a neuron is given by a weight coefficient – *synaptic weight w.*

The process of *learning* or *training* is used to establish necessary knowledge in a neural network. During this process, the synaptic network couplings are modified so that a certain object may be achieved. Generally, the neural network is a means which can learn a non-linear system description from a set of observations. The neural network function is determined by its architecture, by values of individual weights and by functional modes of their elements. The neural network can change its topology continuously.

In research on neural networks, many new network types have been introduced with continuously decreasing emphasis put on the relation to biological neural networks. At present, the most commonly used neural networks have only few aspects in common with the brain.

10.3.2 Neuron Models and Their Properties

The fundamental element or *model neural network unit* is the neuron, the diagram of which is presented in Fig. 10.4. In the left part, the weight parts $w_{i,j}$, which can be adapted to synaptic learning models, enter into the cell from an outer source over dendrites. Such a unit counts certain functions *f* of a weighted sum of these inputs, which is formulated as

$$y_i = f\left(\sum_j w_{i,j} y_j\right), \tag{10.2}$$

where usually the *weighted sum* $\Sigma_j w_{i,j}\, y_j$ is called the input of the network into the unit *i* and is designated as net_i. The unit outputs serve as inputs of other units. In another form, the following is valid for the whole neural network

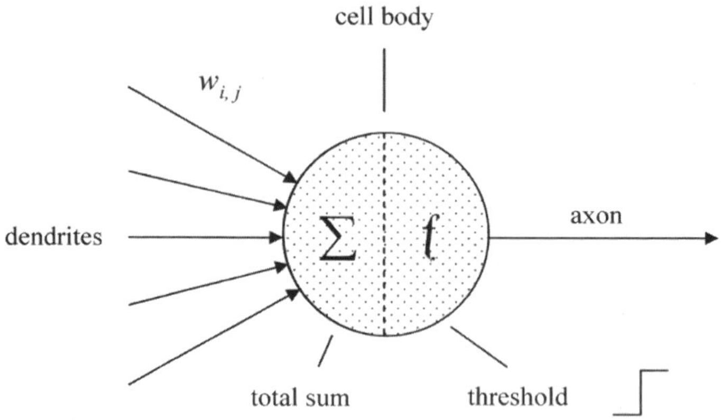

Fig. 10.4. Diagram of a neuron model.

$$y_i = f\left(net_i\right). \tag{10.3}$$

It should be noted that $w_{i,j}$ puts the weight of the unit j into the unit i. The function f is the *unity activating function.* In the simplest case, it represents a linear unity in which f is an identical function and the joint output is the neural network input.

Hodgkin–Huxley model

To be able to compare the biophysical neuron functions referentially, the approximate Hodgkin–Huxley model [10.9] is used. It proceeds from experiments made on a giant axon of an octopus. In these experiments, the authors found three ion currents and described their dynamic behaviour. The result is a mathematical model, having the form of a differential equation, and action potentials passing through ion channels in a cellular diaphragm.

The classic Hodgkin–Huxley model is presented in Fig. 10.5. It consists of a sodium (Na), a potassium (K), and a leakage channel with resistance R. The voltage-dependent ion channels govern the ion current through the cellular diaphragm. The leakage current controls other kinds of channels. The semi-permeable cellular diaphragm separates the inner part from the extracellular fluid and acts as a condenser C. With the input current $I(\tau)$ led into the cell, it causes a subsequent charge increment in the condenser or an escape through the channels in the cellular diaphragm. Because the active ions pass through the cellular diaphragm, an ion concentration, different from that of the extracellular

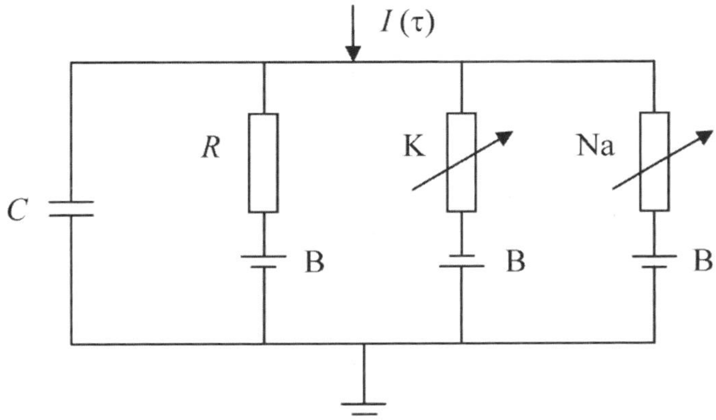

Fig. 10.5. Circuit diagram of the Hodgkin–Huxley model [10.9].

fluid, originates inside the cell. Due to the ion concentration difference, a Nernst potential arises which is expressed as the battery B in the figure.

The mathematical description of the Hodgkin–Huxley model, representing the most fundamental neuron model, can be expressed by two equations

$$I(\tau) = I_C(\tau) + \sum_k I_k(\tau), \tag{10.4}$$

$$C\frac{\mathrm{d}u}{\mathrm{d}\tau} = -\sum_k I_k(\tau) + I(\tau). \tag{10.5}$$

Equation (10.4) expresses the electric charge retained on a part of the diaphragm and it is assumed that the applied current $I(\tau)$ can be divided into the capacitance current $I_C(\tau)$, charging the condenser C, and the component $I_k(\tau)$, passing through the ion channels. The Σ_k represents the sum over all ion channels. Subsequent Eq. (10.5) represents the charge current I_C and proceeds from the capacity definition $C = Q/u$, where Q is the charge and u is the voltage over the condenser. From the biological point of view, it represents the voltage across the diaphragm, and $\Sigma_k I_k$ is the sum of ion currents going through the cellular diaphragm.

From the basic Hodgkin–Huxley model and equations, one proceeds gradually to create more complicated neural models with more numerous ion channels, various kinds of synapses, and various space geometries.

Complicated neural networks and models

The power of models based on the neural networks is in their in-parallel distributed structure and in their ability to learn and generalize afterwards. Here, generalization means that the neural network generates significant results also for input data not used in learning. With both of these abilities being used, complexly structured problems, otherwise difficult to solve, can be solved by means of the neural networks. In practice, the models, based on the neural network principle, are not usually used to solve a total problem individually but form a part of the total solution thereof.

Figure 10.6 shows a simultaneous model composed of neurons (here in cubic shape) arranged into a space network. The neuron consists of an operational element surrounded by a field of interconnecting elements enabling connection (1–2) to a neighbouring neuron. The neuron (2) can have more inputs. The network enables the making of direct connections (2–3–4) and also those (4–5) to the most distant network elements. Then in those networks, neural filaments or possibly the neuron layers can be modelled, and in the layers the information can be handed over in several ways, for example, from each higher-lying neuron into a lower-lying one. The branching off can occur in every layer and so a part of the network is separated in which one neuron from the upper layer is connected to all neurons of the lower layer. On the contrary, the paths can be interconnected so that the information is being concentrated in a funnel form. Especially, the concentrated information is led to the brain. The ways mentioned can be combined. In the structure, there can be positive and even negative feed-backs and closed circular structures, representing a short-term memory.

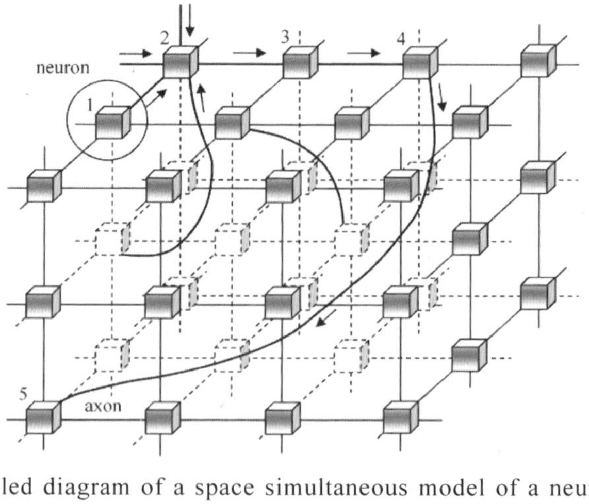

Fig. 10.6. Detailed diagram of a space simultaneous model of a neural network with indicated interconnections of distant places.

The object of the neural network control is to establish a self-setting structure which can collaborate with the external environment and respond to external stimulants. To adapt the modeled structure to external conditions and to be able to learn at least the simplest reactions to external environment, a forgetting function must be built into a model. With the external signal not appearing for a long time, the connections collapse. On the contrary, the connections become stronger for an often appearing signal.

In most fundamental functions, neural networks enable the modeling of the operation of the nervous system and the brain as it corresponds to our contemporary knowledge and ideas about this complicated biological system. They bridge over a period of some disillusionment when the abilities of numerical computers appeared to be limited in this respect.

In addition to the neural network applications in medicine, presented in the introduction of this part, biology can also be mentioned, for example, to model multicellular organisms or parts thereof. Biological cellular systems consist of cells which have many common properties. The most significant couplings are between neighbouring cells. The cell properties can change in time and are dependent on external conditions. The cells can change their position and can die. The neural networks are very important in studying and modeling purposeful behavior, in information processing and transmitting, in pattern classification and recognition, in the synthesis of adaptive, self-adaptable and self-learning systems in robotics, and in many other fields as well. Their significance increases in other engineering branches too, such as in mechanics [10.3] and others.

Practically, neural networks or similar more complex systems can be modelled with simultaneous computers only. With the present development of parallel computers and modeling, accompanied with the integration density, some neural network functions can be modelled in which the neuron internal structure changes randomly and the connections between them are established and can die accidentally. The development aims to higher order cybernetic model creation where a simultaneous sub-system model becomes an operational element of a simultaneous computer. In designing simultaneous models, processor fields are being made, of which various kinds of simultaneous models can be made. As compared with using a numerical computer, the simulation velocity rises by several orders regardless of the number of possible solutions of quantitatively new task types.

With fast-progressing abstraction development of engineering methods in modeling, the solver's ability and readiness to formulate problems and hypotheses on a much higher level, as based on deeper natural biophysics research, becomes a limiting factor. The cybernetic models, especially those of neural networks, have crucial significance in solving the most complicated problems, particularly those which are posed in front of us by present and future realities.

10.4 Applications

The three following examples concern the shortened procedure of simultaneous modeling of three physical fields described by three different differential 2nd-order and 4th-order partial equations. Three other examples concern not only the procedure but also the modeling results obtained using with the cybernetic black-box model. In these examples, both similarity theory and mathematical model transformation theory are applied. Particularly, the last example shows the practical procedure to transform the thermal field in a multilayer system to a thermal circuit and to solve the heat transfer problem in such a system.

Example 10.1 *Unsteady physical field*

Object of the solution To suggest the solution procedure of a physical field.

Method used Simultaneous model.

Task characterization Unstable diffusion field described by the Fourier equation with 1st-kind boundary condition.

The basic equation has the following shape

$$\frac{\partial^2 u}{\partial x^2} + \frac{\partial^2 u}{\partial y^2} = \frac{1}{a}\frac{\partial u}{\partial \tau}, \tag{10.6}$$

or shortened in the following dimensionless form

$$\nabla^2 U = \frac{\partial U}{\partial Fo}, \tag{10.7}$$

where for dimensionless diffusion $U = \frac{u}{u_{ref}}$, and for time the Fourier number $Fo = \alpha\tau/L^2$ are valid, whereas a ($m^2\,s^{-1}$) is the diffusion coefficient, τ (s) is time, and L (m) is the characteristic length.

Solution procedure

For example, a rectangular squared net with geometric step Δx is laid on the definition area. The simultaneous model elements are assigned to individual nodes.

Every element, except elements on the area border, will be joined, with its inputs and outputs, to neighbouring elements, as shown in Fig. 10.7.

The inner area elements are programmed to solve the following relations

$$\Delta v_{i,j,k+1} = u_{i,j,k} \frac{a\Delta\tau}{(\Delta x)^2}, \tag{10.8}$$

$$\Delta u_{i,j,k+1} = \Delta v_{i+1,j,k+1} + \Delta v_{i-1,j,k+1} + \Delta v_{i,j+1,k+1} + + \Delta v_{i,j-1,k+1} - 4\Delta v_{i,j,k+1}, \tag{10.9}$$

$$u_{i,j,k+1} = u_{i,j,k} + \Delta u_{i,j,k+1}. \tag{10.10}$$

At the same time, the stability condition must be fulfilled for numerical solution in the shape of the differential Fourier number which expresses the time step size

$$\Delta Fo = \frac{a\Delta\tau}{(\Delta x)^2} < \frac{1}{4}. \tag{10.11}$$

The boundary conditions are inserted into boundary elements as known values of the potential function. At the solution beginning, the values of $u_{i,j,1}$ are entered into all elements as an initial condition.

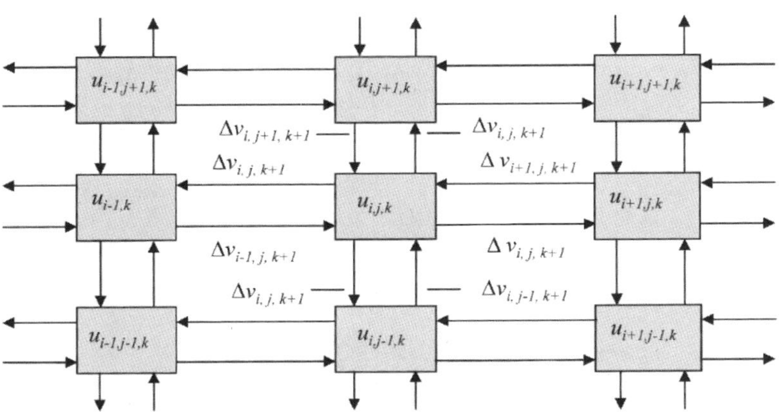

Fig. 10.7. Detail of the operational elements arrangement in a simultaneous unsteady physical field simulation.

Example 10.2 *Physical moving boundary field*

Object of the solution To show, with another equation type, the shortened solution procedure for a two-dimensional task with a moving member.

Method used Simultaneous model.

Task characterization The task suggests the solution of a steady and unsteady task, e.g. of a temperature or another physical field.

For a steady field with a moving member, expressing the movement in the x-axis direction, the following equation holds

$$\frac{\partial^2 u}{\partial x^2}+\frac{\partial^2 u}{\partial y^2}+\frac{w}{a}\frac{\partial u}{\partial x}=0, \tag{10.12}$$

and for an unsteady field

$$\frac{\partial^2 u}{\partial x^2}+\frac{\partial^2 u}{\partial y^2}+\frac{w}{a}\frac{\partial u}{\partial x}=\frac{\partial u}{\partial \tau}. \tag{10.13}$$

With the variable u representing temperature in both equations, it represents equations describing, for example, a temperature field in steady and unsteady material rolling in a rolling mill. In the equations, w (m s^{-1}) denotes the rolling-product movement velocity; a (m^2 s^{-1}) is material thermal diffusivity. Similar equations are valid even for temperature field modeling in operating rolls of a rolling mill.

Solution procedure

This is similar to the previous example. The inner elements solve the equation

$$\Delta u_{i,j,k+1}=\Delta v_{i+1,j,k+1}+\Delta v_{i-1,j,k+1}+\Delta v_{i,j+1,k+1}+\Delta v_{i,j-1,k+1}+ \\ +4\Delta v_{i,j,k+1}+\left(\Delta v_{i+1,j,k+1}-\Delta v_{i,j,k+1}\right)\frac{w\Delta x}{a} \tag{10.14}$$

The initial conditions can be arbitrary. The boundary conditions are entered into boundary elements which can be programmed for various boundary condition types.

Example 10.3 *Outline of generating a simultaneous dynamic process model expressed by a biharmonic equation*

Object of the solution To suggest only a simultaneous model example of a dynamic physical field as described by a biharmonic equation.
Method used Simultaneous model.
Task characterization Various problems of fluid dynamics, solid body dynamics vibration of plates, etc.

As an example, a part of a simultaneous model is suggested in Fig. 10.8 to simulate a one-dimensional dynamic physical field described by the following 4th-order differential equation

$$\frac{\partial^4 u}{\partial x^4} + K_1 \frac{\partial^2 u}{\partial x^2} + K_2 \frac{\partial u}{\partial x} + K_3 u - K_4 \frac{\partial u}{\partial \tau} = 0. \quad (10.15)$$

Every operational element is interconnected with two left-neighbouring and two right-neighbouring elements. In a similar way, a multi-dimensional model can be established.

The simultaneous model can be programmed at the solution beginning as it corresponds to initial conditions of the task. During the solution, the interconnecting paths can be changed according to the solution results. This can be applied in solving indirect tasks and those with area shape changing, etc. The path realization in the model can be deterministic or stochastic.

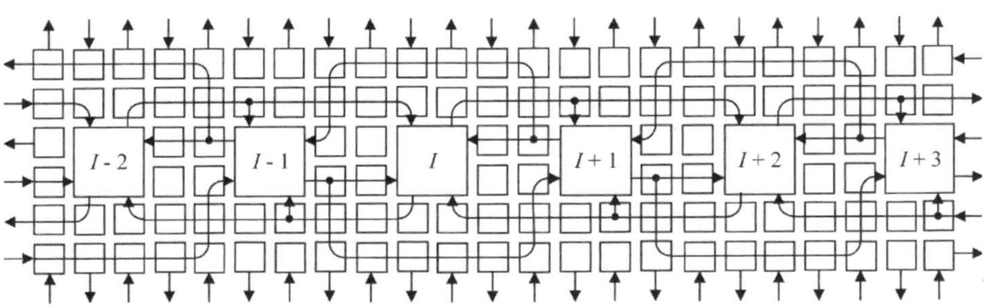

Fig. 10.8. Simultaneous model diagram for the physical field described by a partial 4th-order differential equation.

Example 10.4 *Model of fluid passage through a pipeline*

Object of the solution In a simple example of fluid passage through a pipeline, to show the black-box model function and to determine the outlet hydrodynamic functional.

Method used Functional black-box model.

Task characterization A system S represented by a pipeline in which the viscous fluid flows uniformly at the velocity w (m s^{-1}) and which is L (m) long, with fluid dynamic viscosity η (Pa s) and fluid density ρ (kg m^{-3}) is considered.

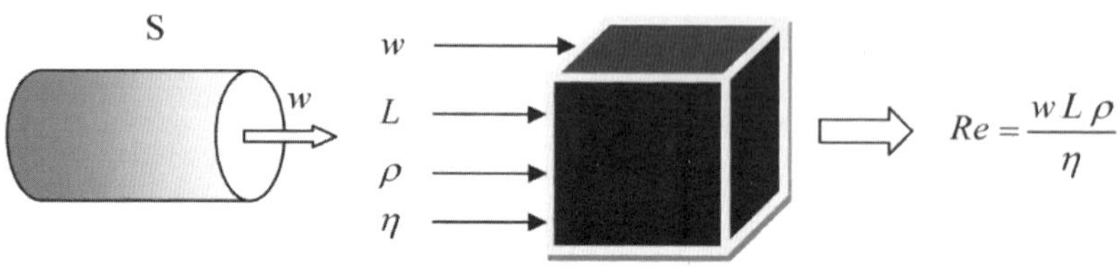

Fig. 10.9. The functional black-box model for viscous fluid passage through a pipeline and resulting hydrodynamic functional *Re*.

Procedure and results of the solution

Four characteristic physical quantities enter the model M. At the outlet, there is a single dimensionless hydrodynamic functional expressing a general viscous fluid passage through a pipeline. It is represented by the known Reynolds number, the size of which represents three basic flow areas, namely the laminar, transient, and turbulent ones. To transform the four inlet quantities to a single functional, some similarity theory methods can be used.

A simple example shows possible application of the functional black-box model and the similarity theory method to determine the hydrodynamic functional which is the well-known Reynolds number.

Example 10.5 *Temperature response model of a sensor for fluid passage*

Object of the solution To determine the temperature response of a thermoelectric thermometer (thermocouple) located in flowing fluid.

Method used Functional black-box model.

Task characterization Into steadily flowing fluid with velocity

w (m s^{-1}) and temperature T_∞(K) in an undisturbed stream, a thermocouple probe was inserted at initial temperature T_{ini} and with local heat transfer coefficient α (W m^{-2} K^{-1}). The thermophysical properties of a sensor are thermal conductivity λ (W m^{-1} K^{-1}), specific thermal capacity c (J kg^{-1} K^{-1}) and density ρ (kg m^{-3}) (Fig. 10.10a).

Procedure and results of the solution

The five above-mentioned quantities and also thermally physical properties of the sensor, thermally active sensor surface, and characteristic length for heat transfer enter the model (Fig. 10.10b). The case does not consider the high flow velocity at which the kinetic energy on the sensor surface is converted to heat increasing the surface temperature.

For the thermocouple temperature response, the following relation is valid in the dimensionless form

$$\Theta_{ef} = \frac{T_{ef} - T_\infty}{T_{ini} - T_\infty} = \exp(-BiFo), \tag{10.16}$$

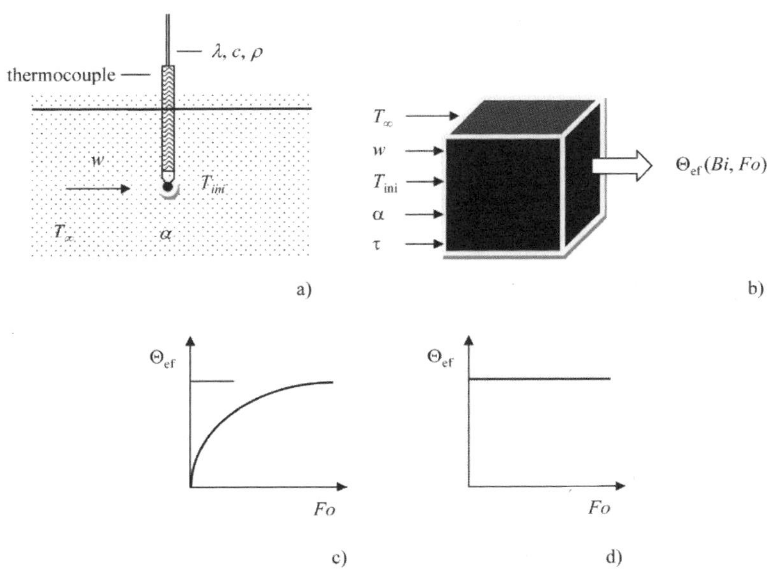

Fig. 10.10. Functional black-box model to determine the thermal response of the thermocouple located in a flowing fluid.

where Bi denotes the *Biot number* and Fo is the *Fourier number* given by relations

$$Bi = \frac{\alpha L}{\lambda}, \quad Fo = \frac{a\tau}{L^2}, \tag{10.17}$$

and the said numbers are the dimensionlessly expressed heat transfer from the fluid into the sensor (Bi) and temperature response (Fo) proceeding in time. In them, a (W m^{-2}K^{-1}) is the heat transfer coefficient, L (m) is the characteristic heat transfer length, λ (W m^{-1} K^{-1}) is the thermal conductivity of the sensor, and a (m^2 s^{-1}) is its thermal diffusivity.

The output response represents the progress of the effective temperature Θ_{ef} for the final Biot number $Bi \ll \infty$ (Fig. 10.10c) and for $Bi \to \infty$ (Fig. 10.10d). The first case (c) corresponds to usual conditions, whereas the second one (d) corresponds to a contact task occurring, for example, in surface temperature measurements with a welded-on thermocouple. With the similarity theory method and the functional black-box model being used, the solution results show that the effective output temperature Θ_{ef} depends on two dimensionless parameters only, which are the Biot Bi and Fourier Fo numbers. At high flow velocities (high Mach numbers M_a), another dimensionless number becomes valid, i.e. the coefficient of conversion of the current kinetic energy into heat.

Example 10.6 *Thermal transfer modeling in high-temperature thermocouple probes.*

Object of the solution To consider the dynamic properties of four different high-temperature thermocouple probes from the point of view of heat transfer, short-time measurement accuracy, life time, and reliability.

Method used Theoretical and experimental methods. The Laplace integral transform of a continuous system to a dynamic thermal circuit is used to determine the thermal response of individual probes. With the considered probes, not only the thermal responses to impact heating but also other dynamic and mechanical properties thereof at high temperatures are established.

Task characterization The task complexity is rooted in molten steel measurement at the temperature of 1500°C and in requirements on measuring probes. The measuring probes represent a composed thermal system having various thermphysical properties. Due to the main direction of the thermal flux into the thermocouple sensor, the task can be solved as one-dimensional and constant.

Procedure and results of the solution

The thermocouple probes are intended for short-term measurements of molten metal at a temperature of about 1500°C. They are subject to high requirements on both the measurement accuracy and thermodynamic and thermomechanical properties.

The ends of the four considered probes are presented in Fig. 10.11, where T_a is the measured fluid temperature, R_3, R_2, R_1 are thermal resistances of the receiver ends with a boundary layer, of an air gap, and of a thermocouple end, C_3, C_2, C_1 are thermal capacities of the receiver end, the grease layer and the thermocouple end. In case (a), the measuring sensor is a PtRh10-Pt thermocouple located in an oxal insertion piece and a protective steel receiver. The measuring connection is covered with a grease layer of Al_2O_3. The sensor does not contact the inner receiver surface where an air gap appears. In contrast to the previous case, in case (b) there is direct contact with the inner receiver surface ($R_2 = 0$). In the subsequent two cases, the probes have no protective receiver and serve primarily for surface temperature measurement, e.g. of castings during solidification in a mould, etc. Case (c) represents a probe used nowadays. The measuring

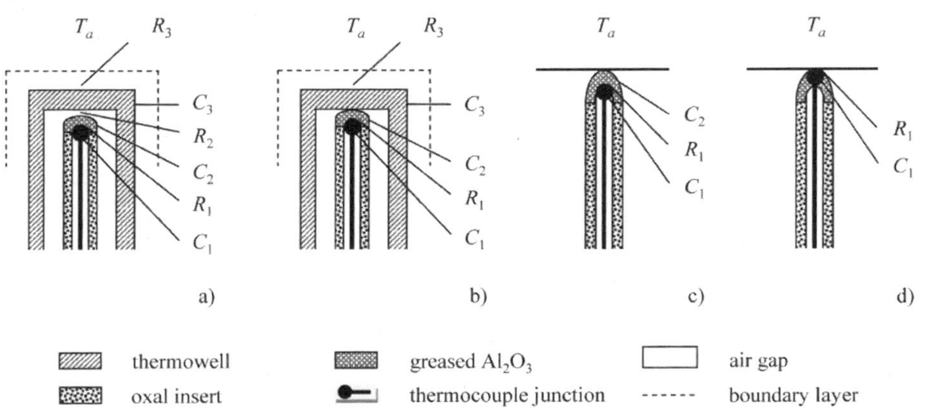

Fig. 10.11. Various arrangements of measuring ends of four verified high-temperature thermoelectric probes.

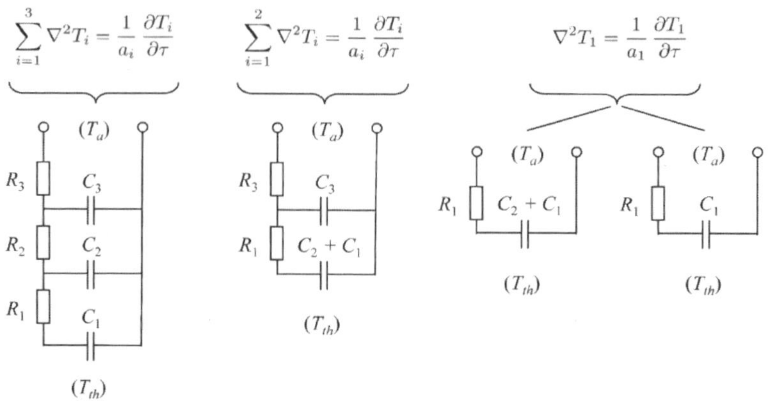

Fig. 10.12. Mathematical models for high-temperature thermocouple probes and their equivalent dynamic thermal circuits obtained after transformation.

end of the probe is protected by a grease layer of Al_2O_3 or possibly by an addition of TiO_2.The frontal layer is 0.5 to 1 mm thick usually. To provide more reliable contact with the casting surface, the method used in case (d), with frontally uncovered ball of the measuring thermocouple end, is more suitable.

Figure 10.12 shows the basic mathematical equations which describe the non-standard heat transfer in differential form in a system of thermocouple probes. In case (a), a three-layer structure in which an imperfect contact of the sensor to the receiver is considered. In the next case (b), there is a two-layer structure with perfect contact of the sensor with the receiver. In cases (c) and (d), one-layer structures are considered. Together with initial and contact boundary conditions and physical parameters of the system, these Fourier equations represent a mathematical model. This can be converted by the Laplace transform to equivalent equations of thermal circuits composed of thermal resistances R and capacities C. With it, the real variable functions $f(\tau)$ are converted to the complex variable function $F(p)$.

The following function is valid for the thermal transfer for the thermocouple probe according to (a), where there is no direct sensor-receiver contact

$$F(p) = \frac{1}{(1+pT_1)(1+pT_2)(1+pT_3)} = \frac{1}{Ap^3 + Bp^2 + Dp + 1}, \quad (10.18)$$

where

$$\begin{gathered} A = R_1C_1R_2C_2R_3C_3, \\ B = R_1C_1R_2C_2 + R_1C_1R_3C_3 + R_2C_2R_3C_3 + \end{gathered} \quad (10.19)$$

$$+R_1C_1R_2C_3 + R_1C_2R_3C_3, \tag{10.20}$$

$$D = R_1C_1 + R_2C_2 + R_3C_3 + R_1C_3 + R_2C_3. \tag{10.21}$$

In case (b), the direct sensor–receiver contact ($R_2 = 0$) is provided and the heat transfer is described by the following function

$$F(p) = \frac{(1+pT)}{(1+pT_1)(1+pT_3)} = \frac{E_p}{Bp^2 + D_p + 1}, \tag{10.22}$$

where

$$B = R_1C_1R_3C_3, \tag{10.23}$$

$$D = R_1C_1 + R_3C_3 + R_1C_3, \qquad E = R_3C_3, \tag{10.24}$$

and $T_1 > T > T_2$ is valid.

For the frontally uncovered thermocouple connection (d), a simple inertial 1st-order member holds

$$F(p) = \frac{1}{(1+pT_1)} = \frac{1}{R_1C_1p}. \tag{10.25}$$

Figure 10.13 shows the functional black-box-type models for the four considered thermocouple probes. T_α is the measured environmental temperature and T_{th} is the temperature measured with a thermocouple sensor. For every observed thermocouple probe the resulting heat transfer is denoted as $F(p)$. Partial transfers in individual layers are $F_1(p)$ to $F_3(p)$.

The experimentally determined dynamic characterizations of the four investigated probes are presented in Fig. 10.14. They were determined by measuring the thermocouple probe response on a single thermal step in submerging into molten steel. The measurement results show

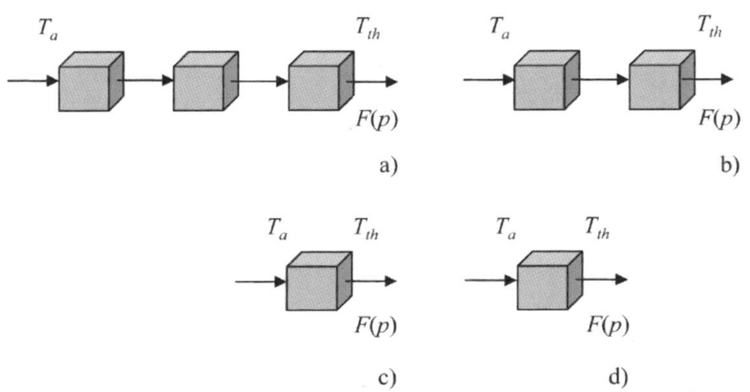

Fig. 10.13. Functional black-box models for four high-temperature thermocouple probes.

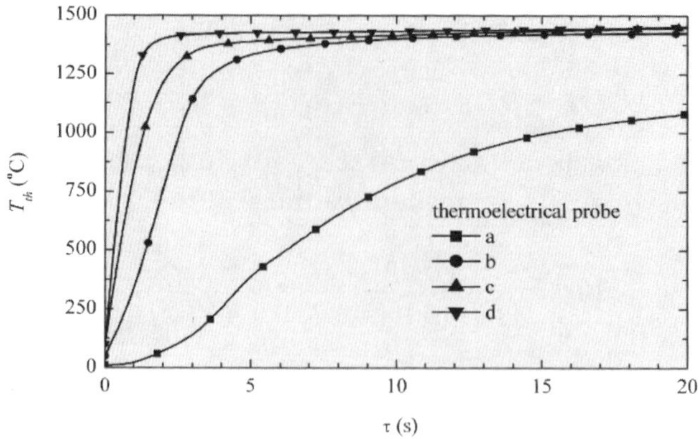

Fig. 10.14. The experimentally determined temperature responses of the four investigated thermocouple probes suddenly submerged into molten steel (1500 °C).

that the considerable influence of imperfect contact of the sensor with the protective receiver and with the measured environment affects the dynamic properties of the thermocouple probe (case a). Direct sensor-receiver contact (b) improves considerably the thermal characteristics. Other characteristics (c, d) relate to the probes for measurement of the surface temperature of castings during solidification in a mould especially. The probe (c), used at present with a protective well-conducting sensor end covering, proved its efficiency. The probe (d), with an uncovered measuring end, has the best dynamic parameters in measuring the surface temperature of a solidifying casting. If slightly inserted into the mould inner space, the end is joined with the solidifying metal surface and produces a perfect contact to the material, due to which the surface temperature can be measured accurately.

APPENDIX

J. Kuneš, *Similarity and Modeling in Science and Engineering*,
DOI: 10.1007/978-1-907343-78-0, © Cambridge International Science Publishing 2012

Appendix A

Physical Quantities and Constants

A.1 Mechanical and Acoustic Quantities

Tab. A.1 Dimensional Mechanical and Acoustic Quantities

Symbol	Quantity, Name	Units	SI Units	SI Dimensions
a	acceleration normal (a_n), tangential (a_t), Coriolis (a_c), local (a_l), drift (a_v), normal gravitational (g_n), gravitational (g)	m s^{-2}	m s^{-2}	$L T^{-2}$
a, c, u, v	speed, velocity	m s^{-1}	m s^{-1}	$L T^{-1}$
w	peripheral (v, u), resultant (w), sonic (a, c), absolute (c), relative (w, v), phase velocity of waves (c)			
a_m	mass diffusivity	m^2 s^{-1}	m^2 s^{-1}	L^2T^{-1}
a_{mp}	pressure mass diffusivity	m^2 s^{-1} Pa^{-1}	m^3 kg^{-1} s	$L^3M^{-1}T$
a_p	pressure filtration diffusivity	m^2 s^{-1}	m^2 s^{-1}	L^2T^{-1}
A, S	area, surface area, cross–section, surface	m^2	m^2	L^2
A, W	work	J	m^2 kg s^{-2}	L^2MT^{-2}
b	width	m	m	L
b, L	angular momentum	kg m^2 s^{-1}	m^2 kg s^{-1}	$L^2 MT^{-1}$
c, δ	linear compliance	m N^{-1}	kg^{-1} s^2	$M^{-1}T^2$
c	velocity of light	m s^{-1}	m s^{-1}	LT^{-1}
c_i	molar concentration	mol m^{-3}	m^{-3} mol	$L^{-3}N$
c_m	mass ratio	-	-	-
c_{mp}	pressure mass capacity	Pa^{-1}	m kg^{-1} s^2	$LM^{-1}T^2$

Tab. A.1 Dimensional Mechanical and Acoustic Quantities (continued)

Symbol	Quantity, Name	Units	SI Units	SI Dimensions
c_t, δ_t	torsional compliance	rad N^{-1} m^{-1}	m^{-2} kg^{-1} s^2 rad	$L^{-2}M^{-1}T^2\alpha$
C_a	acoustic compliance	m^5 N^{-1}	m^4 kg^{-1} s^2	$L^4M^{-1}T^{-2}$
C_i	numerical molecule concentration of i-th part component	m^{-3}	m^{-3}	L^{-3}
d, δ	thickness, width	m	m	L
D, d	diameter	m	m	L
D	deviance moment of cross–section	m^4	m^4	L^4
D_{KA}	Knudsen diffusion coefficient	m^2 s^{-1}	m^2 s^{-1}	L^2T^{-1}
D_{xy}...	deviance moment of body	kg m^2	m^2 kg	L^2M
e, Y	specific energy	J kg^{-1}	m^2 s^{-2}	L^2T^{-2}
	potential (e_p, Y_p), pressure (e_s, Y_s), kinetic (e_k, Y_k), total (e_c, Y_c), frictional (e_t, Y_t), accelerating (e_a, Y_a)			
E	energy	J	m^2 kg s^{-2}	L^2MT^{-2}
E	modulus of elasticity, Young's modulus of elasticity	Pa	m^{-1} kg s^{-2}	$L^{-1}MT^{-2}$
f	frequency	Hz	s^{-1}	T^{-1}
F	force	N	m kg s^{-2}	LMT^{-2}
	volume (F_v), surface (F_A), normal (F_n, N), tangential (F_t, T), resultant (F_R, R), reaction (F_r, A...), frictional (F_t, T), inertial (F_s, D), centrifugal (F_o, D_o), Coriolis (F_c), buoyancy (F_v, ...), weight (G), generalized (Q_i)			

Tab. A.1 Dimensional Mechanical and Acoustic Quantities (continued)

Symbol	Quantity, Name	Units	SI Units	SI Dimensions
F	function of thermoelastic stress	N	m kg s^{-2}	LMT^{-2}
F	complex potential	m^2 s^{-1}	m^2 s^{-1}	L^2T^{-1}
g	gravitational acceleration	m s^{-2}	m s^{-2}	LT^{-2}
G	relaxation modulus of linear viscoelasticity	Pa	m^{-1} kg s^{-2}	$L^{-1}MT^{-2}$
G	shear modulus	Pa	m^{-1} kg s^{-2}	$L^{-1}MT^{-2}$
h	hydrostatic height, depth potential (h_p), pressure (h_s), velocity (h_k), total (h_c), frictional (h_z), accelerating (h_a)	m	m	L
H, p	momentum	kg m s^{-1}	m kg s^{-1}	$L M T^{-1}$
H	momentum flux	N	m kg s^{-2}	LMT^{-2}
I	moment of inertia of the body to axis (I_x,...), to area (I_{xy} ,...), polar (I_p)	kg m^2	m^2 kg	L^2M
I, I_F	force impulse	N s	m kg s^{-1}	LMT^{-1}
I	acoustic intensity	W m^{-2}	kg s^{-3}	MT^{-3}
I'	spectral density of acoustic intensity	W Hz^{-1} m^{-2}	kg s^{-2}	MT^{-2}
I_e	invariant of the stress tensor variation	s^{-2}	s^{-2}	T^{-2}
I_m	force moment impulse	N m s	m^2 kg s^{-1}	L^2MT^{-1}
I_v	invariant of the turbulence tensor	s^{-2}	s^{-2}	T^{-2}
J	square moment of cross–section	m^4	m^4	L^4

Tab. A.1 Dimensional Mechanical and Acoustic Quantities (continued)

Symbol	Quantity, Name	Units	SI Units	SI Dimensions
J_k	torsional stiffness modulus	m^4	m^4	L^4
J_p	polar moment of cross–section	m^4	m^4	L^4
k, s	linear stiffness	$N\ m^{-1}$	$kg\ s^{-2}$	MT^{-2}
k	filtration coefficient	$m\ s^{-1}$	$m\ s^{-1}$	LT^{-1}
k	drying coefficient, velocity constant	s^{-1}	s^{-1}	T^{-1}
k_n	velocity constant of n-order chemical reaction	$m^{3(n-1)}$ $kg^{-(n-1)}\ s^{-1}$	$m^{3(n-1)}\ k\ g^{-(n-1)}\ s^{-1}$	$L^{3(n-1)}M^{-(n-1)}T^{-1}$
k_t, s_t	torsional stiffness	$N\ m\ rad^{-1}$	$m^2\ kg\ s^{-2}\ rad^{-1}$	$L^2MT^{-2}\alpha^{-1}$
K	bulk modulus, volume modulus of elasticity	Pa	$m^{-1}\ kg\ s^{-2}$	$L^{-1}MT^{-2}$
l, s, b L	length, depth, width characteristic length	m	m	L
m	mass	kg	kg	M
M	molar mass	$kg\ mol^{-1}$	$kg\ mol^{-1}$	MN^{-1}
M	force moment inertial (M_s), frictional (M_f) force couple	N m	$m^2\ kg\ s^{-2}$	L^2MT^{-2}
M_H	moment of momentum flux	N m	$m^2\ kg\ s^{-2}$	L^2MT^{-2}
n	revolutions per second, rpm, rotational frequency	s^{-1}	s^{-1}	T^{-1}
n	numerical density of elements	m^{-3}	m^{-3}	L^{-3}
n	amount of substance	mol	mol	N

Tab. A.1 Dimensional Mechanical and Acoustic Quantities (continued)

Symbol	Quantity, Name	Units	SI Units	SI Dimensions
N	number of elements	–	–	–
N	specific sound power	$W\ m^{-2}$	$kg\ s^{-3}$	MT^{-3}
O	perimeter	m	m	L
p	pressure atmospheric (p_a, b), normal (p_n), external (p_e), hydrostatic (p_h), capillary (p_k), static in flowing liquid (p_s, p), kinetic (q), dynamic (p_d, q_s), total (p_c)	Pa	$m^{-1}\ kg\ s^{-2}$	$L^{-1}MT^{-2}$
P	power, power drain	W	$m^2\ kg\ s^{-3}$	L^2MT^{-3}
q	volume acoustic speed	$m^3\ s^{-1}$	$m^3\ s^{-1}$	L^3T^{-1}
q	length loading	$N\ m^{-1}$	$kg\ s^{-2}$	MT^{-2}
q	specific volume flow	$m^3\ s^{-1}\ m^{-3}$	s^{-1}	T^{-1}
q_A	surface loading	Pa	$m^{-1}\ kg\ s^{-2}$	$L^{-1}MT^{-2}$
q_m	mass flow density, mass flow intensity	$kg\ m^{-2}\ s^{-1}$	$m^{-2}\ kg\ s^{-1}$	$L^{-2}MT^{-1}$
q_{mV}	volume mass flow density	$kg\ s^{-1}\ m^{-3}$	$m^{-3}\ kg\ s^{-1}$	$L^{-3}MT^{-1}$
q_V	volume flow density, volume flow intensity	$m\ s^{-1}$	$m\ s^{-1}$	LT^{-1}
Q	loading limit, tonnage	kg	kg	M
Q_A	areal tonnage	$kg\ m^{-2}$	$m^{-2}\ kg$	$L^{-2}M$
Q_m	mass flux	$kg\ s^{-1}$	$kg\ s^{-1}$	MT^{-1}

Tab. A. 1 Dimensional Mechanical and Acoustic Quantities (continued)

Symbol	Quantity, Name	Units	SI Units	SI Dimensions
Q_t	length tonnage	kg m^{-1}	m^{-1} kg	$L^{-1}M$
Q_V, V	volume flow, source spread, underflow absorptivity	m^3 s^{-1}	m^3 s^{-1}	L^3T^{-1}
R, r	radius, semi–diameter	m	m	L
R_a	acoustic resistance	N s m^{-5}	m^{-4} kg s^{-1}	$L^{-4}MT^{-1}$
R_m	mechanical resistance	N s m^{-1}	kg s^{-1}	MT^{-1}
R_S	specific acoustic resistance	N s m^{-3}	m^{-2} kg s^{-1}	$L^{-2}MT^{-1}$
S	static moment of the body to area or to point	kg m	m kg	$L\,M$
S	spectral density of acoustic power	W Hz^{-1}	m^2 kg s^{-2}	L^2MT^{-2}
S, U	linear moment of cross–section	m^3	m^3	L^3
t, τ	time constant, time	s	s	T
T	period	s	s	T
u	moisture content	-	-	-
u	displacement	m	m	L
u	chemical reaction rate	kg m^{-3} s^{-1}	m^{-3} kg s^{-1}	$L^{-3}MT^{-1}$
u_m	molar chemical reaction rate	mol m^{-3} s^{-1}	m^{-3} s^{-1} mol	$L^{-3}T^{-1}N$
U	deformation energy, internal energy	J	m^2 kg s^{-2}	L^2MT^{-2}
v	specific volume	m^3 kg^{-1}	m^3 kg^{-1}	L^3M^{-1}
V	volume	m^3	m^3	L^3

Tab. A.1 Dimensional Mechanical and Acoustic Quantities (continued)

Symbol	Quantity, Name	Units	SI Units	SI Dimensions
V_m	molar volume	$m^3\ mol^{-1}$	$m^3\ mol^{-1}$	L^3N^{-1}
w	acoustic energy volume density	$J\ m^{-3}$	$m^{-1}\ kg\ s^{-2}$	$L^{-1}MT^{-2}$
w_{ch}	chemical reaction displacement velocity	$m\ s^{-1}$	$m\ s^{-1}$	LT^{-1}
w_k	torsional section modulus	m^3	m^3	L^3
w_o	sheer section modulus	m^3	m^3	L^3
x, y, z	coordinate	m	m	L
X_a	acoustic reactance	$N\ s\ m^{-5}$	$m^{-4}\ kg\ s^{-1}$	$L^{-4}MT^{-1}$
Y_a	acoustic admittance	$m^5\ N^{-1}\ s^{-1}$	$m^4\ kg^{-1}\ s$	$L^4M^{-1}T$
Y_m	mechanical admittance	$m\ N^{-1}\ s^{-1}$	$kg^{-1}\ s$	$M^{-1}T$
Z_a	acoustic impedance	$N\ s\ m^{-5}$	$m^{-4}\ kg\ s^{-1}$	$L^{-4}MT^{-1}$
Z_S	specific acoustic impedance	$N\ s\ m^{-3}$	$m^{-2}\ kg\ s^{-1}$	$L^{-2}MT^{-1}$
α	angular acceleration	$rad\ s^{-2}$	$rad\ s^{-2}$	$T^{-2}\alpha$
α, β, γ	plane angle	rad	rad	α
α_i	stechiometric number of the *i*-th component conversion	-	-	-
α_m	specific mass transfer	$kg\ m^{-2}\ s^{-1}$	$m^{-2}\ kg\ s^{-1}$	$L^{-2}MT^{-1}$
α_{mp}	pressure specific mass transfer	$kg\ m^{-2}\ s^{-1}\ Pa^{-1}$	$m^{-1}\ s$	$L^{-1}T$
β	mass transfer coefficient	$m\ s^{-1}$	$m\ s^{-1}$	LT^{-1}
Γ	circulation of velocity	$m^2\ s^{-1}$	$m^2\ s^{-1}$	L^2T^{-1}

Tab. A. 1 Dimensional Mechanical and Acoustic Quantities (continued)

Symbol	Quantity, Name	Units	SI Units	SI Dimensions
δ_p	Soret pressure coefficient, pressure diffusion coefficient	Pa^{-1}	$m\ kg^{-1}\ s^2$	$LM^{-1}T^2$
ε	relative elongation	-	-	-
ε_D	vortex mass diffusivity	$m^2\ s^{-1}$	$m^2\ s^{-1}$	L^2T^{-1}
ε_M	vortex momentum diffusivity	$m^2\ s^{-1}$	$m^2\ s^{-1}$	L^2T^{-1}
ζ	diffusion tortuosity	-	-	-
ζ	permeability of porous material	m^2	m^2	L^2
η	dynamic viscosity, absolute viscosity	Pa s	$m^{-1}\ kg\ s^{-1}$	$L^{-1}MT^{-1}$
ϑ	torsion	$rad\ m^{-1}$	$m^{-1}\ rad$	$L^{-1}\alpha$
ϑ	mass (moisture) transfer potential	K	K	Θ
λ	wavelength	m	m	L
λ, μ	Lamé coefficients, Lamé functions	Pa	$m^{-1}\ kg\ s^{-2}$	$L^{-1}MT^{-2}$
λ	volume density of deformation energy	$J\ m^{-3}$	$m^{-1}\ kg\ s^{-2}$	$L^{-1}MT^{-2}$
λ_m	mass conduction	$kg\ m^{-1}\ s^{-1}$	$m^{-1}\ kg\ s^{-1}$	$L^{-1}MT^{-2}$
λ_{mp}	pressure mass conduction, filtration conduction	$kg\ m^{-1}\ s^{-1}\ Pa^{-1}s$	s	T
μ	porosity	-	-	-
μ, ν	Poisson number, Poisson ratio	-	-	-
ν	kinematic viscosity	$m^2\ s^{-1}$	$m^2\ s^{-1}$	L^2T^{-1}
ρ	density, mass density	$kg\ m^{-3}$	$m^{-3}\ kg$	$L^{-3}M$

Tab. A.1 Dimensional Mechanical and Acoustic Quantities (continued)

Symbol	Quantity, Name	Units	SI Units	SI Dimensions
ρ_A	surface mass density	kg m^{-2}	m^{-2} kg	$L^{-2}M$
ρ_i	mass concentration	kg m^{-3}	m^{-3} kg	$L^{-3}M$
ρ_L	linear mass density	kg m^{-1}	m^{-1} kg	$L^{-1}M$
ρ_l	length mass density	kg m^{-1}	m^{-1} kg	$L^{-1}M$
ρ_v	volume mass density	kg m^{-3}	m^{-3} kg	$L^{-3}M$
σ	surface tension	N m^{-1}	kg s^{-2}	MT^{-2}
σ, τ	stress, shear stress	Pa	m^{-1} kg s^{-2}	$L^{-1}MT^{-2}$
	principal (σ_1, σ_2, σ_3), reduced (σ_{red}), admissable (σ_D, τ_D), proportional limit (σ_U, σ_U), elastic limit (σ_E, τ_E), yield limit (σ_K, τ_K), strength limit (σ_p), fatigue limit (σ_C)			
φ	volume concentration	-	-	-
φ	relative gas moisture	-	-	-
Φ	absolute humidity	kg m^{-3}	kg m^{-3}	$L^{-3}M$
Φ	velocity potential	m^2 s^{-1}	m^2 s^{-1}	L^2T^{-1}
Φ	thermoelastic displacement potential	m^2	m^2	L^2
Ψ	stream function	m^2 s^{-1}	m^2 s^{-1}	L^2T^{-1}
ω	angular frequency	Hz	s^{-1}	T^{-1}
ω	angular speed, angular velocity	rad s^{-1}	rad s^{-1}	$T^{-1}\alpha$

A.2 Thermal and Optical Quantities

Tab. A.2 Dimensional Thermal and Optical Quantities

Symbol	Quantity, Name	Units	SI Units	SI Dimensions
a	thermal diffusivity	$m^2\ s^{-1}$	$m^2\ s^{-1}$	L^2T^{-1}
a_{mt}	thermal mass diffusivity	$m^2\ s^{-1}\ K^{-1}$	$m^2\ s^{-1}\ K^{-1}$	$L^2T^{-1}\Theta^{-1}$
A, W	work	J	$m^2\ kg\ s^{-2}$	L^2MT^{-2}
b	temperature gradient of vapour concentration	$kg\ m^{-3}\ K^{-1}$	$m^{-3}\ kg\ K^{-1}$	$L^{-3}M\Theta^{-1}$
c	specific heat capacity c_p (p = const), c_v (v = const), $\overline{c}$ (median)	$J\ kg^{-1}\ K^{-1}$	$m^2s^{-2}\ K^{-1}$	$L^2T^{-2}\Theta^{-1}$
c_{mt}	mass heat capacity	K^{-1}	K^{-1}	Θ^{-1}
$c\rho$	specific volume heat capacity (specific volume heat)	$J\ m^{-3}\ K^{-1}$	$m^{-1}\ kg\ s^{-2}\ K^{-1}$	$L^{-1}MT^{-2}\Theta^{-1}$
C, K	heat capacity	$J\ K^{-1}$	$m^2\ kg\ s^{-2}\ K^{-1}$	$L^2MT^{-2}\Theta^{-1}$
C_m	molar heat capacity	$J\ mol^{-1}\ K^{-1}$	$m^2\ kg\ s^{-2}\ K^{-1}\ mol^{-1}$	$L^2MT^{-2}\Theta^{-1}N^{-1}$
D	diffusion coefficient	$m^2\ s^{-1}$	$m^2\ s^{-1}$	L^2T^{-1}
E	energy	J	$m^2\ kg\ s^{-2}$	L^2MT^{-2}
E	activation molar energy	$J\ mol^{-1}$	$m^2\ kg\ s^{-2}\ mol^{-1}$	$L^2MT^{-2}N^{-1}$
E	lighting, illumination	lx	$m^{-2}\ cd$	$L^{-2}J$
F	free energy	J	$m^2\ kg\ s^{-2}$	L^2MT^{-2}
G	free enthalpy	J	$m^2\ kg\ s^{-2}$	L^2MT^{-2}
h	specific enthalpy	$J\ kg^{-1}$	$m^2\ s^{-2}$	L^2T^{-2}
h, h_V	specific volume enthalny	$J\ m^{-3}$	$m^{-1}\ kg\ s^{-2}$	$L^{-1}MT^{-2}$

Tab. A.2 Dimensional Thermal and Optical Quantities (continued)

Symbol	Quantity, Name	Units	SI Units	SI Dimensions
H	exposure	lx s	m^{-2} s cd	$L^{-2}TJ$
H, I	enthalpy	J	m^2 kg s^{-2}	L^2MT^{-2}
I_e	irradiance	W sr^{-1}	m^2 kg s^{-3} sr^{-1}	$L^2MT^{-3}\Omega^{-1}$
k	heat passage coefficient	W m^{-2} K^{-1}	kg s^{-3} K^{-1}	$MT^{-3}\Theta^{-1}$
k	specific combustion heat	J kg^{-1}	m^2 s^{-2}	L^2T^{-2}
K	laminous efficiency	lm W^{-1}	m^{-2} kg^{-1} s^3 cd	$L^2M^{-1}T^3J$
l	specific latent heat	J kg^{-1}	m^2 s^{-2}	L^2T^{-2}
L	latent heat	J	m^2 kg s^{-2}	L^2MT^{-2}
L	brightness	nt	m^{-2} cd	$L^{-2}J$
L_e	specific irradiance	W m^{-2} sr^{-1}	kg s^{-3} sr^{-1}	$MT^{-3}\Omega^{-1}$
m	unit cooling rate	s^{-1}	s^{-1}	T^{-1}
M_e	radiant intensity	W m^{-2}	kg s^{-3}	MT^{-3}
M_λ	spectral density of radiant intensity	W m^{-3}	m^{-1} kg s^{-3}	$L^{-1}MT^{-3}$
P	power, power output, power drain	W	m^2 kg s^{-3}	L^2MT^{-3}
P, Φ_e	radiant flux	W	m^2 kg s^{-3}	L^2MT^{-3}
P_λ, Φ_λ	spectral density of radiant flux	W m^{-1}	m kg s^{-3}	LMT^{-3}
q', q_C	linear heat flux density	W m^{-1}	m kg s^{-3}	LMT^{-3}
q'', q_A	surface heat flux density	W m^{-2}	kg s^{-3}	MT^{-3}

Tab. A.2 Dimensional Thermal and Optical Quantities (continued)

Symbol	Quantity, Name	Units	SI Units	SI Dimensions
q''', q_V	volume density of heat flux	W m^{-3}	m^{-1} kg s^{-3}	$L^{-1}MT^{-3}$
q_B	point density of heat flux	W	m^2 kg s^{-3}	L^2MT^{-3}
q_k	specific liquid heat	J kg^{1}	m^2 s^{2}	L^2T^{2}
q_n	heating power	J kg^{-1}	m^2 s^{-2}	L^2T^{-2}
q_{sp}	specific heating power	J kg^{1}	m^2 s^{-2}	L^2T^{-2}
q_v	combustion heat	J kg^{1}	m^2 s^{2}	L^2T^{-2}
Q	heat	J	m^2 kg s^{2}	L^2MT^{2}
$\dot{Q}, \Phi$	heat flux	W	m^2 kg s^{-3}	L^2MT^{-3}
r	specific gas constant	J kg^{-1} K^{-1}	m^2 s^{-2} K^{-1}	$L^2T^{-2}\Theta^{-1}$
r, l_{23}	specific heat of evaporation	J kg^{-1}	m^2 s^{-2}	L^2T^{-2}
r_u	internal specific heat of evaporation	J kg^{-1}	m^2 s^{2}	L^2T^{-2}
R_T	thermal resistance	K W^{1}	m^{2} kg^{1} s^3 K	$L^{2}M^{1}T^3\Theta$
s	specific entropy	J kg^{1} K^{1}	m^2 s^{2} K^{1}	$L^2T^{2}\Theta^{-1}$
S	entropy	J K^{1}	m^2 kg s^{2} K^{-1}	$L^2MT^{2}\Theta^{1}$
t, T	Celsius temperature	°C	K	Θ
T	absolute temperature, thermodynamic temperature	K	K	Θ
∇t	temperature gradient	K m^{1}	m^{1} K	$L^{1}\Theta$
u	specific internal energy	J kg^{1}	m^2 s^{-2}	L^2T^{2}

Tab. A.2. Dimensional Thermal and Optical Quantities (continued)

Symbol	Quantity, Name	Units	SI Units	SI Dimensions
U	internal energy	J	m^2 kg s^{-2}	L^2MT^{-2}
w	cooling rate, heating rate	K s^{-1}	s^{-1} K	$T^{-1}\Theta$
w_A	surface density of energy	J m^{-2}	kg s^{-2}	MT^{-2}
α	heat transfer coefficient	W m^{-2} K^{-1}	kg s^{-3} K^{-1}	$MT^{-3}\Theta^{-1}$
α, γ	linear thermal expansion coefficient	K^{-1}	K^{-1}	Θ^{-1}
α_i, β_i	temperature coefficient of quantity i	K^{-1}	K^{-1}	Θ^{-1}
β	volume thermal expansion coefficient of solids and liquids	K^{-1}	K^{-1}	Θ^{-1}
β^*	Dufour coefficient	K	K	Θ
γ	volume thermal expansion coefficient of gases	K^{-1}	K^{-1}	Θ^{-1}
δ_t	Soret thermal coefficient (thermogradient coefficient)	K^{-1}	K^{-1}	Θ^{-1}
ε	emissivity or absorptivity	-	-	-
ε_T	vortex thermal diffusivity	m^2 s^{-1}	m^2 s^{-1}	L^2T^{-1}
K	attenuation coefficient	m^{-1}	m^{-1}	L^{-1}
λ	wavelength	m	m	L
λ	thermal conductivity	W m^{-1} K^{-1}	m kg s^{-3} K^{-1}	$LMT^{-3}\Theta^{-1}$
λ_{mt}	thermal mass (moisture) conductivity	kg m^{-1} s^{-1} K^{-1}	m^{-1} kg s^{-1} K^{-1}	$L^{-1}MT^{-1}\Theta^{-1}$
λ_t	vortex thermal conductivity	W m^{-1} K^{-1}	m kg s^{-3} K^{-1}	$LMT^{-3}\Theta^{-1}$

Tab. A.2 Dimensional Thermal and Optical Quantities (continued)

Symbol	Quantity, Name	Units	SI Units	SI Dimensions
Λ	thermal conductivity	W K^{-1}	m^2 kg s^{-3} K^{-1}	$L^2MT^{-3}\Theta^{-1}$
ρ_t	specific thermal resistance	m K W^{-1}	m^{-1} kg^{-1} s^3 K	$L^{-1}M^{-1}T^3\Theta$
σ	absorptance (absorption capacity, radiating capacity)	W m^{-2} K^{-4}	kg s^{-3} K^{-4}	$MT^{-3}\Theta^{-4}$
τ_{rx}	relaxation time	s	s	T
φ	optical power, vergency	D	m^{-1}	L^{-1}
Φ	light flux, luminous flux	lm	cd	J
Φ, Ψ	radiant flux density	W m^{-2}	kg s^{-3}	MT^{-3}

A.3 Electrical and Magnetic Quantities

Tab. A.3 Dimensional Electrical and Magnetic Quantities

Symbol	Quantity, Name	Units	SI Units	SI Dimensions
A	magnetic potential vector	Wb m^{-1}	m kg s^{-2} A^{-1}	$LMT^{-2}I^{-1}$
B	magnetic induction	T	kg s^{-2} A^{-1}	$MT^{-2}I^{-1}$
B_i	magnetic polarization	T	kg s^{-2} A^{-1}	$MT^{-2}I^{-1}$
C	capacitance	F	m^{-2} kg^{-1} s^{4} A^{2}	$L^{-2}M^{-1}T^{4}I^{2}$
D	electric displacement field	C m^{-2}	m^{-2} s A	$L^{-2}TI$
D	electric induction	C m^{-1}	m^{-1} s A	$L^{-1}TI$
E	electric field intensity	V m^{-1}	m kg s^{-3} A^{-1}	$LMT^{-3}I^{-1}$
G	electrical conductivity	S	m^{-2} kg^{-1} s^{3} A^{2}	$L^{-2}M^{-1}T^{3}I^{2}$
G_m, Λ	permeance	H	m^{2} kg s^{-2} A^{-2}	$L^{2}MT^{-2}I^{-2}$
H	magnetic field intensity	A m^{-1}	m^{-1} A	$L^{-1}I$
H_i	magnetization	A m^{-1}	m^{-1} A	$L^{-1}I$
I	electric current	A	A	I
J', J_L	linear density of electric current	A m^{-1}	m^{-1} A	$L^{-1}I$
J'', J_A	surface density of electric current	A m^{-2}	m^{-2} A	$L^{-2}I$
J''', J_V	volume density of electric current	A m^{-3}	m^{-3} A	$L^{-3}I$
k	charge carrier mobility	A s^{2} kg^{-1}	kg^{-1} s^{2} A	$M^{-1}T^{2}I$
L	inductance	H	m^{2} kg s^{-2} A^{-2}	$L^{2}MT^{-2}I^{-2}$
L_{12}, M	mutual inductance	H	m^{2} kg s^{-2} A^{-2}	$L^{2}MT^{-2}I^{-2}$

Tab. A.3 Dimensional Electrical and Magnetic Quantities (continued)

Symbol	Quantity, Name	Units	SI Units	SI Dimensions
m	magnetic moment	A m^2	m^2 A	L^2I
M	magnetization	A m^{-1}	m^{-1} A	$L^{-1}I$
p	electric moment	C m	m s A	LTI
P	power, power drain, active power	W	m^2 kg s^{-3}	L^2MT^{-3}
P	electrical polarization	C m^{-2}	m^{-2} s A	$L^{-2}TI$
Q	electric charge	C	s A	TI
R	electrical resistance	Ω	m^2 kg s^{-3} A^{-2}	$L^2MT^{-3}I^{-2}$
R_m	magnetic resistance	H^{-1}	m^{-2} kg^{-1} s^2 A^2	$L^{-2}M^{-1}T^2I^2$
S	Poynting vector	W m^{-2}	kg s^{-3}	MT^{-3}
S	elastance	F^{-1}	m^2 kg s^{-4} A^{-2}	$L^2MT^{-4}I^{-2}$
U	voltage	V	m^2 kg s^{-3} A^{-1}	$L^2MT^{-3}I^{-1}$
U_m	magnetic potential difference	A	A	I
V	electric potential	V	m^2 kg s^{-3} A^{-1}	$L^2MT^{-3}I^{-1}$
V_m	magnetic potential	A	A	I
w	volume energy density of an electromagnetic field	J m^{-3}	m^{-1} kg s^{-2}	$L^{-1}MT^{-2}$
X, Z	reactance (X), impedance (Z)	Ω	m^2 kg s^{-3} A^{-2}	$L^2MT^{-3}I^{-2}$
γ	specific electrical conductance	S m^{-1}	m^{-3} kg^{-1} s^3 A^2	$L^{-3}M^{-1}T^3I^2$
ε	permittivity	F m^{-1}	m^{-3} kg^{-1} s^4 A^2	$L^{-3}M^{-1}T^4I^2$

Tab. A.3 Dimensional Electrical and Magnetic Quantities (continued'

Symbol	Quantity, Name	Units	SI Units	SI Dimensions
ε_0	vacuum permittivity	$F\ m^{-1}$	$m^{-3}\ kg^{-1}\ s^4\ A^2$	$L^{-3}M^{-1}T^4I^2$
μ	permeability	$H\ m^{-1}$	$m\ kg\ s^{-2}\ A^{-2}$	$LMT^{-2}I^{-2}$
μ_0	vacuum permeability	$H\ m^{-1}$	$m\ kg\ s^{-2}\ A^{-2}$	$LMT^{-2}I^{-2}$
ρ	volume density of an electric charge	$C\ m^{-3}$	$m^{-3}\ s\ A$	$L^{-3}TI$
ρ	specific electrical resistance	$\Omega\ m$	$m^3\ kg\ s^{-3}\ A^{-2}$	$L^3MT^{-3}I^{-2}$
σ	surface density of an electric charge	$C\ m^{-2}$	$m^{-2}\ s\ A$	$L^{-2}TI$
τ	linear density of an electric charge	$C\ m^{-1}$	$m^{-1}\ s\ A$	$L^{-1}TI$
Φ, Ψ	magnetic flux	Wb	$m^2\ kg\ s^{-2}\ A^{-1}$	$L^2MT^{-2}I^{-1}$
Φ, Ψ	electric flux	C	$s\ A$	TI

A.4 Subscripts

Tab. A.4 Subscripts

Symbol	Meaning
a	air
ad	adiabatic
A	area quantity
B	point quantity
c	coolant
$crit$	critical
D	object, original
ef	effective
ekv	equivalent
E, e	electrical
f	frictional, phase
g	gas
i, ini	initial
l	liquid
loc	local
L	linear quantity
m	mass
mag, mg	magnetic
med	medium
M	model
n	saturated vapour
p	pressure
par	partial
P	ambient
ref	characteristic, reference
rel	relative
rot	rotational
R, r	radiation
s	solid
st	stagnation
S	surface
T, t	thermal
v	vapour
V	volume quantity
0, 1, 2	space points
1, 2	inner and outer radius
∞	free condition, undisturbed media

Tab. A.5 Physical Constants

Symbol	Quantity, Name	Value (SI Units)	Units	SI Units
$a_0 = \frac{\alpha}{4\pi R\infty} = \frac{4\pi\varepsilon_0 \hbar^2}{m_e e^2}$	Bohr radius	$(5.291\ 772\ 108 \pm 0.000\ 000\ 018) \times 10^{-11}$	m	m
$b = \frac{\frac{hc}{k}}{4{,}965\ 114\ 231\ ...}$	Wien displacement law constant	$(2.897\ 768\ 5 \pm 0.000\ 005\ 1) \times 10^{-3}$	m K	m K
c, c_0	speed of light in vacuum	$2.997\ 924\ 58 \times 10^{8}$ (exact)	m s^{-1}	m s^{-1}
$c_1 = 2\pi\ hc^2$	first radiation constant	$(3.741\ 771\ 38 \pm 0.000\ 000\ 64) \times 10^{-16}$	W m^2	m^4 kg s^{-3}
c_{1L}	first radiation constant for spherical radiance	$(1.191\ 042\ 82 \pm 0.000\ 000\ 20) \times 10^{-16}$	W m^2 sr^{-1}	m^4 kg s^{-3} sr^{-1}
$c_2 = \frac{hc}{k}$	second radiation constant	$(1.438\ 775\ 2 \pm 0.000\ 002\ 5) \times 10^{-2}$	m K	m K
e	elementary charge	$(1.602\ 176\ 53 \pm 0.000\ 000\ 14) \times 10^{-19}$	C	s A
$\frac{-e}{m_e}$	electron charge to mass quotient	$-(1.758\ 820\ 12 \pm 0.000\ 000\ 15) \times 10^{11}$	C kg^{-1}	kg^{-1} s A
$\frac{e}{m_p}$	proton charge to mass quotient	$(9.578\ 833\ 76 \pm 0.000\ 000\ 82) \times 10^{7}$	C kg^{-1}	kg^{-1} s A
$E_h = \frac{e^2}{4\pi\varepsilon_0 a_0} = 2R_\infty hc = \alpha^2 m_e c^2$	Hartree energy	$(4.359\ 744\ 17 \pm 0.000\ 000\ 75) \times 10^{-18}$	J	m^2 kg s^{-2}
$F = N_A e$	Faraday constant	$(9.648\ 533\ 83 \pm 0.000\ 000\ 83) \times 10^{4}$	C mol^{-1}	s A mol^{-1}

Tab. A.5 Physical Constants (continued)

g_n	standard acceleration of gravity	9.806 65 (exact)	m s^{-2}	m s^{-2}
G	Newtonian gravitation constant	$(6.674\ 2 \pm 0.001\ 0) \times 10^{-11}$	$m^3\ kg^{-1}\ s^{-2}$	$m^3\ kg^{-1}\ s^{-2}$
$G_0 = \frac{2e^2}{h}$	conductance quantum	$(7.748\ 091\ 733 \pm 0.000\ 000\ 026) \times 10^{-5}$	S	$m^{-2}\ kg^{-1}\ s^3\ A^2$
$G_0^{-1} = \frac{h}{2e^2}$	inverse of conductance quantum, resistance quantum	$(1.290\ 640\ 372\ 5 \pm 0.000\ 000\ 004\ 3) \times 10^4$	Ω	$m^2\ kg^1.s^{-3}\ A^{-2}$
h	Planck constant	$(6.626\ 069\ 3 \pm 0.000\ 001\ 1) \times 10^{-34}$	J s	$m^2\ kg\ s^{-1}$
$\hbar = \frac{h}{2\pi}$	Planck constant over 2π	$(1.054\ 571\ 68 \pm 0.000\ 000\ 18) \times 10^{-34}$	J s	$m^2\ kg\ s^{-1}$
$h = \frac{h}{2m_e}$	quantum of circulation	$(3.636\ 947\ 550 \pm 0.000\ 000\ 024) \times 10^{-4}$	$m^2\ s^{-1}$	$m^2\ s^{-1}$
$h = \frac{h}{2m_e}$	Boltzmann constant	$(1.380\ 650\ 5 \pm 0.000\ 002\ 4) \times 10^{-23}$	$J\ K^{-1}$	$m^2\ kg.s^{-2}\ K^{-1}$
$K_J = \frac{2_e}{h}$	Josephson constant	$(4.835\ 978\ 79 \pm 0.000\ 000\ 41) \times 10^{14}$	$Hz\ V^{-1}$	$m^{-2}\ kg^{-1}.s^2\ A$
$l_p = \frac{\hbar}{m_p c} = \sqrt{\frac{\hbar G}{c^3}}$	Planck length	$(1.616\ 24 \pm 0.000\ 12) \times 10^{-35}$	m	m
m_d	deuteron mass	$(3.343\ 583\ 35 \pm 0.000\ 000\ 57) \times 10^{-27}$	kg	kg
m_e	electron mass	$(9.109\ 382\ 6 \pm 0.000\ 001\ 6) \times 10^{-31}$	kg	kg

Tab. A.5 Physical Constants (continued)

m_h	helion mass	$(5.006\ 412\ 14 \pm 0.000\ 000\ 86) \times 10^{-27}$	kg	kg
m_n	neutron mass	$(1.674\ 927\ 28 \pm 0.000\ 000\ 29) \times 10^{-27}$	kg	kg
m_p	proton mass	$(1.672\ 621\ 71 \pm 0.000\ 000\ 29) \times 10^{-27}$	kg	kg
$m_p = \sqrt{\frac{\hbar c}{G}}$	Planck mass	$(2.176\ 45 \pm 0.000\ 16) \times 10^{-8}$	kg	kg
$m_u = u = \frac{1}{12} m(^{12}C)$	atomic mass unit	$(1.660\ 538\ 86 \pm 0.000\ 000\ 28) \times 10^{-27}$	kg	kg
m_α	alpha–particle mass	$(6.644\ 656\ 5 \pm 0.000\ 001\ 1) \times 10^{-27}$	kg	kg
m_μ	muon mass	$(1.883\ 531\ 40 \pm 0.000\ 000\ 33) \times 10^{-28}$	kg	kg
m_τ	tau mass	$(3.167\ 77 \pm 0.000\ 52) \times 10^{-27}$	kg	kg
$M_u = \frac{1}{12} M(^{12}C)$	molar mass constant	1×10^{-3} (exact)	kg mol^{-1}	kg mol^{-1}
$n_0 = \frac{N_A}{V_m}$	Loschmidt constant (273 15 K; 101 325 kPa)	$(2.686\ 777\ 3 \pm 0.000\ 004\ 7) \times 10^{25}$	m^{-3}	m^{-3}
N_A, L	Avogadro constant	$(6.022\ 141\ 5 \pm 0.000\ 001\ 0) \times 10^{23}$	mol^{-1}	mol^{-1}
p_0	standard atmospheric pressure	$1.013\ 25 \times 10^{5}$ (exact)	Pa	m^{-1} kg s^{-2}
$r_e = \alpha^2 a_0 = \frac{\alpha^3}{4\pi R_\infty}$	classical electron radius	$(2.817\ 940\ 325 \pm 0.000\ 000\ 028) \times 10^{-15}$	m	m
R	molar gas constant	$8.314\ 472 \pm 0.000\ 015$	J mol^{-1}K^{-1}	m^2 kg s^{-2} K^{-1} mol^{-1}

Tab. A.5 Physical Constants (continued)

$R_K = \frac{h}{e^2} = \frac{\mu_0 c}{2\alpha}$	von Klitzing constant	(2.581 280 744 9 ± 0.000 000 008 6) × 10^4	Ω	$m^2\ kg^1\ s^{-3}\ A^{-2}$
$R_\infty = \frac{\alpha^2 m_e c}{2h}$	Rydberg constant	(1.097 373 156 852 5± 0.000 000 000 007 3) × 10^7	m^{-1}	m^{-1}
S	Sun constant	1.325 × 10^3	$W\ m^{-2}$	$kg\ s^{-3}$
$t_P = \frac{l_P}{c} = \sqrt{\frac{\hbar c^5}{G}\frac{1}{k}}$	Planck time	(5.391 21 ± 0.000 40) × 10^{-44}	s	s
T (0 °C)	zero of Celsius temperature scale	273.15	K	K
$T_P = \frac{mpc^2}{k} = \sqrt{\frac{\hbar c^5}{G}\frac{1}{k}}$	Planck temperature	1.416 79 ± 0.000 11) × 10^{32}	K	K
T_{tp}	triple point of water	273.16	K	K
$V_m = \frac{RT_0}{p_0}$	standard molar volume of ideal gas (273 15 K; 101 325 Pa)	(22.413 996 ± 0.000 039) × 10^{-3}	$m^3\ mol^{-1}$	$m^3\ mol^{-1}$
$Z_0 = \sqrt{\frac{\mu_0}{\varepsilon_0}} = \mu_0 c$	characteristic impedance of vacuum	376.730 313 461 (exact)	Ω	$m^2\ kg\ s^{-3}\ A^{-2}$
$\alpha = \frac{\varepsilon^2}{4\pi\varepsilon_0 \hbar c}$	fine-structure constant	(7.297 352 568 ± 0.000 000 024) × 10^{-3}	-	-
$\alpha^{-1} = \frac{4\pi\varepsilon_0 \hbar c}{\varepsilon^2}$	inverse fine-structure constant	(137.035 999 11 ± 0 000 000.46)	-	-

Tab. A.5 Physical Constants (continued)

$\varepsilon_0 = \frac{1}{\mu_0 c^2}$	electric constant, permitivity of vacuum	$8.854\ 187\ 817... \times 10^{-12}$ (exact)	F m^{-1}	m^{-3} kg^{-1} s^4 A^2
$\lambda_C = \frac{h}{m_e c} = 2\pi\alpha a_0 = \frac{\alpha^2}{2R_\infty}$	Compton wavelength	$(2.426\ 310\ 238 \pm 0.000\ 000\ 016) \times 10^{-12}$	m	m
$\mu_0 = \frac{1}{\varepsilon_0 c^2} = 4\pi \times 10^{-7}$	magnetic constant, permeability of vacuum	$12.566\ 370\ 614\ ... \times 10^{-7}$ (exact)	H m^{-1}	m kg s^{-2} A^{-2}
$\mu_B = \frac{e\hbar}{2m_e}$	Bohr magneton	$(9.274\ 009\ 49 \pm 0.000\ 000\ 80) \times 10^{-24}$	J T^{-1}	m^2 A
μ_d	deuteron magnetic moment	$(0.433\ 073\ 482 \pm 0.000\ 000\ 038) \times 10^{-26}$	J T^{-1}	m^2 A
μ_e	electron magnetic moment	$(-9.284\ 764\ 12 \pm 0.000\ 000\ 80) \times 10^{-24}$	J T^{-1}	m^2 A
μ_n	neutron magnetic moment	$(-0.966\ 236\ 45 \pm 0.000\ 000\ 24) \times 10^{-26}$	J T^{-1}	m^2 A
$\mu_N = \frac{e\hbar}{2m_p}$	nuclear magneton	$(5.050\ 783\ 43 \pm 0.000\ 000\ 43) \times 10^{-27}$	J T^{-1}	m^2 A
μ_p	proton magnetic moment	$(1.410\ 606\ 71 \pm 0.000\ 000\ 12) \times 10^{-26}$	J T^{-1}	m^2 A

Tab. A.5 Physical Constants (continued)

$\sigma_e = \frac{8\pi}{3} r_e^2 = \frac{8\pi}{3}\alpha^4 a_0^2 = \frac{\alpha^6}{6\pi R_\infty^2}$	Thomson cross–section	$(0.665\ 245\ 873 \pm 0.000\ 000\ 013) \times 10^{-28}$	m^{-2}	m^{-2}
$\Phi_0 = \frac{h}{2e}$	magnetic flux quantum	$(2.067\ 833\ 72 \pm 0.000\ 000\ 18) \times 10^{-15}$	Wb	$m^2\ kg\ s^{-2}\ A^{-1}$
$1[eV] = \frac{e}{C}[J]$	electron volt	$(1.602\ 176\ 53 \pm 0.000\ 000\ 14) \times 10^{-19}$	J	$m^2\ kg\ s^{-2}$

References and additional literature

Chapter 1

[1.1] ANTONOV J. G.: *Modeling of biological systems* (in Russ). Naukova dumka, Kiev, 1977, 260 p.

[1.2] BASMADIAN D.: *The art of modeling in science and engineering*. Chapman and Hall/CRC, London, 1999, 654 p.

[1.3] BELLOMO N. (Ed.): *Modeling and simulation in science, engineering and technology*. Springer, 2008, 324 p.

[1.4] BRAKEL J.: *Modeling in chemical engineering*. International Journal for Philosophy of Chemistry. 2000, Vol. 6, No. 2, pp. 101–116.

[1.5] COBELLI C.:, CARSON E. E.: *Introduction to modeling in physiology and medicine*. Academic Press, 2008, 324 p.

[1.6] JEFFERS J. N. R.: *An introduction to systeme analysis with ecological applications*. Edward Arnold, London, 1978, 198 p.

[1.7] KUNEŠ J.: *Modeling of thermal processes*. (in Czech) SNTL, Prague 1989, 424 p.

[1.8] TOSUN I.: *Modelling in transport phenomena*. Elsevier, 2002, 590 p.

Chapter 2

[2.1] CARSON E. R., COBELLI C.: *Modelling methodology for physiology and medicine*. Academic Press, 2001, 421 p.

[2.2] CHALOUPKA L., KRÁL J., KUNEŠ J.: *Measurement and identification of thermal processes in metallurgical systems*. 1st Symposium of the IMECO on Temperature measurement in industry and science. Karlovy Vary 14–16. 10. 1981, DT CVTS-UVZU, Skoda Plzen 1981, pp. 285–293.

[2.3] ERLYKIN A. D, LEBEDEV P. N.: *Models should match phenomenology* (Presentation), 2004. http://www.fi.infn.it/conferenze/ecrs2004/Pages/Presentazioni/08Erlyki n.pdf

[2.4] KUNEŠ J., FRANTA V.: *Some aspects in modelling and systems simulation*. In: Proceedings of the 9th IMACS Congress, Simulation of systems 79. Sorrento 1979, North-Holland Publ. Comp. Amsterdam 1979, pp. 7–13.

[2.5] KUNEŠ J., VAVROCH O., FRANTA V.: *Comparison of various simulation methods by solution of thermomechanic problems*. In: Proceedings of the 11th World Congress IMACS System simulation and scientific computation. Oslo 5–9. 8. 1985, Vol. 5, pp. 69–71.

[2.6] KUNEŠ J., VAVROCH O., FRANTA V.: *Fundamentals of Modeling*. (in Czech), SNTL, Prague 1989, 264 p.

[2.7] VITKOVITCH D.: *Field analysis. Experimental and computtioal methods*. Van Nostrand, London, 1966, 503 p.

J. Kuneš, *Similarity and Modeling in Science and Engineering*,
DOI: 10.1007/978-1-907343-78-0, © Cambridge International Science Publishing 2012

Chapter 3

[3.1] BAKER, J. E., WESTINE P. S., DODGE F. T.: *Similarity methods in engineering dynamics.* Elsevier, Amsterdam, 1991, 384 p.

[3.2] BARENBLATT G. I.: *Dimensional analysis.* Gordon and Breach Science Publisher. New York, 1987, 135 p.

[3.3] BARENBLATT G. I.: *Scaling, self similarity, and intermediate asymptotics*, Cambridge University Press, 2000, 412 p.

[3.4] BARENBLATT G. I.: *Scaling*. Cambridge University Press, 2001, 214 p.

[3.5] BATZIAS F. A., NIKOLOU N. P., KAKOS A. S.: *Computer aided dimensional analysis for knowledge in chemical engineering processe*s. In: Proc. Internat. Conference of Computational Methods in Science and Engineering (IICCMSE 2003), Sept. 12–16, 2003, Kastora, Greece, 2003. pp. 74–82.

[3.6] BIRKHOFF G.: *Hydrodynamics. A study in logic, fact and similitude.* Princeton University Press, Princeton, 151 p.

[3.7] BIRKHOFF G.: *Similarity in two-phase fluid flows.* In: *Group theoretical methods in mechanics*. In: Proc. of the point IUTAM/AMU symposium. USSR Academy of Science. Siberian Branch. Novosibirsk, 1978, pp. 58–79.

[3.8] BRIDGMAN, P. W.: *Dimensional analysis.* New York, AMS Press, 1978, 120 p.

[3.9] BRÜCKNER S. *The dimensional analysis for Matlab.* In: Proc. 5th International Workshop on Similarity Methods, 4–5. Nov. 2002, Stuttgart University, 2002.

[3.10] BRÜCKNER S., RUDOLPH S.: *Hierarchical modeling using dimensional analysis.* In: 3rd International Workshop on similarity methods, 13–14 Nov. 2000, Universität Stuttgart, 16 p.

[3.11] BUCKINGHAM E.: *On physically similar systems: Illustrations of the use of dimensional analysis.* Physical Review, 4 (1914), pp. 345–376.

[3.12] BUCKINGHAM E.: *The principle of similitude.* Nature, 96 (1915), p. 396–397.

[3.13] BUTTERFIELD R.: *Dimensional analysis for geotechnical engineers.* Geotechnique, 49 (1999), No. 3, pp. 357–366.

[3.14] CALCHEMY Software Inc.: *Automatic units conversion by dimensional Analysis* http://calchemy.com/uclive.htm> [cit. 2008–05–23].

[3.15] CARMONA R., TEHRANCHI M. R.: *Interest rate models: An infinite dimensional stochastic analysis perspective.* Springer 2006, 235 p.

[3.16] CRAIG E.: *Dimensional analysis in applied psychological research.* Journal of Psychology, 55 (1963), p. 223.

[3.17] CRAIG GLORIA P.: *Clinical calculation using dimensional analysis: Instructors manual.* Lippincott Williams and Wilkins, 1996.

[3.18] CRAIG GLORIA P.: *Clinical calculations made easy: Solving problems using dimensional analysis.* Lippincott Williams and Wilkins, 2004.

[3.19] CIZIKOV J. M.: *Theorie and modeling of metall cutting processes.*(in Russ.) Metallurgia, Moscow, 1970, 296 p.

[3.20] CMELIK R. F., GEHANI N. H. *Dimensional analysis with C++.* IEEE Software. 1988, Vol. XX, pp. 21–27.

[3.21] DE JONG F. J.: *Dimensional analysis for economists.* North Holland Publishing Co., Amsterdam, 1967, 220 p.

[3.22] DJAKONOV G. K.: *Questions of theory similarity of physico–chemical processes,* Moscou, Izdat. AN SSSR, 1956, 206 p.

[3.23] DOUGLAS J. F.: *An introduction to dimensional analysis for engineers.* Pitman and sons, London, 1969, 137 p.

[3.24] DUNCAN W. J.: *Physical similarity and dimensional analysis: An elementary treatise.* Edward Arnold, London, 1953, 154 p.

[3.25] EIGENSON L. S.: *Modeling* (in Russ) Sovetskaia nauka, Moscow, 1952, 372 p.

[3.26] RENAULT-PELTERIE R.: A*nalyse dimensionnelle et metrologie*. F.Rouge,Lausanne, 1950, 112 p.

[3.27] FLEMING R. (ed.): *Scale-up in practice*. Reinhold Publ., New York, 1958, 134 p.

[3.28] FOCKEN C. M.: *Dimensional methods and their applications*. Edward Arnold, London, 1953, 224 p.

[3.29] GLOCKNER P. G., SIHGH M. C. (ed*.): Symmetry, similarity and group theoretic methods in mechanics*. Calgary. The University of Calgary, 1974, 640 p.

[3.30] GÖRTLER H.: *Dimensionsanalyse: Teorie der Physikalischen Dimensionen mit Anwendungen*. Berlin-Heidelberg, Springer, 1975, 247 p.

[3.31] GUKHMAN A. A.: *Introduction to the theory of similarity*. Academic Press, 1965.

[3.32] HACKESCHMIDT, M.: *Strömungstechnik. Ähnlichkeit-Analogie-Model. Wissenschaftliche Grundlagen des Experimentierens*. Leipzig, VEB Deutcher Verlag für Grundstoffindustrie, 1972, 425 p.

[3.33] HANSEN A. G.: *Similarity analysis of boundary value problems in engineering*. Prentice-Hall, New Jersey, 114 p.

[3.34] HART G. W.: *Multidimensional analysis: Algebra and systems for science and engineering*. Springer 1995.

[3.35] HAYNES R.: *Introduction of dimensional analysis for geographers. Essay 33*, In: Concepts and techniques in modern geography. Norwich (UK), Geo Abstracts, 1982.

[3.36] HILFINGER P. N.: *An data package for dimensional analysis*. ACM Transactions on programming languages and systems. 1988, Vol. 10, No. 2, pp. 189–203.

[3.37] HORNUNG H. G.: *Dimensional analysis: Examples of the use of symmetry*. Dover Publications. 2006, 80 p.

[3.38] HUNTLEY H. E.:*Dimensional analysis*. New York, Dover Publications, 1967, 158 p.

[3.39] HUNTLEY H. E.: *The divine proportion – A sudy in mathematical beauty*. Dover Pub lications Inc., New York, 1970.

[3.40] ISAACSON E., ISAACSON M.: *Dimensional methods in engineering and physics*. Edward Arnold, London, 1975, 220 p.

[3.41] JACOBY W. G.: *Data theory and dimensional analysis. Quantitative applications in the social sciences*. Sage University Paper No.78, 1991, 96 p.

[3.42] JUPP E. W.: *An introduction to dimensional method*. Cleaver-Hume Press, London, 1962, 89 p.

[3.43] KASPRZAK W., LYSIK B., RYBACZUK M.: *Dimensional analysis in the identification of mathematical models*. World Scientific Pub., 1990.

[3.44] KATTANEK S., GRÖBER R., BODE C.: *Ähnlichkeitstheorie*. VEB Deutsches Verlag für Grundstoffindustrie, 1967, 173 p.

[3.45] KIRPICHEV M. V.: *Theory of similarity and heat modeling*. (in Russ.), Nauka, Moscow, 1987, 167 p.

[3.46] KLINE S. L.: *Similitude and approximation theory*. McGraw-Hill, New York, 1965, 228 p.

[3.47] KREITH F.: *Fluid Mechanics.(3th chap.:Simulitude-dimensional analysis and date correlation., 14.chap.: Micro/Nanotribology)*. CRC Press 1999, 224 p.

[3.48] KOŽEŠNIK J.: *Theorie of similarity and modeling*. (in Czech), Academia, Prague, 1983, 264 p.

[3.49] KREITH F.: *Fluid Mechanics. Chapter 3.: Similitude-dimensional analysis and date correlation. Chapter 14.: Micro/Nanotribology*. CRC Press 1999, 224 p.

[3.50] KUNEŠ J., TRKOVSKÝ V.: *Physical similarity of thermal processes in machine systems.* Strojnický časopis 47 (1996), No.4, pp. 254–262.
[3.51] KURTH R.: *Dimensional analysis and group theory in astrophysics.* Pergamon Press, Oxford 1972, 230 p.
[3.52] KUTATELADZE S.S.: *Similarity analyse in thermo-physics.* (in Russ), Nauka-Sibirskoe otdelenije. Novosibirsk, 1982, 280 p.
[3.53] KUTATELADZE S. S: *Similarity analysis and physical models.* (in Russ), Nauka-Sibirskoe otdelenije. Novosibirsk, 1986, 296 p.
[3.54] LANGHAAR H.L.: *Dimensional analysis and theory of models.* Wiley, New York, 1951, 166 p.
[3.55] LEHMAN P.J.: *Dimensional analysis in applied psychological research.* J. Psychology 55 (1963), pp. 223–226.
[3.56] MASSEY B. S.: *Units, dimensional analysis and physical similarity.* London, Van Nostrand Reinhold, 1971, 140 s.
[3.57] MURPHY G.: *Similitude in engineering.* Ronald Press, New York, 1950, 302 p.
[3.58] NOVIKOV I. I., BORIŠANSKIJ V. M.: *Theorie of similarity in thermodynamics and heat transfer.* Atomizdat, Moscow, 1979, 184 p.
[3.59] PALACIOS J.: *Dimensional analysis.* MacMillian, London, 1964, 236 p. PANKHURST R. C.: *Dimensional analysis and scale factors.* Chapman and Hall, London, 1964, 151 s.
[3.60] PAWLOWSKI J.: *Ähnlichkeitstheorie in der physikalisch-technischen Forschung. Grundlagen und Anwendung.* Springer, Berlin, 1971, 172 p.
[3.61] PETTY G. W.: *Automated computation and consistency checking of physical dimensions and units in scientific programs.* Software – Practice and Experience. 2001, Vol. 31, pp. 1067–1076.
[3.62] REZNIAKOV A. B.: *Similarity methods.* (in Russ), Akademia nauk Kaz. SSR, Alma Ata, 1959, 152 p.
[3.63] SEDOV L. I.: *Similarity and dimensional methods in mechanics.* Academic Press, New York, 1959, 479 p.
[3.64] SENA L. A.: *Units of physical quantities and their dimensions.* Mir Publishers, Moscow, 1972.
[3.65] SHIVE J. N., WEBER R. L.: *Similarities in physics.* Adam Hilger, Bristol, 1982, 275 p.
[3.66] SCHEPARTZ B.: *Dimensional analysis in the biomedical science.* Thomas, 1980, 172 p.
[3.67] SHIRER D.L., SQUARE W.: A computer program for dimensional analysis. American Journal of Physics. 1972, Vol. 40, No. 3, pp. 489–490.
[3.68] SCHURING D. J.: *Scale models in engineering.* Pergamon Press,New York, 1977, 299 p.
[3.69] SKOGLUND V. J.: *Similitude. Theory and applications. International Textbook,* Scranton (PA), 1967, 320 p.
[3.70] SLOAN A. D., HAPP W. W.: *Computer programm for dimensional analysis.* NASA TND 5165, 1969, 29 p.
[3.71] SMIRNOV V. S.: *Similarity method in the theorie of rolling process.*(in Russ.) Nauka, Leningrad, 1971, 179 p.
[3.72] SONIN A.A.: *The physical basis of dimensional analysis.* Massachusetts Institute of Technology, Cambridge, MA, 2001, 56 p.
[3.73] STAICU C. I.: *Restricted and general dimensional analysis: Treatment of experimental data.* Abacus Press, Tunbridge Wells, Kent, UK, 1982, 304 p.
[3.74] SZÜCS E.: *Similitude and modelling.* Elsevier, Amsterdam, 1980, 335 p.

[3.75] SZIRTES T.: *Applied dimensional analysis and modeling*. McGraw Hill, New York, 1997, 790 p.
[3.76] TAYLOR E. S.: *Dimensional analysis for engineers*. Clarendon Press, Oxford, 1974, 162 p.
[3.77] THOMAS P.H.: *Dimensional analysis: A magic art in fire research?* Lund University, Rap. 3094, 46 p.
[3.78] TRAUSTEL S.: *Modellgesetze der Vergasung und Verhüttung*. Akademie Verlag Berlin, 1949, 89 p.
[3.79] UMRIGAR Z. D.: *Fully static dimensional analysis with C++*. ACM SIGPLAN Notices. 1994, Vol. 29, No. 9, pp. 135–139.
[3.80] VENIKOV V. A.: *Theory of similarity and simulation*. MacDonald, Technical and Scientific, London, 1969.
[3.81] ZIEREP J.: *Ähnlichkeitsgesetze und Modellregeln der Strömungslehre*. G.Braun Verlag, Karlsruhe, 1972, 139 p.
[3.82] ZLOKARNIK M.: Scale-up. Modellübertragung in der Verfahrenstechnik. Wiley-VCH, Weinheim, 2000, 226 p.
[3.83] ZLOKARNIK M.: *Dimensional analysis and scale- up in chemical engineering*. Springer, 1991.
[3.84] WIKIPEDIA.: http://en.wikipedia.org/wiki/Dimensional analysis

Chapter 4

[4.1] DARUKA I.: *A phenomenological model for the collective landing of bird flocks*. Proc. Royal Soc. B, Vol. 276, 2009, No.1658, pp. 911– 917.
[4.2] HYUNHO SHIN, JONG-BONG KIM: *A phenomenological constitutive equation to describe various flow stress behaviors of materials in wide strain rate and temperature regimes*. J. Eng. Materials and Technology. Vol.132, 2010, Issue 2, (6 pages).
[4.3] KUNEŠ J., VESELÝ Z., HONNER M.: *Thermal Barriers*. (in Czech) Academia, Prague, 2003, 302 p.
[4.4] MARTAN J.: *Thermo-kinetic model of laser-material interaction in the form of criterial equations*. Ph.D. Thesis. University of West Bohemia, Faculty of Applied Sciences, Plzen, Czech Republic, University of Orleans, Ecole Polytechnique, Orleans, France, 2005, 153 p.
[4.5] TAPATI DUTTA, PRABIR KUMAR BISWAS: *Phenomenological model for the unified studies of different physical properties of polymers*. Die Angewandte Makromolekulare Chemie, Vol. 207, 1993, Issue 1, pp. 9–21.
[4.6] TRKOVSKÝ V.: *Modelling and control of thermal deformations of large machining tools* (in Czech). Ph.D. Thesis. Czech Technical University in Prague, Faculty of Mechanical Engineering, Prague 1988, 119 p.

Chapter 5

[5.1] ACKERMAN E., GATEWOOD L. C.: *Mathematical models in the health science. A computer aires approach*. University of Minnesota Press. 1979, 358 p.
[5.2] BASARAB M., KRAVČENKO V. F., MATVEJEV V. A.: *Mathematical modeling of physical processes in giroskopy*. Radiotekhnika, Moscow, 2005, 175 p.
[5.3] BARDZOKAS D. I., ZOBIN A. I: *Mathematical modeling of physical processes in composite materials with the periodic structure*. Fizmatkniga, URSS, Kiev 2003, 376 p.

[5.4] BUTUC M. C., BARATA DA ROCHA A., GRACIER J.J., FERREIRA DUARTE J.: *Study on forming limit diagrams prediction using a phenomenological and a physical approach of plasticity theory.* Journal of Physics IV France, 2003, Vol. 105, pp.73–80.

[5.5] CHANIN M. A., ELKIN L. I.: *Mathematical model of micro-bacteria population.* (in Russ.), Nauka, Moscow, 1976, 152 p.

[5.6] CIBULKA O. E.: *Analysis of dynamics of pulsed laser induced phase changes.* Ph.D. Thesis.University of West Bohemia, Plzen, 2008, 70 p.

[5.7] FINKELSTEIN L., CARSON E. R.: *Mathematical modelling of dynamic biological systeme.* Research Studies Press. Forest Grove (Kreton), 1979, 329 p.

[5.8] FRAUENTHAL J. C.: *Mathematical modelling in epidemiology.* Springer, 1980, 118 p.

[5.9] HONNER M.: *Heat waves simulation.* Intern. Journal Computers & Mathematics with Applications. 38 (1999),pp. 234 – 243.

[5.10] HONNER M., KUNEŠ J.: *On the wave diffusion and parallel nonequilibrium heat conduction,* Trans. ASME, Journal of Heat Transfer, Vol. 121, 1999, pp. 702 – 707.

[5.11] HONNER M.: *Micro-thermomechanics in advanced technologies.* Ph.D. Thesis. University of West Bohemia, 1999, Plzen (in Czech).

[5.12] HONNER M., KUNEŠ J.: *Microscale nonequilibrium heat transfer in technologies using intensive heat sources.* Thermal Science 2000 – Proceedings of the ASME – ZSITS International Thermal Science Seminar, Bled, Slovenia, June 11 – 14, 2000, pp. 113 – 119.

[5.13] HONNER M., KUNEŠ J.: *Microscale nonequilibrium heating of materials in advanced physical technologies.* Engineering mechanics, Vol. 9, 2000, No. 5, pp. 297–308.

[5.14] HVYNHO SHIN: *A phenomenological constitutive equation to describe* various flow stress behaviors of materials in wide strain rate and temperature *regimes.* Journal of Engineering Materials and Technology., Vol. 132, 2010, Issue 2, (6 pages).

[5.15] KASPRZAK W., LYSIK B., RYBACZUK M.: *Dimensional analysis in the identification of mathematical models.* World Scientific Pub. Co. Inc, 1990.

[5.16] KUNEŠ J.,TRKOVSKÝ V.: *Wärmeverhalten von Getriebekästen bei extremen Witterungsverhältnissen.* Tribologie und Schmierungstechnik. 38 (1991), No.4.

[5.17] KUNEŠ J., KREJČÍ L., HONNER M.: *Wave heat conduction in technologies.* Strojnický časopis (Mechanical Engineering), Vol. 47 (1996), No. 6, pp. 346 – 362.

[5.18] MARCHUK G. I.: *Mathematical models in immunology.* Nauka, Moscow 1980, 264 p.

[5.19] TIKHONOV A.N., SAMARSKI A. A.: *Equations of mathematical physics.* Nauka, Moscow, 19, 736 p.

[5.20] TRKOVSKÝ V., KUNEŠ J.: *Mathematisches Modell der thermoelastischen Verformungen von Werkzeugmaschinen.* Konstruktion 45 (1993), pp. 66–72.

[5.21] TZOU D. Y.: *Macro- to microscale heat transfer: The Lagging behaviour.* Taylor & Francis, New York 1997,

[5.22] VESELÝ Z.: *Thermomechanical processes in heteregeneous layered structure of TBC during thermal shock.* Ph. D. Thesis. University of West Bohemia, Plzen, 2002, 171 p. (in Czech)

[5.23] VESELÝ Z., KUNEŠ J., HONNER M., MARTAN J.: *TB dynamic behaviour during thermal shocks – simulation and experiment.* In: Advanced Computational Methods in Heat Transfer. Heat Transfer VII. WIT Press, 2002, pp. 503–512.

[5.24] TAYLOR S., COBURN S. P., TOWNSEND D.W.: *Mathematical modeling in experimental nutrition: vitamins, proteins, methods.* Academic Press, 1996, 362 p.

[5.25] ZARUBIN V. S., KUVYRKIN G. N.: *Matematičeskije modeli termomechaniky.* Fizmatlit, 2002, 168 p. WIKIPEDIA: http://en.wikipedia.org/wiki/Phenomenology_(science)

Chapter 6

[6.1] ALLEN F. H.: *Hydraulic model techniques.* Longman 1952, London.

[6.2] FÄRBER A. M.: *Hydraulic modelling.* Universtät Stuttgart – Institut für Wasserbau, 2003, 81p.

[6.3] FLEMING R.(ed.): *Scale-up in praktice.* Reinhold Publ. Corp., New York, 1958,134 p.

[6.4] HUTTER K, JOHNK K.: *Continuum methods of physical modeling. Continuum mechanics, dimensional analysis, turbulence.* Springer, 2004,

[6.5] KUNEŠ J., CHALOUPKA L., TRKOVSKÝ J., ZUZAŇÁK A.: *Heat pipe – an active chill in the thermal "steel-sand-mould system"* In: 3IHPS – International heat-pipe symposium, Tsukuba (Japan), Sept. 12–14, 1988, pp. 211–218.(also AFG Transactions, USA, pp. 559–563)

[6.6] LEVI I.I.: *Modeling of hydraulic processes.* (in Russ.) Gosenergoizdat, Moscow, 1960, 211 p.

[6.7] SHARP J. J.: *Hydraulic modeling. Butterworth*, London, 1981.

[6.8] WAUGH J. G. , STUBSTAD G. W.: *Hydrobalistics modeling.* Naval Undersea Center, San Diego, 1972, 345 p.

[6.9] YALIN M. S.: *Theory of hydraulic models.* McMillan, London, 1971. New York, 1997, 266 p. WIKIPEDIA: http://en.wikipedia.org/wiki/Physical_model

Chapter 7

[7.1] BEKEY G. A., KARPLUS W. J.: *Hybrid computation.* John Wiley, New York, 1968.

[7.2] CHARLTON T. M.: *Model analysis of plane structures.* Pergamon Press, Oxford 1966,

[7.3] FRANKE O. L., PINDER G. F., PATTEN E.P.: *An electric-analog simulation of elliptic partial differential equations using finite element theory.* Mathenatic and Computers in Simulation. 24 (1982), No. 1, pp. 65–71:

[7.4] GUTTENMACHER L. I.: *Electrical models.* Technika, Kiev, 1975, 175 p.

[7.5] KARPLUS J. W.: *Analog simulation.* McGraw.Hill, NewYork 1958, 434 p.

[7.6] KARPLUS J. W., SOROKA W. W.: *Analogs methods. Computation and simulation.* McGraw-Hill, New York 1958, 483 p.

[7.7] KOVAŘÍK J., KUNEŠ J.: *Thermal stresses in sparking plugs.* Technical Digest 6 (1964), No.2, pp. 21 – 27.

[7.8] KOZDOBA L. A.: *Electrical modeling in heat and mass transfer.* (Russ.) Energia, Moscow, 1972, 296 p.

[7.9] KUNEŠ J., ULRYCH B.: *An analogue study of reactor vessel heating.* Technical Digest 8 (1966), No.1, pp. 24–33.

[7.10] KUNEŠ J., ULRYCH B.: *Analogue method for modelling the fields of gradients.* Technical Digest 8 (1966), No.6, pp. 403–412.

[7.11] KUNEŠ J., ULRYCH B.: *Analogue – digital analysis of thermal shock in high-pressure valve body.* Technical Digest 9 (1967), No.11, pp. 776–783.

[7.12] KUNEŠ J., ULRYCH.: *An analogue study of turbine blade cooling*. Technical Digest 2 (1968), No.11, pp. 4–11.

[7.13] KUNEŠ J., ŠVÍGLER V., VAVROCH O,: *Thermal stress in reactor outlet piping transition part under thermal shocks*. Skoda Concern, Nuclear Power Plants Division. Information Centre Plzen, 1970, 22 p.

[7.14] KUNEŠ J.,: *Neue Generation der Beukenmodelle. Elektrowärme* International 39 (1981), No. 2, pp. 192–203.

[7.15] MATSEVITY J. M.: *Electrical modeling of non-linear problems of engineering thermal physics*. (in Russ.) Naukova dumka, Kiev, 1977, 254 p

[7.16] MATSEVITYJ JU. M., KUNEŠ J.: *Gibridnoje modelirovanije teplovych processov*. Naukova dumka, Kiev, 1987, 268 p.

[7.17] MURÉNY G., SHUPLY D. J., LUO H. L.: *Engineering Analogies*. Iowa State University, Ames, Iowa, USA, 1963, 255 p.

[7.18] OLSON H. F.: *Solution of engineering problems by dynamical analogie*. Van Nostrand, New York, 1966, 278 p.

Chapter 8

[8.1] ATLURI S.N.: *Methods of computer modeling in engineering and the sciences*. Vol. I, Tech Science Press, 2005, 560 p.

[8.2] DONEA J., HUERTA A. *Finite element methods for flow problems*. Chichester, Wiley, 2003. 350 p.

[8.3] FORSYTHE G. E.,WASOW W.R.: *Finite-difference methods for partial differential equations*. Dover Publications, 2004. 464 p.

[8.4] GAUL L., KÖGL M., WAGNER M. *Boundary element methods for engineers and scientists*. Springer, 2003. 488 p.

[8.5] KANTHA L. H., CLAYSON C. A.: *Numerical models of oceans and oceanic*. Academic Press, 2000, 940 p.

[8.6] KLEIBER M. *Handbook of computational solid mechanics: survey and comparison of contemporary methods*. Springer, 1998. 763 p.

[8.7] KUNEŠ J.: *Models and simulations of controlled thermal processes*. In: Proceeding 'Simulation 77', Montreux, Switzerland, 1977.

[8.8] KUNEŠ J.: *Models and simulation of biothermal process in the human body*. In: Modelowanie zagadniej brzegowych: In: Proceedings of the 4th. Conference of Inst. Biocybernetiki. Pol. Ak. Nauk, 1978, pp. 138–152.

[8.9] KUNEŠ J., LEDVINOVÁ A.: *Simulation of the biothermal proces in the body of a foundry workman*. In: Proceeding Simulation of systems in biology and medicine, Prague 1987. DT CSVTS Prague, 1978, pp. 118–125.

[8.10] KUNEŠ J., LEDVINOVÁ A., VAVROCH O.: *Simulation of the thermal barrier influence on the thermal proces in human body*. Proceedings Simulation of systems in biology and medicine, Prague 1980. DT CSVTS Prague, 1980, pp. 219–228.

[8.11] KUNEŠ J., BARTÍK J.: *Special hybrid simulation system and simulators of industrial thermal processes*. Intern. Symposium 'Systems Analysis and Simulation', Berlin 1–5.9. 1980, In: Mathematische Forschung, Vol. 5, Akademie Verlag Berlin 1980, pp. 342–346.

[8.12] KUNEŠ J. VAVROCH O., FRANTA V.: *Hybrid simulation of non-linear heat transfer in metallurgical systems*. 1st Intern. Conf. Applied Modelling and Simulation. Lyon, France 7–11. 9. 1981, AMSE 1981, Vol, 3, pp. 153–156.

[8.13] KUNEŠ J.,VAVROCH O., FRANTA V.: *Influence of nonlinearities and recrystallization upon heat conduction*. In: 7th Intern. Heat Transfer Conference, München 1982, Hemisphere Publ. Corp., Washington 1982, Vol. 2, pp. 59–62.

[8.14] KUNEŠ J., FRANTA V., VAVROCH O., SEDLAČEK J.: *Comparison of various simulation methods involved in the solution of thermechanical problems*. In: Proceedings of 11th IMACS World Congress, Oslo, August 5–9, 1985, Vol. 5, pp. 59–71.
[8.15] KUNEŠ J., SEDLÁČEK J.: *Hybrid simulation of solidification processes*. (in Czech). Strojnický časopis (Mechanical Engineering), 38 (1987), No. 2,
[8.16] KUNEŠ J., SEDLÁČEK J.: *Simulation of an optimally controlled solidification process*. In: Proceedings of 12th IMACS World Congress, Paris, July 18–22, 1988, 3 p.
[8.17] KUNEŠ J., VESELÝ Z.: *Numerical solution of the thermomechanical proces in the thin layer structure of thermal barrier*. Mechanics and Engineering Sciences, 7 (2000), pp. 207–218.
[8.18] LANG V.: *Pulse method solution of thermal properties of multi-layer material structure*. Ph.D. Thesis. University of West Bohemia, Plzen, 2007, 100 p.
[8.19] LEVEQUE R. J. *Finite volume methods for hyperbolic problems*. Cambridge: Cambridge, University Press, 2002. 558 p.
[8.20] LUCQUIN B., PIRONNEAU O., Kern M.: *Introduction to scientific computing*. Chichester, Wiley,1998. 361 p.
[8.21] MATSEVITY J. M., KUNEŠ J.: *Hybrid modeling of thermal processes*. (in Russ.) Naukova dumka, Kiev, 1987, 268 p.
[8.22] OZISIK N. *Finite difference methods in heat transfer*. Boca Raton, CRC Press, 1994. 432 p.
[8.23] PRINCE M. R., RAPPAZ M., BELLET M.: *Numerical modelling in materials science and engineering*. Springer, 2004, 552 p.
[8.24] REDDY J.N. *An introduction to the finite element method*. McGraw-Hill, 2005, 912 p.
[8.25] SAMARSKI A.A., VULKOV L.G., VABISHCHEVICH P.N.: *Finite difference methods: Theory and applications*. Nova Science Publishers, 1999. 278 p.
[8.26] TRKOVSKÝ V., KUNEŠ J., VAVROCH O.: *Die Ursache der Mikrooberflächenrissen beim Induktionharten des Radzahnes*. In: 37. Internat. Wiss. Koll. 21-24.9.1992. Ilmenau, TU Ilmenau, Band I, pp.66–72.
[8.27] ŠVANTNER M.: *Determination of residual stresses in multi-layer structures*. Ph.D. Thesis. University of West Bohemia, Plzen, 2004, 117 p.
[8.28] WROBEL L., ALIABADI M. *The boundary element method: Applications in thermo-fluids and acoustics*. Vol. 1. Chichester, Wiley, 2002. 468 p.
[8.29] ZIENKIEWICZ O. C., TAYLOR R. L., ZHU J. Z.: *The finite element method: Its basis and fundamentals*. Elsevier, Butterworth-Heinemann, 2005, 733 p.

Chapter 9

[9.1] BARTHLOMEW D. J.: *Stochastic models for social progress*. John Wiley, New York 1973.
[9.2] BHAT N. U.: *Elements of applied stochastic processes*. John Wiley, 1998.
[9.3] BINDER K. HERMAN D. W.: *Monte Carlo simulation in statistical physics*. Springer 1992, 129 p.
[9.4] CHUKWUDI IBE O.: *Markov processes for stochastic modeling*. Academic Press, 2008, 490 p.
[9.5] EMERY A. F., CARSON W. W.: *A modification to the Monte Carlo method – The Exodus method*. Transaction of the ASME, ser. C, Journal of Heat Transfer, 90 (1968), p. 328–332.
[9.6] FISHMAN G. S.: *Monte Carlo: Concept, algorithmus and applications*. Springer, 1995.

[9.7] FRANTA V.: *Hybrid stochastic model of multidimensional physical fields*. In: 3st Modelowanie zagadnien brzegowych. Wyd. Pol. Akad.Nauk, Warszawa 1975, pp. 113–122.

[9.8] FRANTA V.: *Probability numerical solution of the field theory indirect problem*. In: 4th. Modelowanie zagadnien brzegowych, Jablonna 1978. Wyd. Pol. Akad. Nauk, Warszawa 1978, pp. 36–44.

[9.9] FRANTA V., KUNEŠ J., VAVROCH O.: *Hybrid Monte Carlo modeling method of heat conduction solution*.(in Russ.), Elektronnoe modelirovanije, 6 (1984), No.3, pp. 84–88.

[9.10] FRANTA V., VAVROCH O.: *Stochastic Exodus method in thermomechanics*. (in Czech), Strojnický časopis (Mechanical Engineering), Vol. 38 (1987), No. 2, pp. 207–219.

[9.11] FRANTA V., VAVROCH O.: *Exodus simulation of indirect and optimization problems*. In: Proceedings of the IASTED International Symposium Modelling, Identification and Control. Gridelwald, Switzerland, February 17–20, 1987, p. 425–428.

[9.12] HAAS P. J.: *Stochasic Petri nets. Modelling, stability, simulation*. Springer 2006,

[9.13] HAJI-SHEIKH A., SPARROW E. M.: *The solution of heat conduction probleme by probability methods*. Trans. of the ASME. J. of Heat Transfer ,ser. C, 89 (1967), pp. 121–131.

[9.14] HAMMOND B. L., LESTER W. A., REYNOLDS P. J.: *Monte Carlo methods in ab initio quantum chemistry: Quantum Monte Carlo for molecules*. World Scientific Publishing,1994.

[9.15] HANDLER H.: Monte Carlo solution of partial differencial equations using a hybrid computer. *IEEE Trans. of Electronic Computers*, EC-16, 1967, p. 603-610.

[9.16] HONNER M., FRANTA V.: *PVD technology simulation system*. Engineering Mechanics, Vol. 5, No. 1, pp. 21–30.

[9.17] HONNER M., ČERVENÝ P., FRANTA V., ČEJKA F. : *Heat transfer during HVOF deposition*. Surface and Coating Technology, 1998, Vol. 106, pp. 94–99.

[9.18] HONNER M., VESELÝ Z., ŠVANTER M.,: *Exdodus stochastic method application in the optimal heating control system for continuous furnances*. Scandinavian Journal of Metallurgy, Vol. 33, No. 6, 2004, pp. 328–337.

[9.19] HOWELL J. R.: *Application of Monte Carlo to heat transfer problems*. Advances in Heat Transfer, 5 (1968), pp. 1–54.

[9.20] JELEPOV B. S.: *Monte Carlo methods solution of boundary problems*. (in Russ) Nauka, Novosibirsk 1980, 176 p.

[9.21] JERMAKOV S. M.: *Monte-Carlo method and another questions*. (in Russ.) Nauka, Moscou 1975, 472 p.

[9.22] KULKARNI V. G.: *Modeling, analysis, design and control of stochastic systems*. Springer, 2000.

[9.23] NELSON B. L.: *Stochastic modelling. Analysis and Simulation*. McGraw Hill, 2001.

[9.24] PRIOR D.V.,BURNS P. J.: *A paralel Monte Carlo model for radiative heat transfer*. SIAM Meeting, Boston, MA, July 21–25, 1986.

[9.25] RUBENSTEIN R. Y.: *Simulation and the Monte Carlo method*, Wiley, New York, 1981.

[9.26] SABELFELD K. K.: *Monte Carlo methods in boundary value problems*. Springer, Berlin, 283 p.

[9.27] ŠROUB J.: *Stochastic methods in new physical technologies*. Ph.D. Thesis University of West Bohemia, Plzen, 2010, 100 p

[9.28] TICHONOV V. I., MIRONOV M. A.: *Markov's processes*. (in Russ.), Moscow, Sovetskoe radio, 1977, 488 p.

[9.29] WIKIPEDIA: *Monte Carlo method*. http://en.wikipedia.org/wiki/Monte Carlo method [cit. 2008-05-23].

Chapter 10

[10.1] ANDERSON, J. A., ROSENFELD, E. (editors): *Neurocomputing: Foundations of research*. MIT-Press, Cambridge, Mass, 1988.

[10.2] BHADESHIA H. K. D. H. (1992): *Neural networks in Materials Science*. ISIJ International Vol. 39, No.10, pp. 966–979.

[10.3] BRÜCKNER S., RUDOLF S.: *Neural Networks applied to smart structure control*. In: SPIE Aerosense, 2000, Orlando, FL, 11 p.

[10.4] GERSTNER W., KISTLER W.: *Spiking Neuron Models*. Cambridge University Press, 2002.

[10.5] HAYKIN, S.: *Neural Networks. A comprehensive foundation*. 2. Ed., Prentice Hall, NJ, 1999, 842 p.

[10.6] HERTZ, J., KROGH A., PALMER, R. G.: *Introduction to the theory of neural computation*. Addison-Wesley, Redwood City, CA,1991.

[10.7] NORGAARD M., RAVN O., POULSEN N. K., HANSEN L. K. *Neural networks for modelling and control of dynamic systems. A practitioner's handbook*. Springer, 2000, 246 p.

[10.8] PRING M. J.: *Breaking the Black Box*, McGraw-Hill, 2002,

[10.9] WIKIPEDIA: http://en.wikipedia.org/wiki/Black_box http://en.wikipedia.org/wiki/Neural_network

Index

J. Kuneš, *Similarity and Modeling in Science and Engineering*,
DOI: 10.1007/978-1-907343-78-0, © Cambridge International Science Publishing 2012

N

P

R

S

T

U

V

Printed by Printforce, the Netherlands